HIMALAYAN ECOLOGY AND ENVIRONMENT

HIMALAYAN ECOLOGY AND ENVIRONMENT

By

A.B. Chaudhuri, IFS (Retd.)

Consultant, Environment and Forests

Ex-Director, Forest Survey of India

ASHISH PUBLISHING HOUSE

8/81, Punjabi Bagh, New Delhi - 110 026

Published by :
S.B.Nangia
for Ashish Publishing House
8/81, Punjabi Bagh
New Delhi - 110 026
☎ *50 0581*
541 0924

Show Room
5 Ansari Road
Darya Ganj
New Delhi-110002
☎ *3274050*

ISBN 81-7024-434-X

1992

Typeset at
APCON
'Aashirwad', Green Park
☎ *686 3359*
666 847

Printed in India at
Efficient Offset Printers
New Delhi - 110 028

By the Same Author

- Resources of Tropical and Temperate Forest
- Sunderbans Mangrove - Ecology and Wildlife
- Environment and Ecology of Shrubs and Herbs
- Plant, Wildife and Man
- Mine Environment and Management

This work is dedicated to the memory of the foresters and forest workers but for whose dedicated and sincere efforts the flora, fauna and environment of the Himalayas would not have been saved.

Preface

No single geographical feature has had greater influence on the life, culture and history of the people of Indian Sub-continent than the Himalaya spreading over 2500 km in a huge crescent and embracing about 500,000 sq. km. in its confines. It has been a route as well as a barrier to the migration of plants and animals. The Himalayas are a vast repository of natural resources: soil, water, minerals, forest, plants and animals, but it has felt the impact of human activities for millennia. The increasing pace of unplanned development for historical, demographical and cultural reasons has led to serious ecological repercussions. This has been caused also due to diverse pattern of land utilization, deforestation for agronomic and pastoral activities, fuelwood requirement, shifting cultivation, indiscriminate hunting and in recent years tourism.

The National Committee on Environmental Planning and Co-ordination (NCEPC) held a Seminar on Resources Development and Environment in the Himalayan Region in April, 1978. This committee constituted a Task Force on the development of Western Ghats which in its report alerted the scientific community on the initial importance of protection and preservation of Silent valley. The recommendation of NCEPC on the Himalayan Environment resulted in establishment of the Institute of Himalayan studies.

The Himalayas are the loftiest chain of mountains in the world, with peaks piercing the sky, has been the source of inspiration to many ages and philosophers have sought refuge in its recesses to meditate on higher divine values; poets and artists have derived inspiration and made magnificent contributions to spiritual upliftment of the people in very many ways. The Himalayas have influenced the life culture and history of this sub-continent.

In the mighty sweep of the Himalayas the western end has the vale of Kashmir, famous for its attractive flowers and gardens, and in Garhwal Himalayas the Valley of flowers. In the eastern region the great explorers e.g. Kingdom ward, Burkhill, Buchanan-Himilton, Griffith, Duthie, Gammie, Bor and others collected myriads of species of plants which are priceless gems for the gardens and laboratories of the world.

The vast expanse of glaciers cover 33,000 sq. km. and their melting snows feed numerous rivers which sustain life in the plains in India and other countries. The vast complex of hills and valleys nourished agriculture, horticulture, forestry, animal husbandry, and have a vast wealth of mineral resources.

Gigantic hydroelectric power projects, various developmental activities and increase of cattle and human population have led to increased soil erosion, disintegrated hillsides, and at places, even to the subsidence of mountain slopes. More trees are being felled to meet the human need and unauthorized cutting of trees go unabated. Overgrazing and fire have caused erosion and eradication of very many species of plants and animals. All these have resulted in silting of rivers and consequent flood, drought and landslides. In recent years the Central and State governments have given comprehensive and integrated look, at the whole Himalayan system in all ramifications, from Kashmir in the west to Arunachal Pradesh and other Hill States in the east and various strategies have been adopted to protect the vast resources.

The problems of this vast hill ranges and of the people living in the mountains are vast and some attention has been paid to protect the resources and improve the living conditions of the people. The NCEPC in their seminar and lately the Planning Commission in their draft resolution of Eight Five Year Plan on the development of the hill regions of the country have considered various problems of this region and suggested remedial measures.

The subject is vast and covers wide ranges of features viz., geology, geomorphology, soil, agricultural system, water resources, plant resources, animal resources, crop resources plant genetic resources, grassland resources, fresh water resources in highland, aquatic and terrestrial ecology, mineral resources, tourism resources, forest resources, live stock resources, glacier resources and many others. Obviously, it is impossible to deal with all these aspects in a volume. The author has, therefore, selected three very important subjects viz., (i) land use and forest cover from the study of aerial photographs and satellite imagery (or state of environment in the Himalayas), (ii) biogeography including floral and faunal resources of eastern Himalayas with a case study of Arunachal Pradesh, (iii) indepth qualitative and quantitative studies of the ecology of vegetation of eastern Himalayas with a case studies of vegetation of Darjeeling Himalayas which is a representative area for entire N.E. Himalayan forest flora.

The author has been personally associated with all these works which have been distinctly brought out in three parts. The Land use status, the richness of flora and fauna of the Himalayas in general and eastern Himalayas in particular have been clearly presented along with analysis and distribution of avifauna, insects, amphibia and reptiles of the area. Various maps on migration presented in Part II is from the compiled work of M.S. Mani in 'Ecology and Biogeography of India (1974)'.

It is sincerely hoped that the book will be useful to those who want to know about land use type, flora and fauna and the balance in nature owing to rich assemblage of herb, shrub, tree flora and the interactions of animals, birds, insects, reptiles within themselves and man.

A. B. Chaudhuri

131, Netaji Subhas Chandra Bose Road
Block 10 Flat 4
Calcutta-40

Foreword

The forests of the Himalayan region have felt the impact of intense human activities in recent years. Clearing of forests, overgrazing, fire, shifting cultivation, indiscriminate felling and other unplanned activities have changed the physiography of the Himalayas. Besides, deforestation for agronomic and pastoral need has further reduced the forest cover to a precariously low level. Cumulative effect of all these factors have brought about deleterious ecological changes in vast floral and faunal resources. In this context Shri A. B. Chaudhuri's present voluminous and diverse work on landuse type by use of aerial photographs and Satellite imageries, on biography and conservation and on qualitative and quantitative eco-botany of Himalayan landuse type, flora and fauna is not only imperative but also very appropriate and pragmatic.

The present work has been accomplished on sound statistical methodology. It encompasses the use of latest techniques on the use and interpretation of Satellite imageries and supplemented by checking of ground truth, a broad spectrum of forest types met within vast stretches of tropical, subtropical, temperate and subalpine forests. Though the foresters have done ecological work on tree flora to some extent, practically very meagre work has been done on herb, shrub and other life forms of vegetation. The findings which have been presented in an interesting pattern expose a clear picture of the composition, structure, density, frequency, canopy class and various aspects of the vegetation of the Himalayas. Listing of vegetation has not been the aim as in a flora; but plants falling inside the quadrats and transects have been recorded and a short description of the species has been provided.

The author has brought forth his wide research data and comprehensive facts to bear on this most important aspects of vegetation hitherto unnoticed. All these facts have been supported by graphs, diagrams and plant distribution and successional charts.

Shri Chaudhuri who was a brilliant scholar and rank holder of Calcutta University in the discipline of Botany and opted the adventurous career in Indian Forest Service devoted most of his opportune time as a keen observor and ardent researcher in pursuing this original study on quantitative ecology of herbs, shrubs and trees of Himalayan forests.

The work deserves to find its valuable use with the conservationists, wildlife managers, botanists, ecologists, biologists and environmentalists. The vast data presented should prove to be a valuable aid for future researcher for basic and applied work in the field. It is expected to meet the long felt needs of reliable quantitative data on Himalayan vegetation. The diversity of facts presented by the Author would surely be a welcome addition to quantitative ecological literature which is too scanty in this country. It would give me immense pleasure if the work finds its right place and fulfils the objectives aimed at.

I congratulate Shri Chaudhuri for this endeavour.

T.M. Das
M.Sc., Ph.D. (Lond.) D.I.C.

Life Science Centre
Calcutta University
5, Ballygunge Circular Road
Calcutta-700019

Contents

PART III

Qualitative and Quantitative Ecology and Environment of the Vegetation of Eastern Himalayas

PART I

STATE OF ENVIRONMENT IN THE HIMALAYAS

Contents

Part -I

State of Environment in the Himalayas

1

The Himalayan Influence

From time immemoral the mighty Himalayan mountains (6,10,200 sq. km.) with their snow-capped peaks and extremely rich forest zones have been excerting a tremendous influence on the vast majority of India's flora, fauna and ecology and have also been shaping and determining the course of the country's unique civilisation of "Unity in Diversity", - not barring, however, the economic, religious and cultural lives of the Indian people. For this alone and on account of their sweeping panorama of majestic beauty and grandeur which inspired awe and reverance, the ancient Indians considered the Himalayas to be the "Castle of Gods" and truly, the cradel of Indian civilisation and culture.

As a matter of fact, the Himalayas with their snows, glaciers, forests, rivers and diverse types of animals, birds and other avifauna etc. have been the source and inspiration of great Indian thinkers and even to-day, they stand there with their majestic and massive stature to attract the attention of the whole world to their numerous appeals. Indeed they serve as the nerve centre of Indian existence. India can, therefore, ill afford to under-rate the Himalayan influence on its climates. rainfall patterns and on the entire economic and cultural life of her people. Every effort should, therefore, be made to save the Himalayas from degradation and the whole nation must, with a single minded dedication strive to conserve their natural resources and eco-system by all possible means.

The Denudation and the Crisis:

Due to the population-explosion and growing pressure on natural resources from human and animal populations - resulting in over-exploitation of forest resources in some areas, much of the Himalayan wealth has been depleted. And as the matters look to-day, the ecology, which is already disturbed, stands the ominous risk of degradation. the mountains which were once green and thickly forested with extremely rich floral and faunal zones look bleak and barren. Frequent occurrence of floods, droughts and near-desert conditions obtaining in many parts of the country are but Nature's signals for greater disasters to overtake the country if we fail to arrest further denudation of the Himalayan ranges and improve upon the already critical position.

Central as well as all the State Governments have been emphasising time and again the paramount need for conserving our forests and also for promoting ecological security. While speaking at the function to launch the world Conservation Strategy on March 6, 1980, the then Prime Minister went so far as to say that:

"The interest in conservation is not a sentimental one, but the rediscovery of a truth well known to our ancient sages. The Indian Tradition teaches us that all forms of life— human, animal, plant are so closely iterlinked that the disturbance in one gives rise to imbalance in the other."

Possible Steps to Avert the Crisis :

Amongst other things, the Indian constitution enjoins as follows :

"It shall be the duty of every citizen of India to protect and improve the natural environment including forests, lakes, rivers, and wildlife and to have compassion for the wildlife."

We should, therefore, get our inspiration and guidelines from this article to conserve and improve our forest resources, particularly those of the Himalayan region as there is practically no difference of opinion in this regard among scholars, scientists, politicians ad a considerable section of the enlightened public. The case for conserving the Himalayan resources thus comes in here. It should, therefore, be our duty to tackle the problem from right now so that the degradation once started may not accelerate further and becomes irreversible.

As indicated already, the then Prime Minister, while expressing her great anxiety over large-scale felling of trees had rightly stressed the view that there should be no project in the Himalayas at the cost of ecology. Dr. M.S. Swaminathan, on various occasions, has highlighted his impression that Indo-gangetic agriculture is in danger owing to indiscriminate felling and shifting cultivation in the hills.

The Central Soil and Water Conservation Engineering Institute indicated that at least half of the Himalayan slopes of 306 million ha. needs soil and water conservation measures. The Uttar Pradesh Government has already taken up integrated soil and water conservation and soil survey in river valley projects. The Chipco movement enthusiasts wanted a blanket ban on all feelings for 10-25 years. The Siwalik regions are glaring example of reckless destruction of forests in the outer Himalayas where 'Chaos' have expanded manifold. Profitability from high altitude forests has to be evaluated in consideration of loss of soil, disturbance of vegetative layers, elimination of certain species, unsetting of vertical and horizontal layers of vegetation and imbalance in the hydrological regime. Due to these compelling reasons and due to the fact that the bulk of Indian forests is situated in the sensitive catchment areas of innumerable rivers, the need for a ban on felling of trees in the Himalayas above 2100 metres is justifiable. The urgency should be realised and members of the public alerted so that the proposed move may get the peoples' support and co-operation at all levels.

To be able to appreciate the magnitude of the problem and to evaluate and assess the Himalayan resources from a proper perspective, some descriptive but important information have been furnished in the succeeding paragraphs of this study under different captions, such as, Physiography of the Himalayas, The rivers, Geology and soil, Drainage, Soil and Water erosion. The slopes, Forests, Diverse ecologically and biologically rich niches, (a) High-altitude animals, (b) Moths, butterflies and other insects (c) Avifauna (d) Amphibia and reptiles (e) Ethnobotany, Migration of fauna, Plant-animal interaction.

Some relevant figures, diagrams and maps have also been added to identify and locate the different Himalayan Forest zones with a view to realising the situation in its entirity.

Physiography :

India has the privilege and blessings of three mountain systems, viz., the Himalayas, the Vindyas and the Coastal ranges. Of these three systems again the Himalayas happen to be the greatest and most massive in extent and height.

The great Himalayan ranges are 2400 km. long and 240 km to 400 km wide, with a number of peaks rising well over 7,200 metres constitute the most striking feature in the geography of the world. The terrain ranges all the way down from lofty snow-clad peaks to deep valleys and detritus accumulations with base of the foot hills at an altitude of less than 300 metres.

Longitudinally this mountain system is classified into three parallel zones with well-marked orthographical, hydrological and vegetative features - (i) The Great Himalayas, (ii) The middle Himalayas and (iii) the Outer Himalayas or the Siwaliks. The entire 2400 km. length has been named from west to east as Punjab Himalayas (500 km.), Kumaon Himalayas (300 km.) Nepal Himalayas (800 km), Bhutan Himalayas (400 km.) and Arunachal Himalayas (400 km). The western Himalayas (from Western Nepal west wards) differ from the Eastern in orientation, in greater length, higher latitude. cooler and drier climate and in far greater width of the mountain west of the Sutlej.

A transverse section drawn through, the valley of Kashmir from the plains of the Punjab to the Karakoram ranges is three times as long as one drawn any where, transversely across the Eastern Himalayas.

Total forest area (officially acknowledged) in the hills in the States of Jammu and Kashmir, Uttar Pradesh, Himachal Pradesh, Sikkim. West

Bengal and Arunachal Pradesh covers 1,29,870 sq. km. Bhutan has an area of about 21,000 sq. km. while Nepal has a forest area of about 15,000 sq. km. This brings the total recorded (official) forested area in the Himalayas to 1,65,000 sq. km. of the total geographical area of 6,10,000 sq. km.

Studies made from aerial photographs and contour surveys show that the states of Jammu and Kashmir, Uttar Pradesh and Himachal Pradesh have about 21,200 sq. km. above 2100 metres. Taking Nepal, Bhutan, West Bengal, Sikkim and Arunachal forested areas together area above 2100 metres would be about 56,200sq.km. 21,200 + 35,000 sq. km.).

A typical altitudinal distribution of area under forests, specifically relating to Uttar Pradesh is shown below as an example:

Metres	Percentage
Below 300	10.0
300 - 600	9.8
600 - 900	6.3
900 - 1200	6.7
1200 - 1500	8.3
1500 - 18000	9.4
1800 - 2100	7.8
2100 - 3000	15.0
3000 - 3450	5.9
3450 and above	20.8
Total :	100.0

Thus the higher ranges supporting scanty to little forest growth account for more than one fifth of the area. This trend holds good for other states also.

The Rivers :

Many of the great Himalayan rivers are the Indus, the Ganga, the Sutlej, the Bhagirathi, the Jamuna, the Brahamaputra and a large number of their tributaries. The Himalayan rivers are antecedent and have consequently eroded deep gorges through the Himalayan chains and ranges. The peculiar configuration of the Himalayan river valleys is chiefly due to transverse gorges, the nature of rocks and the amount of rainfall in the region. The dynamics of rainfall plays a critical role in agriculture, forestry, flood and silt discharge conditions of the big rivers.

Geology and Soil :

The outer hills are composed of shales, phyllites, schists and sandstone and are very unstable. These areas are subject to landslips and erosions. Towards the middle more stable rocks (usually gnessis and sometimes granites) occur, but the inner and higher hills are mostly composed of primary rocks.

The major soil groups found in the region are palehumuths (brown-hills soils), Hapludalfs (sub-montane soils), Cryoborolls (mountain meadow soils) Lithic entisols (skeletal soils), red loamy soils and associations thereof.

The main problem associated with the soils, is high intensity of water-erosion, owing to steep slopes, erratic land use, indiscriminate grazing, lopping and felling of trees at random. thus, the problem of soils erosion has become very acute and has turned out to be an environmental liability of the region.

The complex geological history of the Himalayas and the tectonic stresses which heaved up this range - geologiacally the youngest in the earth's history - have contributed to the aggravating of the earth's dynamic processes of weathering, erosion, mass eastage, seismicity, etc., which affect and modify the natural environment.

The Indus has a catchment area of 468,000 sq. km. upto Indo-Pakistan border, of which area, that drained in India is 321,000 sq. km. The total catchment of Ganga and its tributaries is over 511,000 sq. km. of which 300,000 lies in India. The Brahmaputra drains a total catchment of 580,000 sq. km. of which 187,000 sq. km. lies in India (Source C.W.C.).

Drainage :

It will be seen that the total catchment and drainage area of the major river systems springing from the Himalayas is about 10,72,064 sq. km.

The Indus river system (the rivers Jhelum, Ravi and Sutlej) has a length of 38.25 km. which drains an area of 1,11,184 sq. km. with a maximum discharge of 81,500 m^3 per sec. The Ganga river system (the Ganga, Yamuna, Brahmaputra etc.) has a length of 4797 km. which drains an area of 18,89,500 sq. km. with a maximum discharge of 1,64,520 m^2 per sec.

Some tributaries like the Ghagra, the Gandak and the Koshi drains 1,90,000 sq. km. areas in Nepal. The Ramganga basin covers an area of 32,493 sq. km. The Gomti river drains an area of 30,437 sq. km. The Ghagra has a basin of 1,27,950 sq. km. of which 45% is in India.

Soil and Water Erosion:

India with a very high human population of about 800 million and a grazing animal population of 350 million is striving hard for rapid industrialisation. The meagre forest area is dwindling rapidly and about 5 million hectares of forests have been deforested during last 25 years.

Bulk areas of the country which are arid and semi-arid suffer from wind erosion; humid and semi-humid areas suffer from water erosion. Ravines have eaten up 3.7 million ha. of land. More than 4.8 million ha. have been cleared for river valley projects. The damage caused by erosion to soil fertility is estimated at 53,70,000 tonnes per year costing about 700 crores of rupees. The situation was otherwise even 50 years back when the vast country-side was sylvan, fertile and the rivers were perennial with crystal clear water and erosion was practically unknown.

The entire physiography of the country has, therefore, undergone a drastic change and the forest area has dwindled far below from a bare 22.8 percent, which is ominously inadequate. Only 30 million hectares out of 65 million hectares classified under forests have adequate cover. About 65 million hectates represent denuded land surface, which are permanently prone to erosion.

The extent to which damage is done by soil erosion will be evident from the carriage of 36,60,00,000 tonnes of sand and clay to the Bay of Bengal alone.

The Slopes:

Slopes, degrees and percentage of slopes have been studied in Western Nepal, Bhutan, Alaknanda, Kashmir valley and other inventory areas. As table prepared for Bhutan (produced as table-1) may be taken as typical pattern. As a first approximation, these figures may be extended over whole of the Himalayas though there are possibilities that a higher proportion of areas may fall in higher altitudinal zones in the western Himalayas than in the Eastern.

Slopes above 100% to 200% is generally considered steep where no felling of trees should normally be done. In fact all slopes above 60% should be considered as steep in view of extensive grazing, absence of sufficient herbs and shrubs flora and occurrence of forest fire which have brought degeneration to the soil already.

The Forest :

The Himalayan forests have diverse type of vegetation which comprise tropical, sub-tropical, montane temperate (which comprise wet temperate, moist temperate and dry temperate forests) and dry moist alpine forests and scrub land.

Exposed to monsoon currents, the Himalayan range is floristically a very rich area. The altitudinal factors determine zonal distribution of vegetation. But the crops are in various stages of retrogression due to fragmentation on a very large scale for agricultural purposes.

Climatically also, the Himalayan forests are unique in the sense that in a small journey one moves from humid tropical forests with their orchids, liana and wild bananas, rhododendrons, laurels and conifers to the alpine zone i.e. from swamp to snow and from tropical heat to arctic cold.

This juxtaposition of so many biotopes or life zones from tropical heat to alpine cold, all telescoped within a straight line distance of hardly 80 Km (in eastern Himalayas) have given the Himalayas an assemblage of flora and fauna unrivalled anywhere in the world. Dry and wet valleys add to the ecological complexities of the jumbled habitats which are rich in flowering plants, birds, and butterflies.

Broad-leaved forests comprising of Oaks, Maples, Laurels, Rhododendrons etc. represent the Eastern Himalayan forests, whereas conifer forests comprising of Deodar, Chirpine, Bluepine Spruce and Silver fir and little quantity of Walnuts, maples, willows and poplars represent the western Himalayan forests. All these forests are extremely rich in Fungi, Lichens, Mosses, Ferns, Climbers, Flowering herbs and Shrubs of innumerable species of colourful varieties.

Miles after miles of the hill slopes are covered with colourful (red, mauve, white, blue, cream coloured) Rhododendron flowers that offer a visiter of kaleidoscopic view. Extensive grasslands have a pleasant assemblage of herbs and shrubs yielding flowers red, blue, mauve, yellow white and other colours. Many of such flowering species are indigenous and are of great horticultural value. The forest floors are soft where innumerable species of mosses, liver worts, fern allies form inches of thick carpet that harbour innumerable species of insects. In a word, the Himalayas have more than half the total of 20,000 plant species of India (Sikkim Himalayas have more than 4000 species of flowering plants, 300 species of ferns, 300 species of mosses and liverworts in a very small area). Some climbers have fantastic form and they play a considerable role in the ecological complex. Shrubby vegetation forms an essential part of "Synusia" of forest community. Many species have colourful flowers which have wide range of structure to provide food and shelter to birds, insects moths and butterflies.

The herb flora are still richer. They comprise of decorative and medicinal plants of immense horticultural and economic value. On such flora the future economy of the hill people and the country as a whole will primarily lie. Even small plants like ferns play a significant role in the balance of nature. They have value in protecting soil, providing food and shelter to ground dwelling birds, supplying materials for botanical research and also providing picturesque landscape in the undulating hills. Some ferns form special type of ecological niche. Innumerable species of orchids (more than 600 species) which are residents of the Himalayan hills are price of our forests and will, in their protection and culture, attract attention of flower lovers and botanists from all over the world.

Physiological, morphological and biological importance of Epiphytes, which are special type of life form and biotic association, in the

Himalayan hills are of immense economic and biotic value. Special niche which the fern *Asplenium nidus* in their funnel like leaf base provide, litters and detritus collect to form a new biotope where arboreal ant lizard and bird association grow.

In a word, it may emphatically stated that the vast gene pool of the Himalayan plant species will in their protection attract tourists, birds lovers, animal lovers, botanists, zoologists, naturalists and researchers in their thousands and build a stable economy of the himalayan hill and a biologically rich niche.

Diverse Ecological Niche :

According to the ecological classification of forests types by Champion and Seth the Himalayan region can be broadly classified under 80 different ecological niche. Taking herb and shrubs into account there may be more than 100 ecological niche. This forms more than at least double ecotone zones (where one type of vegetation merges into another type) which are generally floristically and faunistically very rich and diverse.

The floral diversity is parallel to the mammalian and other macro and micro-faunal diversity. Quantitative estimation of flora and fauna of the area indicates this parallelism. The study of flowering and fruiting pattern, types of leaves, their basal area and size classes of fruits (phenophase, phenodynamic and phenometry) testify to the diversity and richness of flora. This diverse flora also maintains a diverse avifauna (Diagram 1,2,3 and 4 in Part III)

An analysis of the Himalayan habitat by the author has revealed.

i. Wide range of leaf surface (800 mm to 2500 mm)
ii. Various types leaves ranging from simple alternate (51.9%) to compound opposite (7.6%).
iii. Canopy layers of top storey (25.8% of tree species), middle storey (48.4% of tree species) and lower storey (25.8% of tree species).
iv. Duration of flowering and fruiting all over the year (Diagram 1,2,3,4, in part III)
v. Size of fruits having positive correlation with the size of frugivorous birds (diagram 2 and 3 in part II)
vi. Positive correlation between size of insects and size of insectivonous birds (Diagram 3 of part II)

a. High altitude animals :

The Himalayan region typifies a unique and ideal habitat where faunal and floral elements of west (European, Mediterranean, African) and east (Chinese, Japanese, Malayasian) have intermigrated and adjoining eastern Himalayas. In fact the sikkim and Eastern Nepal in particular and eastern Himalayas as a whole are the meeting zones of such elements and conservation of such zone should be considered essential.

The temperate Eastern Himalayas are presented by Thar, Goral, Serow, Tibetan sheep, Blue sheep, Takin, Chiru, Musk deer, Barking deer, Tibetan wild Ass, Snow leopard, Cat bear and ninety other mammals (India has 500 mammalian species). Besides Thar, Goral, Serow, Musk deer, the Western Himalayas have Markhor, Ovis, Red bear, snow foxes, Marmot etc. Must rat, Tailless rate, Lynx and many other mammalian species.

Conservation of trees, herbs shrubs and grasslands are essential for food and shelter of these animals. These animals would in future be a sustained source of raw materials (animal protein, fur, zoological species etc.) and would add to the wealth of the local people and the country.

b. Moth, Butterflies and other insects :

The insect population in the Himalayas is extremely rich and greatly diversified. Not only the wooded forest, even the barren rocky and glacier areas have abundant insect life. A distributional study in Eastern Himalayas shows abundance of Coleoptera between 1200 and 1900 m. Dermaptera between 1500 and 2000 m, Hymenoptera between 900 and 2200 m. Coleoptera and Lepidoptera groups of insects are concerned with plants and surpass other species numerically (Diagram 4 of part II).

Vast species of insectivorous birds thrive on insects. The nature has provided a balance in the sizes of the insects and the insectivorous birds.

Butterflies and Moths, which are most conspicuous owing to their elegance, colourful pattern and cause wide range of damage of plants (they are economically very important also) have 1443 species in India. The Eastern Himalayan region is rich with 835 species while the western Himalayas have 421 spp. Selected herbs and shrubs of at least eleven plant families are essential for the survival of such insects.

On barren rocks, on alpine meadows, on snows, on aquatic habitat, on damp moss covered rock surface, in rock crevices and in soil thrive innumerable species of insects. They offer, a rare opportunity to the biologist for carrying out multifacet scientific studies. Such areas are biologically very important.

c. Avifauna :

India has about 2000 species of birds of which the Sikkim Himalayas have 600 spp. Arunachal and Bhutan Himalayas have 600 spp. Arunachal and Bhutan Himalayas have about the same, Nepal Himalayas have about 800 species and Western Himalayas have about 400 species. which signify floral diversity and biogeographical importance of the Himalayas.

Abundance of fruits, flowers, colourful flowers, nectars, pollens insects, horizontal and vertical continuous forest corridor, is essential for sustenance and survival of such a treasure of fauna of the Himalayas.

Conservation of leaf-litter in the forest floor is essential for Pheasants, Grouse, Partridge, Fowl and other ground dwelling birds (Grazing and fire will eliminate such species).

The laughing thrush, Finch, Wood peckor, Bunting, Flower pecker, Chloropsis Cuckoo, Barbet, Babbler, Pigeon, Dove, Quail, Pheasant, Chat, Robbin, Partridge, Snow cock, Himalayan ruby throat, Red start, Stone chat, Golden eagle, Bearded eagle and many other species enrich and add colour to the temperate and alpine Himalayas. Many such species are colourful and game birds and are cultivated as farm birds.

Such vast gene-pool of bird varieties are an immense treasure of the Himalayas and their conservation and culture will change the socio economic status of the people of the country.

d. Amphibia and Reptiles :

Diversified plant, bird, insect and animal species in the floral and faunal complex make room for a wide range of species of amphibia and reptiles.

Bufo, Rana, Rhacophorous (amphibia), *Geomyda, Kachau, Testudo, Lissemys, Trionyx, Chitra* etc. and various species of snakes and reptiles should also be considered as treasures of the tropical and temperate hills. These animals are treasures in the entire complex of balance in the Himalayan eco-system.

Ethnobotany :

The Eastern Himalayas have more than 300 species of seeds, fruits and rhizomes bearing plants that provide food to the people. These could be quantitatively estimated, chemically analysed, scientifically established and nutritiously evaluated for pin-pointing their scientific and economic potential.

Migration of Fauna (Insects, Avifauna and Animals):

The migration of Ducks, Geese, Cranes etc. into India occurs through Siberia -China - Over Tibet into Eastern India, Russia over the Pamirs - into North West India and Pakistan and Central Asia - Iran, Afganistan - Pakistan (Sind) - West India. On this long march the vast migratory population need transit facilities in the Himalayas and the maintenance of habitat has a tremendous significance in such migration. Local migration of summer and winter resident visitors for food and breeding takes place and for their movement horizontal and vertical corridors of forests are essential.

Plant - Animal Interaction :

Plants and animals have intimate and reciprocal relation and these are adapted to the local environmental conditions. The physical and chemical factors in environment are responsible for the size and composition of the communities. Colouration of flowers, shape of flowers, scent and size of fruits are correlated to various insect and avifauna species. Bess are sensitive to red, yellow and blue colours ; Hive bees prefer blue and yellow flowers ; Bats prefer drab green colours, Beetles and Moths prefer red, purple, white and pink flowers ; Flies prefer scarlet or yellow or red flowers. All these colours are provided by the Himalayan flowers in the following proportions :

Colour	Trees	Shrubs	Herbs
White	33	33	24
Yellow	23	24	22
Purple	4	18	18
Red	7	10	8
Pink	5	9	13
Blue	8	6	15
	100	100	100

These natural phenomena are well related to each other.

2

Landuse Type from Aerial Photo and Satellite Imagery Interpretation

Aerial Photo Interpretation

An area of 1,00,265.704 sq. k.m. in Jhelum and Chenub Valley in Jammu and Kashmir, Alaknanda Valley in Uttar Pradesh, Rohru, Chopal and Simla of Himachal Pradesh, Western Nepal and Bhutan (part) were analysed and the findings are as under. The area only covers about 16% of the Himalayan scenario. However, a complete picture has been drawn after interpretation of Satellite Imagery described later in this Chapter.

Particulars	*Area in Sq. km.*	*% of total area interpreted*
I. Vegetative area	51,800.879	52.1
(a) Tree Vegetation	47,568.677	47.9
- Chir *(Pinus roxburghii)*	4,942.280	5.0
- Kail *(Pinus smythiana)*	3,471.570	3.5
- Deodar *(Cedrus deodora)*	608.250	0.7
- Kail -deodar *(Pinus-Cedrus)*	1,025.000	1.0
- Fir *(Abies SP)*	4,093.350	4.1
- Fir-spruce *(Abies-Picea)*	1,938.650	2.0
- Mixed confers	4,245.12	4.3
- Conifers & broad-leaved-mixed	2,531.970	2.5
- Mixed hardwoods	24,632.395	24.8
- Degraded vegatation	222.400	0.2
- Plantations	4.440	-
(b) Alpine pastures & scrubs	985.742	1.0
(c) Pasture land	1,917.000	1.9
(d) Blanks	726.230	0.7
(e) Others	376.390	0.4
II. Non-forest land	47,555.635	47.9
(a) Snow-covered areas	3,083.824	3.1

Particulars	Area in Sq. km.	% of total area interpreted
(b) Tarraced cultivation *	1,117.267	1.1
(c) Valley cultivation *	89.640	0.1
(d) Cultivation & habitation **	4,142.740	4.2
(e) Lakes & water spreads	193.870	0.2
(f) Exposed rocks & barren slopes	9,257.683	9.3
(g) Non-forestry plantations	47.360	0.1
(h) Farm wood-land	276.280	0.3
(i) Orchards	28.060	-
(j) Others ***	29,298.911	29.5
Total area interpreted	99,336.514	100
Photo-gaps	836.340	
Clouds	92.850	
Total Geographical area	1,00,265.704	

N.B. * Terraced and valley cultivation have been delineatd separately for U.P. Survey area only.

** For Bhutan orchards have been including in the landuse class-cultivation and habitation.

*** For Chenab valley and Western Nepal survey areas, entire non-forested area have been placed under the category-Others.

JAMMU AND KASHMIR

Area under different landuse types of Chenab and Jhelam valley are as under :-

(A) Chenab valley (J&K)

Landuse Class	Area Km²	%
(i) Forest land	4,036.40	41.0
(a) Vegatation forests	3,744.33	38.0
1.Chir	193.96	2.0
2.Kail	733.69	7.5
3. Deodar	657.79	6.7
4.Fir	1,324.01	13.4
5.Others (mainly broadleaved)	834.88	8.5
(b) Vegetation scrubs etc.	292.07	3.0
(ii) Farm Wood land	276.28	2.8

(iii)	Non-forestry plantations	47.36	0.5
(iv)	Agriculture crop land	718.33	7.3
(v)	Pasture land	726.33	7.3
(vi)	Urban, village & industrial lands	47.36	0.5
(vii)	Barren land	3,710.09	37.7
(viii)	Others	284.18	2.9
	Total Geographical area	9,846.23	
(B)	Jhelum Valley (J&K)		
	Wooded land	2,633	25.8
(i)	Blue-pine/Deodar	1,025	10.0
(ii)	Fir/Spruce	1,336	13.1
(iii)	Other hardwoods	272	2.7
	Non-wooded land	7,563	
(i)	Scrubland	1,543	5.1
(ii)	Others	6,023	69.1
	Total Geographical area	10,196	

Analysis of Results:

(a) Chenab Valley

It is interesting to note that out of 41% of forest area, 38% is under vegetation of different crown densities while about 3% of area is under scrub. Of the non-forested area only about 7.3% is under crop land while about 37.7% of area is barren land. It seems, there is lot of scope for developmental activities in barren areas with the help of social forestry schemes.

(b) Jhelum Catchment

Forestry position in Jhelum Catchment is worse than in Chenab Catchment. There is only 25.8% tree forest of varying capability while 5.1% of project area is under scrub. About 69.1% of area is non forested. There is lot of scope for extending forestry activities.

Analysis of Results:

About 46.3% of area is under vegetation cover. 42.7% is under tree vegetation while 3.6% is either scrub, blanks, plantations or degraded areas. 53.7% of geographical area is non forested. Forestry position as

HIMACHAL PRADESH

The area figures of different landuse types in Rohru, Chopal and Simla Divisions is as under :-

Area in Km²

Particulars	Name of Forest Division						Total Area	%
	Simla Area	%	Rohru Area	%	Chopal Area	%		
1	2	3	4	5	6	7	8	9
FOREST AREA	310.39	44.2	691.32	42.4	467.68	55.8	1469.39	46.3
a) Tree Forest	265.56	37.8*	630.92	38.7	459.65	54.8	1356.13	42.7*
Fir and Spruce	16.34	2.3	103.11	6.3	46.04	5.5	165.49	5.2
Mixed conifers	76.61	10.9	186.19	11.4	141.16	16.8	403.96	12.7
Deodar	19.96	2.8	9.01	0.6	1.49	0.2	30.46	1.0
Conifers mixedwith B.L.	12.50	1.8	68.83	4.2	61.86	7.4	143.19	4.5
Kail	67.74	9.7	165.93	10.2	181.06	21.6	414.73	13.1
Chir	60.50	8.6	14.34	0.9	17.52	2.1	192.36	2.9
Upland hardwoods	8.84	1.3	83.51	5.1	10.52	1.2*	102.87	3.2
Low land hardwoods	3.07	0.4	-	-	-	-	3.07	0.1
b) Non-tree Forest	44.83	6.4	60.40	3.7	8.03	1.0	113.26	3.6
Alpine Pastures & Alpine Scrubs	0.23	-	56.20	3.4	2.79	0.3	59.22	1.9

1	2	3	4	5	6	7	8	9
Forest Blanks	0.20	-	2.83	0.2	1.36	0.2	4.39	0.1
Plantations	-	-	-	-	0.54	0.1	0.54	-
Degraded forests	11.15	1.6	1.26	0.1	2.29	0.3	14.70	0.5
Scrubs	33.25	4.8*	0.11	-	1.05	0.1	34.41	1.1
NON-FORESTED AREA	391.18	55.8	940.62	57.6	370.81	44.2	1702.61	53.7
Permanent Snow & glaciers	-	-	14.05	0.9	-	-	14.05	0.4
Water	1.53	0.2	3.91	0.2	-	-	5.44	0.2
Exposed rocks & barren slopes	80.57	11.5	536.61	32.9	155.26	18.5	772.44	24.4
Agriculture & Habitation	296.32	42.3*	373.72	22.9	214.59	25.6	884.63	27.9
Orchards	12.76	1.8	12.33	0.7	0.96	0.1	26.05	0.8
Total area interpreted	701.57	100	1631.94	100	838.49	100	3172.00	100
Photo gap	812.22	53.7	-	-	-	-	812.22	20.4
Total Geo. area	1513.79	-	1631.94	-	838.49	-	3984.22	

% Figures have been calculated with respect to area interpreted but for gap the % pertains to total geographical area.

* Rounding off adjustments.

such is not very good in this hilly state where at least 60% of geographical area should be under vegetation cover.

UTTAR PRADESH

Area statements of different landuse types of Alaknanda Catchment are as under :-

Area in Km²

Landuse class/ Forest type	*Area under density class*				*%*
	High	*Medium*	*Low*	*Total*	
1	*2*	*3*	*4*	*5*	*6*
FOREST AREA				4263.323	40.8
(a) Tree forest	1824.992	1132.789	647.81	3605.591	34.4
* Chir	298.50	331.47	223.66	853.63	8.2
* Kail	15.08	1.73	1.90	18.69	0.2
* Fir-Spruce	326.86	81.61	28.69	437.16	4.2
* Mixed conifers	70.35	81.226	22.60	174.176	1.7
* Broad leaved mixed with conifers	292.312	160.128	71.11	523.55	5.0
* Upland hardwoods	808.31	443.576	255.22	1507.106	14.4
* Lowland hardwoods	13.60	33.049	44.63	91.279	8.7
(b) Alpine pastures				278.22	2.7
(c) Degraded forest				207.70	2.0
(d) Scrub-steep				19.602	0.2
(e) Scrub-gentle				133.66	1.2
(f) Plantations				1.20	—
(g) Blanks				27.35	0.3
NON-FOREST				6192.264	59.2
(a) Snow covered				2871.754	27.5
(b) Water spreads				38.19	0.4
(c) Exposed rocks				1950.053	18.6
(d) Barren grass land				123.35	1.2
(e) Terraced cultivation				1117.267	10.7
(f) Valley cultivation				89.64	0.8
(g) Orchards				2.01	-

Analysis of Results.

Area under vegetation cover is only 40.8% out of which 34.4% is under tree forest and 6.4% under scrub, pastures, plantations and blanks etc. Only 17.4% of area has crown density of 60% and above, while 10.8% is under medium density and 6.2% under low density below 20%. Non-forest area is very high i.e. of the order of 59.2%.

NEPAL

The area under different landuse and forest type is as under :-

Landuse pattern *forest type*	*Area in sq km under density class*				%
	High 50%	*Medium 25-50%*	*Low 5-25%*	*Total*	
1	2	3	4	5	6
Chir Pine	1003.312	884.128	955.680	2833.120	7.8
Blue pine	1096.048	316.656	165.376	1578.080	4.3
Mixed conifers	1371.120	564.864	320.272	2256.256	6.2
Lowland hardwoods	657.104	638.128	1088.080	2383.312	6.5
Upland hardwoods	2338.976	1366.896	1872.896	5578.768	15.2
Total foresed area	6466.560	3770.672	4402.304	14639.536	40.0
Percentage	17.0	10.3	12.0		
Non-forested				21967.731	60.0
Grand total				36607.267	100

Analysis of Results

40% of project area is under forest cover while 60% is non-forest. Only 17.7% of area is of high forest density and 10.3% area is medium density. 12.0% of area has crown density of less than 20%. Low density areas should be subjected to better stocking by afforestation and protection measures.

Analysis of Results

29.4% of the total area interpreted is non-forested whereas 70.6% is forested. Out of this total 70.6% tree forest area is 63.4%, Alpine pasture cover 6.3% and blanks cover 1.0%. Out of 63.4% area under the forest, 9.7% is of low density cover and needs complete protection, whereas 12.4% of the land area under medium density cover of tree forest requires judicious working. Remaining 41.3% of the land area which is under dense forest cover, will form the backbone for the forest based industrialization of the region which projects quite an encouraging picture of luxuriant forest resources of this part of the country.

BHUTAN (NORTH-CENTRAL)

The area under different landuse and forest type is as under :-

Laduse class	*Area in sq. km under density class*				*%*
	High 60% &	*Medium 20.60%*	*Low 2%*	*Total (2+3+4)*	
1	2	3	4	5	6
FOREST AREA				5909.29	70.6
(a) Tree forest	3276.53	987.78	768.48	5032.79	63.4
(i) Chir	49.80	95.38	53.02	198.20	2.5
(ii) Kail	169.06	111.29	279.88	560.23	7.1
(iii) Fir	845.01	271.22	163.98	1280.21	16.1
(iv) Mixed conifers	711.11	138.29	75.72	925.12	11.7
(v) Broadleaved mixed with conifers	301.80	131.29	54.82	487.91	6.1
(vi) Broad-leaved	1199.75	240.31	141.06	1581.12	19.9
(b) Alpine pastures				497.23	6.3
(c) Blanks				79.27	1.0
NON-FORESTED AREA					
(a) Snow covered				2334.85	29.4
(b) Lakes & Water spreads				187.77	2.4
(c) Exposed rocks &				33.54	0.4
barren slopes				1674.79	21.1
(d) Cultivation & habitation including terraced cultivation and Orchards				438.75	5.5
Total area interpreted				7144.14	
AREA UNDER CLOUDS				16.07	0.2

BHUTAN EASTERN

Landuse class	*Area in sq. km under density class*				*%*
	High	*Medium*	*Low*	*Total*	
1	2	3	4	5	6
FOREST AREA				5691.98	84.2
(a) Tree forest	3467.78	1065.53	649.31	5182.62	76.6
i) Chir	101.90	95.88	62.44	260.22	3.8

1	*2*	*3*	*4*	*5*	*6*
ii) Kail	55.52	52.08	58.55	116.15	2.5
iii) Fir	706.15	115.06	33.58	854.79	12.6
iv) Mixed conifers	310.96	35.13	27.14	373.23	5.5
v) Broad-leaved mixed with conifers.	337.23	187.10	101.53	625.86	9.3
vi) Broad leaved	1956.02	580.28	366.07	2902.37	42.9
(b) Alpine pastures				409.71	6.1
(c) Blanks				9.95	1.5
NON-FORESTED AREA				1068.74	15.8
(a) Snow covered				0.50	-
(b) Lakes & Water Spreads.				15.04	0.2
(c) Exposed rocks & barren slopes.				776.03	10.0
(d) Culti. & habitation including terraced culti. & orchards.				377.17	5.6
Total geog. area				6760.72	100

Analysis of Results :

84.2% of the survey area is forested, out of which 76.6% is tree forest, 6.1% under alpine pastures and 1.5% under blanks. It is encouraging to note that 51.3% of total land area is of high density forest, 15.7% is of medium density tree forest (may be subjected to judicious working) and on 9.6% is of low density class which requires protection. Only 15.8% of project area is non-forested.

BHUTAN (EAST SOUTHERN)

Landuse class	*Area is sq km under density class*				*%*
	High	*Medium*	*Low*	*Total*	
1	*2*	*3*	*4*	*5*	*6*
FOREST AREA				3393.71	83.8
(a) Tree forest	1505.52	908.84	644.28	3058.64	72.5
i. Chir	91.35	199.78	196.94	488.07	12.0
ii. Fir	149.73	33.10	6.56	189.39	4.7
iii. Mixed conifer	55.08	16.72	1.84	73.64	1.8
iv. Broad-leaved mixed with conifers	129.82	180.53	70.76	311.11	9.4

1	2	3	4	5	6
v. Broad-leaved	1079.54	478.71	368.18	1226.43	47.6
(b) Alpine pastures				322.34	8.0
(c) Blanks				12.73	0.3
NON-FORESTED AREA				654.89	16.2
(a) Lakes & water spreads				3.49	0.1
(b) Exposed rocks & barren slopes.				19.86	0.5
(c) Culti. & habitation including terraced culti. and orchards.				631.54	15.6
Total area interpreted				4048.64	100
CLOUDS				16.42	0.4
PHOTO GAPS				72.17	1.7
Total Geog. Area				4137.19	

Analysis of Results :

83.8% of the total area interpreted is forests, only 7.5% being under tree cover. Areas under high, medium and low density tree cover account for 37.2%, 22.4% and 15.4% respectively of the total land area. Non-forested region account for 16.2% of the total area interpreted.

BHUTAN (South Western)

The are under different landuse and forest types is as under :-

Land-use class/ Forest type	*Area in sq. km.under density class*				
	High	*Medium*	*Low*	*Total*	*%*
FOREST AREA				8822.02	85.6
(a) Tree forest	3866.10	2929.32	1520.62	8316.04	80.7
i. Chir	2.15	8.54	2.3	12.72	0.1
ii. Fir	255.67	143.48	45.80	444.95	4.3
iii. Mixed conifers	16.24	16.54	6.05	38.83	0.4
iv. Broadleaved mixed with conifer	123.97	192.83	53.55	370.35	3.6
v. Broadleaved	3468.07	2567.93	1413.19	7449.19	72.3
(b) Alpine pastures				350.28	3.4
(c) Plantations				2.70	-
(d) Blanks				153.00	1.5

NON-FOREST AREA	1483.95	14.4
(a) Snow covered	9.75	0.1
(b) Lakes & Water spread	98.17	1.0
(c) Exposed rocks & barren slopes.	331.07	3.2
(d) Cultivation & habitation including slopes.	1044.96	10.1
Total area interpreted	10305.97	
Clouds	7.70	0.1
Photo gaps	4.61	-
Total Geographical area	10318.28	

Analysis of Results :

85.6% of the interpreted area is forested whereas 80.7% of the area is under tree cover. 3.4% of the total land area is under pasture, whereas 1.5% constitute forest blanks besides a negligible area 2.70 sq. km is under plantations of various spp. Contribution of low, medium and high density forest is upto the level of 14.8%, 28.4%, 37.5% of the interpreted area respectively. Non-forested areas account for 14.4% of the area. Maps for negligible portion of the area 12.31 sq. km accounting approximately 0.1% of the total geographical area only could not be prepared due to clouds and photo gap.

Vegetation Classification :

The Himalayan forests can be classified under ten (10) broad types described by Champion and Seth (1968) which have been enumerated in next paragraph. It has been possible to identify major tree forests from aerial photointerpretation as detailed in a few paragraphs below. From verification of ground truth some additions has been made in the list of associate species and the altitudinal distribution of vegetations.

Broad Outlines of Vegetation of the Himalayas :

The Himalayan forests have diverse types of vegetation which comprise tropical, sub-tropical montane temperate, sub-alpine and alpine.

According to Champion and Seth the following main forest types occur in the Himalayas.

- Tropical Moist Deciduous Forests
- Tropical Dry Deciduous Forests
- Sub-Tropical Pine Forests
- Sub-Tropical Dry Evergreen Forests
- Montane wet-Temperate Forests

* Montane wet-Temperate Forests
* Himalayan Moist Temperate Forests
* Himalayan Dry Temperate Forests
* Sub-alpine forest
* Moist alpine scrub
* Dry alpine scrub.

Exposed to monsoon currents, the Himalayan ranges are floristically very rich. Botanically the Himalayas can be divided into Western Himalayas and Eastern Himalayas. A broad summary of vegetative characters of these two regions is given later in this chapter.

i. Western Himalayas :

The rainfall in this region varies from 1000 mm to 2000 mm per annum, with considerable variation between the outer and inner valleys. The inner valleys get less rainfall and more snow. The climate generally becomes drier and more extreme towards west. According to G.S. Puri, the vegetation can be divided into altitudinal zones, such as, submontane zone upto 1500 m, temperate zone from 1500 m to 3300 m and alpine zone from 3300 or 3600 m to snow line. The altitudinal zonation of vegetation is not very strict and at places geology and soil play a dominent role.

The Siwaliks are covered by *Shorea robusta* forests. These forests are broken by geology, soil as well as biotic interference and accordingly miscellaneous, riverain and swamps etc. are also found. In miscellaneous forests *Terminalia, Anogeissus, Stereospermum, Adina, Ougeinia, Lagerstroemia* and *Acacia* species occur along with sal or without sal. Besides these, *Dendrocalamus strictus* is also found at places as mixed crop. *Acacia catechu* and *Dalbergia* sissoo are the main crop in riverain tract. In Swampy areas *Toona ciliata, Bischofia javanica, Albizzia procera, syzigium cumini* etc. with a thick undergrowth of evergreen bushes, ferns etc. occur.

Pinus roxburghii comes on quarzite and boulder conglomerates on top of hills. In burn and heavily disturbed areas thorny species as well as open and degraded vegetation formations are found. *Pinus roxburgii* is the most important species of sub-tropical pine zone.

In the temperate zone five species of Oaks i.e., *Quercus ilex, Q. glauca, Q. incana, Q.dilatata* and *Q. semicarpifolia* alongwith a number of deciduous broad leaved species such as *Acer, Aesculus, Populus, Ulmus, Alnus, Betula, Prunus, Rhododendron* etc. are found. In the zone of Oaks, conifers forming pure or mixed forest constitute an important feature of forest vegetation. In outer monsoon zone *Pinus excelsa* and *Cedrus deodara* are found while in areas with heavy snow fall at higher altitudes *Picea smithiana*, and *Abies pindrow* occur. *Pinus gerardiana* occurs in inner dry valleys.

In sub-alpine and alpine areas vegetation generally consists of silver fir, *Betula, Juniper, Rhododendrons* with a variety of plants forming cushion like features.

ii. Eastern Himalayas:

The Eastern Himalayas form a humid region and is divided into tropical temperate and alpine zones according to altitude. They are distinguished from Western Himalayas by higher monsoon rainfall, high humidity, milder temperature and less snowfall.

The tropical zone is formed by sal forest with patches of riverain forest, savannah, mixed deciduous forest and swamps mainly in Siwaliks. The undergrowth is more luxuriant than in Western sal forests.

The temperate zone is divided into lower and upper types. The lower type is dominated by broad leaved species while upper zone has more conifers. The Main confier species are *Abies webbiana, Picea spinulosa, Larix griffithii* and *Tsuga dumosa* with Junipers, Willows and *Rhododendrons*. In the alpine zone the most common species are Rhododendrons and Junipers.

The conifers are more abundant in the vegetation of Western Himalayas compared to those of Eastern Himalayas. The broad leaved species along with Orchids and fern are more abundant in vegetation of Eastern Himalayas compared to those of Western Himalayas.

Physiognomy of Vegetation :

The physiognomy of vegetation in Himalayas vary from closed crown formation to scattered trees, high growth to stunted three growth along with shruby and herbal formations pastural lands. The different degraded formations are present side by side with climax vegetation.

Delineation of Landuse and Vegetation Types Using Aerial Photographs of Small Scale :

The general description of different vegetation and landuse classes delineated in Himālayas on aerial photographs of small scale are as under:-

Alpine Pastures and Scrub :

These forests are found in Himalayas above 2900 m. Trees are of stunted growth and height less than 5 m with scrubby vegetation and pasture land. The important shrubs are *Rhododendron, Juniper Viburnum* etc.

Fir/Spruce. *(Abies/Pinus or Tsuga) :*

Such forest occurs above deodar and blue pine zones generally between 2500-3500 m. Fir is often associated with spruce. On cool Northern aspects fir tends to be pure, whereas on warm southern aspects spruce tends to be gregarious.

Blue Pine. *(Pinus smythiana)* :

Blue pine occurs fairly extensively between elevations of about 1500 m and 2500 m though along the spurs and warmer sites it extend upto 3500 m. Such vegetation class is generally pure with some admixture - of deodar and spruce at places. Crown images of blue pine are round, biggest and lighter than fir/spruce or deodar.

Deodar *(Cedrus deodara)* :

Deodar occurs between the altitude of 1500 m to 2500 m. Deodar generally grows as pure patches with some admixture of Kail spurs and spruce on cooler sites. On photo, deodar appears darker in tone chir and blue pine and its crown shadows are conical.

Chir Pine *(Pinus roxburghii)* :

This type generally occurs between 1000 - 2200 m. elevation. In depressions and cooler aspects it has deciduous lower storey association.

In the upper range, chir is gregarious and almost pure. In nullas and depressions some broad leaved species form an intimate mixture.

Fir-Blue Pine *(Abies Sp - Pinus smythiana)*

The vegetation area in which the occurrence of blue pine is more than 50% and fir with minimum of 20% have been put under this category.

Mixed Conifers :

In the transition zone various coniferous species may grow together in varying proportions. Deodar/*Tsuga* may be mixed with Chir pine, blue pine or spruce at different sites and there may be two or three species in a stand. In such cases, identification of individual species becomes difficult. Such stands have been put under the category mixed conifers.

Broad Leaved Mixed with Conifers :

This type has been identified in the State of U.P., Bhutan and Arunachal Pradesh. Line of difference of this type from those of mixed hard woods and mixed conifer is very thin and it requires great deal of caution to distinguish the above three types from each other.

Broad Leaved:

This type commonly occurs over the entire range of altitudes. It is composed of large number of species. On the aerial photographs this type appears in light grey colour. The crowns are generally spreading and irregular. This type has further been sub-classified into two sub-groups viz. upland and low land hardwoods, the criterion for differentiation is only the altitude line (1200 m). But the composition of species as well as their tone on the aerial photographs has no sharp difference as such.

Sal:

Where Sal constitutes over 50% of the crop density, the forest has been recognised as the Sal type. In some survey areas sal with more than 80% of composition is recognised as pure Sal. Miscellaneous species are often found mixed in varying proportions. It is found all along the foot hills of Lower Himalayas. On aerial photographs Sal appears in dark grey tone and uniform in stand height and stocking as compared to mixed hardwoods. Young Sal growth appears lighter in tone and finer in texture than middle aged and mature sal.

Sal Mixed with Broad-leaved :

In this type sal is the predominent species but is generally mixed with other miscellaneous species like *Anogeissus latifolia, Terminalia tomentosa, Adina cordifolia, Buchnania lanzan* etc. The proportion of sal is less than 50% of the total crop density. Small patches of pure sal may often be met within an otherwise Sal mixed hardwood stand.

This type appears in aerial photographs in dark to light grey tone depending on the proportion of sal. Density appears less and ground may often be visible whenever the constituent spp. are leafless.

In most of the survey areas when a species was found to form more then 50% of crop density it was delineated as pure crop. In some survey areas a crop was recognised pure only when species composition was more than 80%. Besides this, following types were identified in non-vegetation.

* Snow cover areas
* Terrraced Cultivation
* Valley cultivation
* Cultivation & habitation
* Lakes & water spread
* Exposed rocks & barren slopes
* Non-forestry plantations
* Farm woodland
* Orchards
* Others.

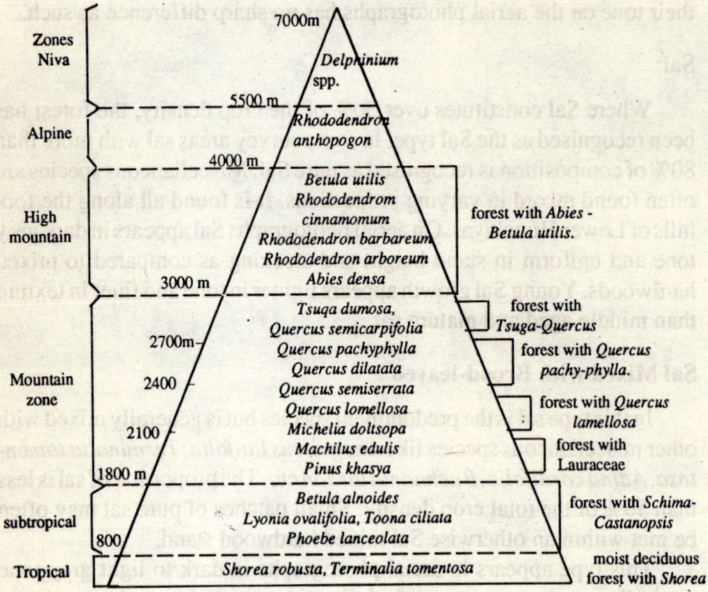

Altitudinal Distribution of vegetation in Eastern Himalayas

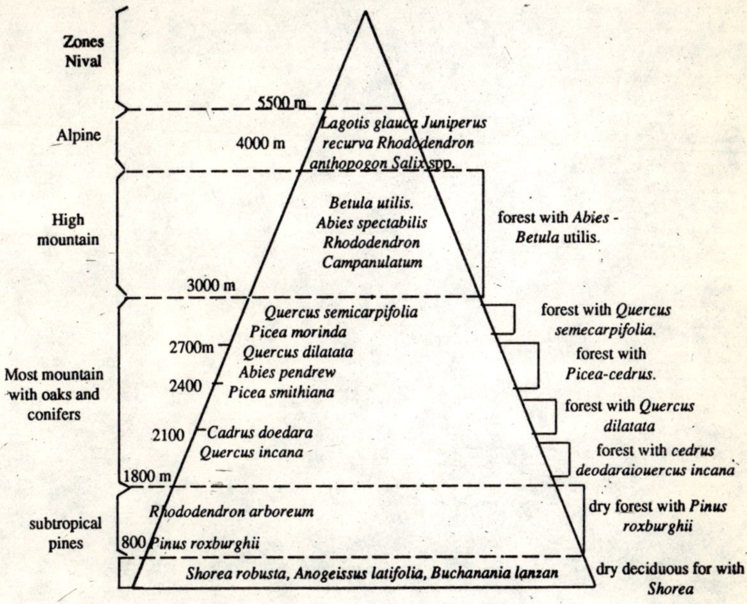

Altitudinal Distribution of vegetation in Western Himalayas

3

Forest Cover

(From Satellite imagery interpretation)

Indian Scenario:

Conservation of the Himalayan forests and various resources has been one of the major issues of the country to-day. Various Government and non-Government organisations are involved in identifying the extent of problems and are suggesting remedial measures about conservation of resources and rehabilitating the loss.

The Department of Forests in various States in the Himalayan regions are in control of about 1,65,000 sq. km. of land. Vast areas of the Himalayan forests are under control of non-Government agencies. Besides, bulk areas are cultivated and grazing lands where Forest Department have no control. Two foreign countries occupy 1,87,600 sq. km. of such Himalayan region. The morphology of land in Indo-Gangetic-Brahmaputra plains are also results of management of land in these countries besides the Indian States in the hilly regions as all the rivers flow southwards in the plains of India.

Various Forest Departments are not posted with accurate information on tree cover and their density; little is known about vast forested tracts of the Himalayan region in Arunachal Pradesh. Some States are not in a position to assess the density and extent of depletion of forest land. Similar situation prevails in other areas outside the control of the Forest departments. Availability of topo sheets and aerial photographs were limited in the present study owing to defence consideration.

The National Remote Sensing Agency (NRSA) of the Department of Space estimated the forest cover of the country using Satellite imagery at two periods of time viz., 1972-75 and 1980-82. The interpretation was supported by limited ground truth verification. Diapositive imagery with false colour composite made from landsat series of Satellites on 1:1 million scale for the two following broad cover types were identified and delineated by visual interpretation of the imagery:-

Forest land	Non-forest land
- Closed forest (areas with more than 30% tree cover)	- Agricultural land
	- Grassland

- Open forest or degraded forest (area with less than 30% tree cover)
- Mangrove forest
- Shrubland
- Non-forest plantation
- Barren land.

(others include areas under snow, cloud, shadow etc.)

The changes in forest cover of the country in the 7-year interval as determined by NRSA are as under:

	1972-75	1980-82	Area in sq. km. Change*
Total Forest Cover.	555.80	4634.70	(-) 191710
Total Forest cover as percentage of total geographical area.	16.89%	14.10%	(-) 2.79%
Closed Cover	464226	360229	(-) 103987
Closed forests as percentage of the total geographical area.	14.12%	10.96%	(-) 3.16
Open/degraded forests	87673	100592	(+) 12919
Open and degraded forests as % of the total geographical area.	2.67%	3.06%	(+) 0.39%
Mangrove forests.	3281	2649	(-) 632
Mangrove forest as % of total geographical area	0.099%	0.081%	(-) 0.018%

* (+) means increase; (-) means decrease.

The Forest Survey of India (FSI) made an assessment of forest cover by visual interpretation of Landsat imagery on 1:1 million scale for the period 1981-83. This was supplemented by intensive ground truth verification and from analysis of aerial photograph. The F.S.I. estimated the forest cover in the country at 19.7 percent of the geographical area. This was different from the findings of NRSA. The two organizations lately reconciled their findings by further examination and the following were the result:

Sl. No.	Category	Forest area as percentage of total geographical area			
		NRSA estimates for		FSI estimated	Reconciled estimates.
		1972-75.	1980-82	1981-83	1981-83
1.	Dense forest	14.12%	10.96%	10.9%	10.88%
2.	Open forest.	2.67%	3.06%	8.6%	8.41%
3.	Mangrove forest.	0.099%	0.081%	0.1%	0.12%
4.	Coffee plantation.	—	—	0.1%%	0.11%
	Total Forest	16.89%	14.01%	19.7%	19.52%

Other areas are:

Scrub area	- 76,796 Sq.Km.	or 2.34%
Uninterpreted area	- 11,524 "	or 0.35%
Non forest (includes tea estate)	25,57,436 "	or 77.79%

These findings, however, relate to the forest cover and non forest areas of the country as a whole. Fortunately the Governments of Bhutan and Nepal got the resource survey of forest wealth done through Forest Survey of India, the former over 42,000 sq.km. (entire Bhutan area) and the latter over 50,000 Sq. Km. (Western Nepal only) by the use of aerial photographs which enabled the author to get maps of entire Himalayas prepared. Interpretation of other areas was done from Satellite imageries.

The findings reveal that closed forests extend over 133,601 sq. km. or 21.89 percent of the total geographical area (610,200 sq.km) of the Himalayas.

Landuse Pattern of the Himalayas Using Landsat Imageries:

Black and white landsat imageries on band 5 of the base year 1975 were used for the study. The field information avaliable in the Forest Survey of India was made use of and no field trip was under-taken. The interpreted landsat imageries were transferred on to base map on 1:1 million scale. The area of different landuse classes was calculated with the help of dot grid templet. The information given in the this report can be taken as first level of information and can be used for planning at macro level.

Landsat Product Used:

Black and White prints of band 5 of the following scenes were used.

For Arunachal Pradesh Black and White imagery (Diapositives) on band 5 & 7 were used.

JAMMU & KASHMIR

Path	Row	Year	Remarks.
162	034	Feb. 1975	
	034	Sept. 1972	
161	034	Nov. 1975	
	035	Oct. 1975	
	036	Nov. 1975	
	037		Gap
160	035	Apr. 1975	
	036	Nov. 1975	
	037	Nov. 1975	
159	035	Mar. 1975	
	036	Feb. 1973	
	037	Nov. 72/75	
158	035	June 1975	
	036	May 1975	
	037	Mar. 1975	
	035	Oct. 1972	
	036	Oct. 1972	
	037		Gap

HIMACHAL PRADESH

Path	Row	Year	Remarks.
159	037	Nov. 72/75	
	038	Nov. 1972	
158	037	Mar. 1975	
	038	Mar. 1975	
	039	Mar. 1973/75	
157	038	Mar. 1975	
	039	Feb/May/June1973/1974	

UTTAR PRADESH

Path	Row	Year	Remarks.
157	038	Mar. 1975	
	039	Feb/May/June/1972/1974	
156	039	Oct. 1972	
	040	Mar. 1975	
155	039	Feb. 1973	
	040	Feb. 1975	

NEPAL

155	039	Feb. 1973	
	040	Feb. 1975	
154	039	Mar. 1975	
	040	Mar 1975	
	041	Mar. 1975	
153	040	Mar. 1975	
	041	Feb. 1975	
152	040	Apr. 1975	
	041	Dec. 1972	
151	041	Nov. 1975	
150	041		Gap
	042	Mar. 1975	

SIKKIM

149	041	Dec. 1975

BHUTAN

149	040	Mar. 1975
	041	Dec. 1975
148	040	Oct. 1972
	041	Dec. 1975
147	041	Dec. 1975

WEST BENGAL

148	041
149	042

ARUNACHAL PRADESH

Path	Row	Year	Remark
147	041	Feb. 1977	
	042	Feb. 1977	
146	040	Jan. 1977	
	041	Jan. 1977	
145	040	Jan. 1977	
	041	Jan. 1977	
144	040	Dec. 1976	
	041	Mar. 1975	

Landuse Classes Delineated:

The following landuse classes were delineated on the imageries:

* Closed forest (F1)
  Well stocked forests having an average crown cover ranging from 50% to 60%. In some cases crown cover is as low as 40% and as high as above 80%.
* Open forest (F2)
  These are forests which have been badly affected due to biotic interference or natural calamities. The tree cover has been depleted to average crown density of about 25% i.e. 10% to 40%. The natural & official regeneration is similar to that of adjoining forest and they have also been included in this category.
* Closed forest with shifting cultivation (F3)
  These are forests with long cycle of shifting cultivation include new and old clearings, secondary growth forests etc.
* Open forests with shifting cultivation (F4).
  Include degrade forests affected by shifting cultivation.
* Water bodies (W):
  Includes rivers, streams, lakes etc.
* Non-Forest - N.
  This includes areas such as agricultural land, cultivable and uncultivable waste land, pasture and grazing land, small tree covers, towns etc.

JAMMU AND KASHMIR

Out of total geographical area of the State about 44% falls in Ladakh which is a cold desert. Area under good vegetation cover is about 6.4% and about 5.5% is highly degraded and open forest, which might have been included under orchards etc. also. Looking at the high demand of fuel wood etc. in the valley concerted efforts are required to meet the demand. Farm forestry is already popular in the valley. Efforts should be made to improve the stocking and area of forests.

HIMACHAL PRADESH

The area under vegetation cover is about 39.3% of geographical area. Good forests i.e. forests with crown density 40% and above cover about 10.1% of geographical area while 29.5% of area is covered either by open forests, thickets or by apple orchards etc. In a hilly state as Himachal Pradesh, this gives a dismal picture and efforts on war footing are required to cover the barren hills with more dense vegetation cover.

UTTAR PRADESH

The study area is confined mainly to eight hill districts only 22.2% of geographical area is under medium to high density forest while about 29.5% of area is either open forest, thickets or orchards etc.

SIKKIM AND WEST BENGAL

In Sikkim and West Bengal most of the area was under clouds and hence the representative picture of vegetation cover was not available.

ARUNACHAL PRADESH

In Arunachal Pradesh about 68.2% of geographical area was under medium to high density forest. The position of vegetation cover comparatively good.

NEPAL

In Nepal about 13,460 sq km area was either in imagery gap or under clouds. Of the remaining area, 15.8% is under medium to high crown density forest while about 17.8% is open forest. The remaining area is non-forested.

BHUTAN

In Bhutan 7138 sq km area was covered with clouds so no break up could be given. Out of the remaining area about 53.7% is under medium to high density forest while about 23% area is under open forest, thickets or under orchards.

With all the limitations of the visual interpretation of satellite imageries on band 5 Black and White it can be said broadly that Eastern Himalayas have a better coverage of vegetation compared to Western Himalayas. In Arunachal Pradesh shifting cultivation is evident and this practice is to be curbed.

In Western Himalayas the condition of vegetation cover is comparatively poor. Effective vegetation cover varies from Jammu & Kashmir at 6.4%, Himachal Pradesh 10.1% and Uttar Pradesh 22.2%. In case of Nepal it is only 15.8%. It is generally recommended that at least 60% of geographical area in hills should be under cover of vegetation. The condition as such is not that hopeful in the Western Himalayas. Two pronged strategy is recommended to stop further deterioration of vegetation cover in Himalayas.

i. There should be restricted felling of timber trees on slope more than 45°

ii. Large scale afforestation works should be taken up on Government and private land.

Table 3.1
Area estimates of Vegetation Types in the Himalayas in different States by Visual Interpretation of Landsat Imagery.

State	Geog. Area in km²	Forested area in km²				Total	Non-forested area in km²				
		F1	F2	F3	F4	Total	N	Cl	W	Gap	Total
J & K	222,236	14,168	12,156	-	-	26,324	190,404	-	3,996	1,512	1,95,912
H. P	55,673	5,632	15,312	-	-	20,944	34,148	-	193	388	34,729
U.P.	51,122	11,365	15,061	-	-	26,426	24,696	-	-	-	24,696
Nepal	141,000	22,340	25,080	-	-	47,420	79,220	2376	900	11084	93,580
Sikkim	7,299	1,164	920	-	-	2,084	148	5067	-	-	5,215
Bhutan	46,600	21,136	9,148	-	-	30,284	9,178	7138	-	-	16,316
W. Bengal (Part)	2,692	772	772	-	-	1,484	36	1172	-	-	1,208
Arunachal	83,578	57,024	11,510	3313	4628	76,475	6,635	-	468	-	7103

Table - 3.2
Actual Forest Cover

(Area in sq. km)

Sl. No.	State/UT	Geog. area	Forest area as official-ally report-ed	Forest area as finally assessed by FSI	Forest area as assessed by NRSA	Finally assessed forest area (Col. 5) as percentage of officially reported area (Col. 4)
1	2	3	4	5	6	7
1.	Andhra Pradesh	276820	63771	50194	40438	78.7
2.	Arunachal Pradesh	83580	51540	60500	521024	117.4
3.	Assam	78520	30708	26386	19790	85.9
4.	Bihar	173880	29230	28748	20139	98.3
5.	Goa (Including Deman & Diu)	3810	1053	1285	1139	122.0
6.	Gujarat	195980	18777	13570	13159	72.3
7.	Haryana	44220	1685	644	401	38.2

Table 3.2—Contd.

1	2	3	4	5	6	7
8.	Himachal Pradesh	55670	21325	12882	9130	60.4
9.	Jammu & Kashmir	222240	20892	20880	14361	99.8
10.	Karnataka	191770	38644	32264	31397	83.5
11.	Kerala	38870	11222	10402	7376	92.7
12.	Madhya Pradesh	442840	155414	127749	90215	82.2
13.	Maharashtra	307760	64055	47416	30350	74.0
14.	Manipur	22360	15155	17679	13572	116.7
15.	Meghalaya	22490	8514	16511	12458	193.9
16.	Mizoram	21090	15935	19092	11971	119.8
17.	Nagaland	16530	8625	14351	8095	166.4
18.	Orissa	155700	59555	53163	39425	89.3
19.	Punjab	50360	2803	766	499	27.3
20.	Rajasthan	342210	31151	12478	5972	40.1
21.	Sikkim	7300	2650	2839	2883	107.1
22.	Tamil Nadu	130070	22319	18380	15884	82.3
23.	Tripura	10480	6280	5743	5138	91.4
24.	Uttar Pradesh	294411	51269	31443	27780	61.3

Table 3.2—Contd.

1	2	3	4	5	6	7
25.	West Bengal	87850	11879	8811	9051	74.2
26.	A & N Islands	8290	7144	7603	7653	106.4
27.	Chandigarh	114	6	2	-	33.3
28.	Dadra & Nagar Haveli	490	203	237	125	116.7
29.	Delhi	1490	42	15	10	35.7
30.	Lakshadweep	30	-	-	-	-
31.	Pondicherry	492	-	8	-	-
Total:		3287797	751896	642041	490515	85.4
Percentage of geog. area.			22.8%	19.52%	14.9%	

A rational landuse planning is called for. Unless it is done immediately the recurring floods cannot be controlled and money spent on engineering works and on other multipurpose River Valley Projects may go waste.

India, Nepal and Bhutan should work together to improve the ecological heritage of Himalayas on which depends the prosperity and well being of entire sub-continent. The Table - 1 gives the summary of visual interpretation. The Tables 3.2, 3.3 and 3.4 (source F.S.I. - State of Environment Report) give the situation in India.

Table - 3.3
Actual Forest Area by Density Classes

(Area in sq. km)

Sl. No.	*State/UT*	*Dense Forest (crown density above 40%)*	*Open Forest (crown density 10-40%)*	*Mangroves*	*Total Tree covered area*
1	*2*	*3*	*4*	*5*	*6*
1.	Andhra Pradesh	28580	21119	495	50194
2.	Arunachal Pradesh	51096	9404	-	60500
3.	Assam	18415	7971	-	26386
4.	Bihar	13490	15258	-	28748
5.	Goa (including Daman & Diu)	763	522	-	1285
6.	Gujarat	7850	5293	427	13570
7.	Haryana	43	601	-	644
8.	Himachal Pradesh	9908	2974	-	12882
9.	Jammu & Kashmir	12978	7902	-	20880
10.	Karnataka	16394	15870	-	32264
11.	Kerala	8569	1833	-	10402
12.	Madhya Pradesh	72174	55575	-	127749
13.	Maharashtra	27244	20032	140	47416
14.	Manipur	4670	13009	-	17679

1	2	3	4	5	6
15.	Meghalaya	5749	10762	-	16511
16.	Mizoram	2938	16154	-	19092
17.	Nagaland	6379	7972	-	14351
18.	Orissa	28573	24391	199	53163
19.	Punjab	96	670	-	766
20.	Rajasthan	3048	9430	-	12478
21.	Sikkim	1867	972	-	2839
22.	Tamil Nadu	10866	7491	23	18380
23.	Tripura	340	5403	-	5743
24.	Uttar Pradesh	18876	12567	-	31443
25.	West Bengal	3512	3223	2076	8811
26.	Andaman & Nicobar Island	6807	110	686	7603
27.	Chandigarh	-	2	-	2
28.	Dadra & Nagar Haveli	187	50	-	237
29.	Delhi	-	15	-	15
30.	Pondicherry	-	8	-	8
31.	Lakshadweep	-	-	-	-
	All India Total	361412	276583	4046	642041

Table- 3.4

Forest Cover by Different Forest Types

(Based on interpretation of Landsat Imageries of 1981-83)

(Rs. in sq. km.)

State/UT	Total	1	2	3	4	5	6	7	8	9	10	11	12	13	14, 15, 16
Andhra Pradesh	50194	-	-	33150	495	11444	4689	416	-	-	-	-	-	-	-
Arunachal Pradesh	60500	13617	-	-	-	-	-	-		18732	-	17492	-	-	10659
Assam	26386	7768	3051	14466	-	-	-	-	1101	-	-	-	-	-	-
Bihar	28748	-	-	3078	-	25670	-	-	-	-	-	-	-	-	-
Gujarat	13570	-	494	2980	427	5727	3942	-	-	-	-	-	-	-	-
Haryana	644	-	-	-		322	322	-	-	-	-	-	-	-	-
Himachal Pradesh	12882	-	-	-	-	2346	43	-	-	3953	512	-	4464	-	1561
Jammu & Kashmir	20880	-	-	-	-	339		-	-	4552	1098	-	11151	312	3428
Karnataka	32264	8439	10730	8020	-	3092	1515	-	-	-	-	468	-	-	-
Kerala	10402	4552	3495	2353	-	-	-	-	-	-	-	-	-	-	-
Madhya Pradesh	127749	-	-	63992	-	63575	182	-	-	-	-	-	-	-	-
Maharashtra	47416	-	5483	19139	140	20149	2193	-	312	-	-	-	-	-	-
Manipur	17679	6146	-	7474	-	-	-	-	-	3591	-	468	-	-	-
Meghalaya	16511	-	-	14860	-	-	-	-	1018	633	-	-	-	-	
Mizoram	19092	-	-	8164	-	-	-	-	-	-	10928	-	-	-	-
Nagaland	14351	5675	1840	848	-	-	-	-	-	2153	-	3079	-	-	756
Orissa	53163	-	243	35603	199	17118	-	-	-	-	-	-	-	-	-
Punjab	766	-	-	-	-	303	383	-	-	-	-	-	-	-	-
Rajasthan	12478	-	-	-	-	11216	1262	-	-	-	-	-	-	-	-
Sikkim	2839	-	-	-	-	-	-	-	-	308	-	937	-	-	1594
Tamil Nadu	18380	1358	224	3022	23	10093	1759	980	-	-	-	921	-	-	-
Tripura	5743	-	-	5743	-	-	-	-	-	-	-	-	-	-	

Table 3.4—Contd.

State/UT	Total	1	2	3	4	5	6	7	8	9	10	11	12	13	14, 15, 16
Uttar Pradesh	31443	-	-	7048	-	8715	201	-	-	8455	-	..	6397	-	627
West Bengal	8811	-	-	2971	2076	3414	-	-	350	-	-	-	-	-	-
Goa (including Daman & Diu	1285	234	53	998	-	-	-	-	-	-	-	-	-	-	-
Union Territories															
Andaman & Nicobar Islands.	7603	3458	811	2648	686	-	-	-	-	-	-	-	-	-	-
Chandigarh		2	-	-	-	-	2	-	-	-	-	-	-	-	-
Dadar & Nagar Haveli.		237	-	-	237	-	-	-	-	-	-	-	-	-	-
Delhi		15	-	-	-	-	15	-	-	-	-	-	-	-	-
Lakshdweep		-	-	-	-	-	-	-	-	-	-	-	-	-	-
Pondicherry		8	-	-	-	-	-	-	8	-	-	-	-	-	-
Total:	642041	51249	26424	236794	4046	183620	16491	1404	2781	42377	12538	23365	22012	312	18628
% of total forest area	100	8.0	4.1	37.0	0.6	79.6	2.6	0.2	0.4	6.6	2.0	3.6	3.4	-	2.9

KEY

1 TROPICAL WET EVERGREEN FOREST
2 TROPICAL SEMI EVERGREEN FOREST
3 TROPICAL MOIST DECIDUOUS FOREST
4 LITTORAL AND SWAMP FOREST
5 TROPICAL DRY DECIDUOUS FOREST
6 TROPICAL THORN FOREST
7 TROPICAL DRY EVERGREEN FOREST
8 SUB TROPICAL BROAD LEAVED HILL FOREST
9 SUB TROPICAL PINE FOREST
10 SUB TROPICAL DRY EVERGREEN FOREST
11 MONTANE WET TEMPERATE FOREST
12 HIMALAYAN MOIST TEMPERATE FOREST
13 HIMALAYAN DRY TEMPERATE FOREST
14, 15, 16 SUB ALPINE AND ALPINE FOREST

BIBLIOGRAPHY

1. Champion, H.G. and Seth. S.K. (1968) - Revised Forest Types of India.
2. Forest Survey of India (1983) Forest cover of India.
3. Forest Survey of India (1988) State of Forests in India.
4. Resources Development and Environment in the Himalayan Region.(1978) - National Committee on Environmental Planning and Coordination.
5. Reports on forest Resources compiled .... On Chenub Valley, Alkananda Valley by reinvestment Survey of Forest Resources Rohru, Chopal and Simla region and other reports.

PART II

HIMALAYAN BIOGEOGRAPHY AND CONSERVATION OF RESOURCES

Contents

IN PRAISE OF ARUNACHAL PRADESH

A Song of Sun

From dark silent
Forests
And the humid heat
of foot hills
where big game dwells under
Canopy of forest home
And the gliding hornbills
Swish the sky
Sky that startles
with shrieks
of bird of prey
Then and now
To the peaks
on the watersheds over the Snows, and
Then the gentle slopes,
The flowery heaven

Of pink yellow
Blue white, purple
Crimson and other hues
As far as eyes can see
Bulbous Shrubs
Give out flower
And then leaves
when snow melts

When snow melts,
Crystal clear, blue
Streams rush
Floating ice blocks
(of continuing thaw
reaching above and above-)
Down to valleys, where
musk deer and mountain goats
feed on
Velvety moss,
Live with Stunted pines
and Rhododendrons.
Lower
They pass
Taller trees
of bamboos, of pines and of
rhododendrons
Lower
They cut
Tropical green
Deep
And join
To foam and to fret
In big waters
Oh! the varied faces
of breath taking beauty
and splendour
From
Steep multicoloured rocks.

(Source : Govt. of Arunachal Pradesh)

Preface

India's complex environment and great biological diversity have guided distribution pattern of individual plant species and communities and also of various animal groups-reptiles, amphibia, birds and mammals. It has, therefore, been always difficult to have an acceptable biogeographic zonation. There are divergent views on the subject. Keeping these factors in view, one has to strive to achieve the objectives and values of conservation and protected areas under discussion in the present study which are :-

i) To enumerate floral and faunal status and safeguard representatives and viable samples of all major biological communities within the country.

ii) To provide long-term and in-situ conservation of genetic and specific diversity.

iii) To promote and facilitate scientific research and recreational benefits.

iv) To provide ecological reference centres so as to monitor environmental changes due to natural and man-made disturbances.

Demarcation of areas for protection of gene pools are guided by endemicity of floral and faunal species and concentration of several species within a small area. Such areas are sites for rapid speciation. These aspects have been discussed with particular stress. Besides the existing ones, new conservation areas have been suggested keeping plant and animal communities in view. Except for special sites like Andaman and Nicobar Islands (Island ecosystem) where isolation of species stand supreme than that of community concept and individual and endemic species the present study area offers one of the best site for biogeographic investigation.

By setting up of more than 60 National Parks, 300 Wildlife Sanctuaries besides many zoological parks and botanical gardens, India has initiated a massive programme to protect its floral and faunal resources. Owing to heavy pressure on land by a huge growing population, India cannot have too extensive protective areas even in respect of animals of low productivity. Moreover, large animals have complex habitat requirements and it is difficult to meet with their requirement.

The migratory and dispersal routes ("Geographical stepping stones") or 'Corridor' concept is an essential item of consideration in the study of flora and fauna of the Himalayas and adjoining plains.

In the present study no analysis or research on Biogeographic classification and other relevant issues has been brought to the sphere of discussion. For identifying protected areas in different biogeographic zones, a classification pattern which has three basic components that corresponds in the broad sense to Udvardy's classification of protected (Map-3) areas have been adopted (Biogeographic zone, Biotic province and Biome).

The auother has gone little further to present various concepts of classification, their merits and demerits and various floral and faunal affinities that the study area has. The study area is so rich in floral and faunal resources and considering the geology migration of species over long long years and complexity of the situation due to confluence of various floral and faunal elements entire subject has only been discussed in a nutshell.

4

Arunachal Pradesh

Arunachal Pradesh is "a land of mighty rocks and luxuriant forests, gentle streams and raging torrents; it presents a breath-taking spectacle of Nature in her glory, beauty, unspoilt and untamed in a profusion of flora and fauna, customs, language and dress." It extends from 27°N to 29°N and 92°E To 97°E latitudes and longitudes respectively.

Arunachal Pradesh-the land of primeval forests and dawn-lit mountains-is a sprawling mountaneous territory spread over 83,743 sq. km. and is habited by more than 6.3 lakh people (1981 Census). Bounded by Bhutan to the west, China to the north and north-east, Burma to the east and the plains of Assam to the south. It is the home of more than twenty major tribes and considered to be one of the most splendid, variegated and multilingual tribal areas of the world.

It is the land of the rising sun that is blessed with rich heritage of forest resources having multitudinal biotopes-flora and fauna-and with wide variety of genetic resources. Here, morning comes with sweet rhythm of birds vibrating through the greeny land stretching to infinity and the forests echo with call of the wild. It is a wonderland of spectacular landscapes and a home of a rich flora and fauna which is paradise for the naturalists, treasure for ecologist, a heaven for tourist and a rich laboratory to the scientists.

The terrain is extremely difficult and cut up by countless streams, the mightest of them being Siang. These turbulent streams and rivers which become raging torrents in the rain.

The weather is capricious. Premonsoon rains occur in March and monsoon losts from May to the end of December. Snowfall occur even above 1500 m. There are many drier hill slopes in Kameng and Tawang districts. The rainfall varies from 1000 mm to 5750 mm. The altitude varies from 100 m to 7000 m.

Significant temperate forests occur surrounding Perila, Singe-Bzond, Thong and Tawang in Kameng and Tawang districts, hills above Apatani valley in Lower Subansiri valley districts, higher slopes of Bori above inner valley of Siang district. Alpine forests occur in Sela area, Jhang valley and above Tawang and Mego in Kameng district, some forests in Tale valley in Subansiri district, above Walong in Lohit district.

The geology is fragile and may be perused from the following :-

In Kameng, Subansiri and Siang districts (old) :-

1. Mid Miocene	-	Siwalik belt (Sand stone & Clay)
2. Permian	-	Gondwana belt (hard sand stone, slate and shale)
3. Pre Permian	-	Bicham group (Mini quartzite, Carbonates)
4. Early Palaeozoic	-	Bomdila group (Greiss & Schist)
5. Pre Cambrian with younger granite	-	Sela group (Granite & Schist)

In Lohit district :-

1. Lower Palaeozoic	-	Crystalline limestone, Schist, Schistose meta sediments, Serpentirites
2. Pre Cambrian	-	Para metamorphites
3. Low Palaeozoic	-	Grandiarile
4. Pre Cambrian	-	Diorite Complex

In Tirap district :-

1. Alluvium		
2. Dehing Series. Tertiary	-	Barail Series, Dasang Series, Para metamorphiles.

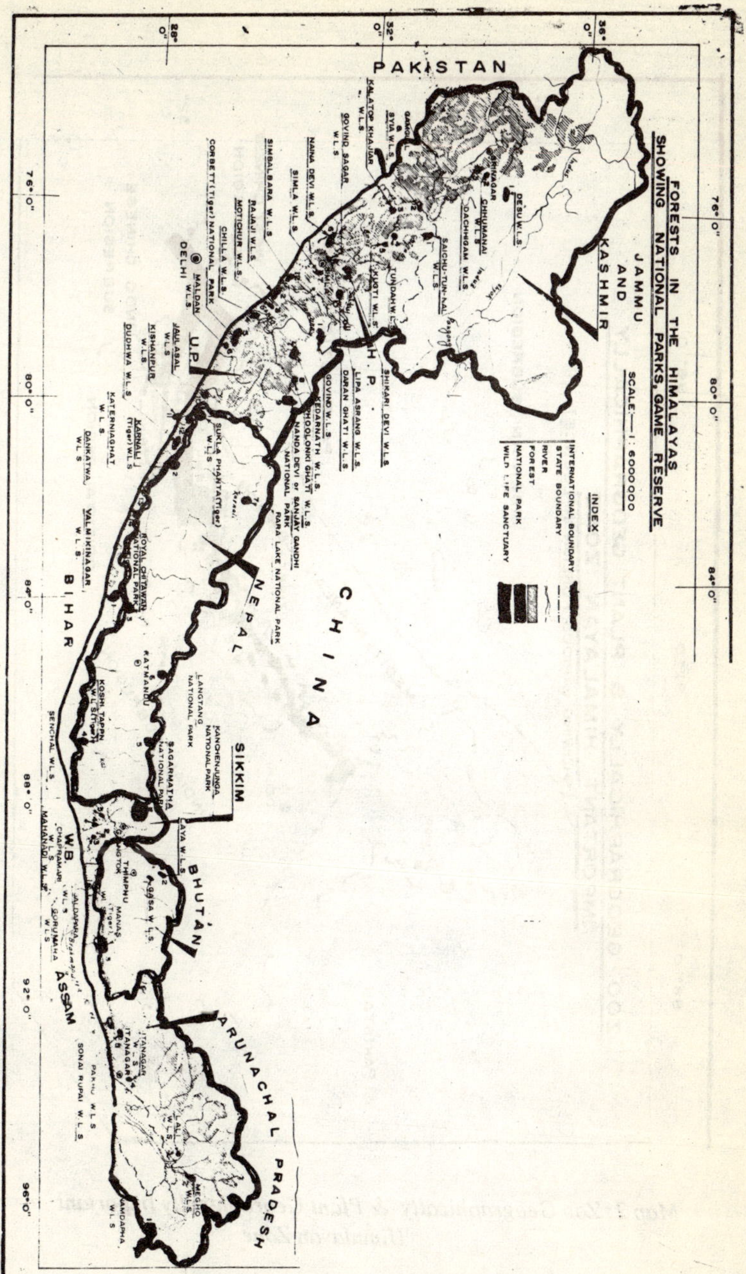

Map 1: Forests in the Himalayas Showing National Parks, Game Reserve Jammu and Kashmir

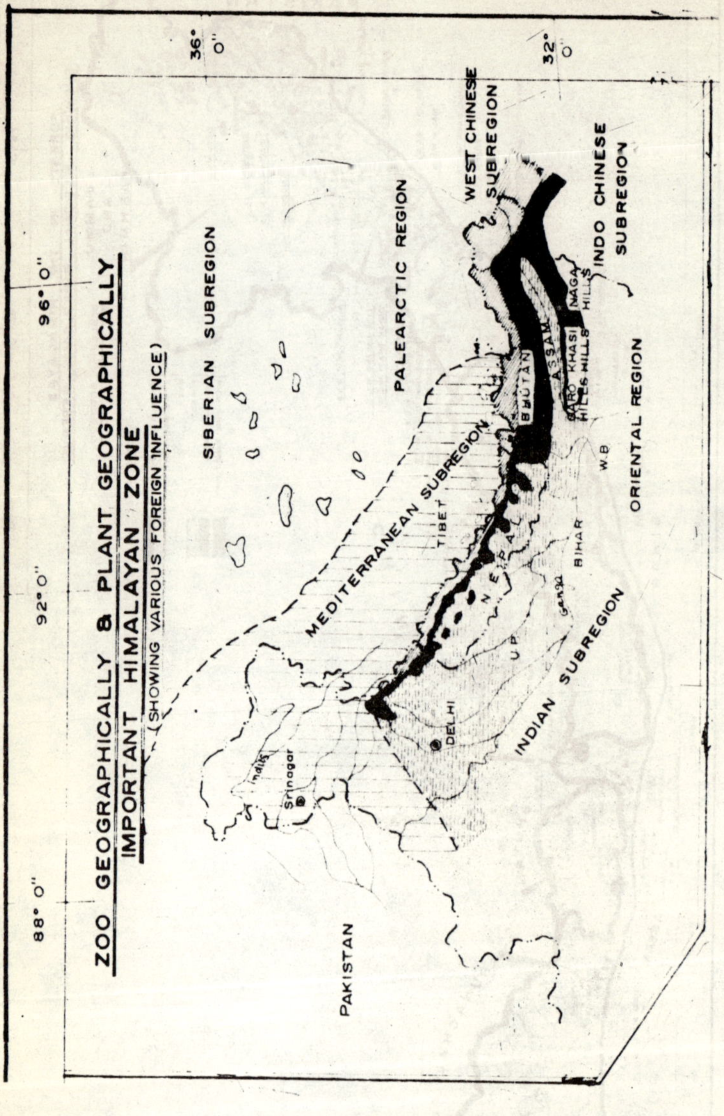

Map 2: Zoo Geographically & Plant Georaphically Important Himalayan Zone

5

Quality of Arunachal Pradesh as a Rich and Unique

(Biogeographic and Conservation Zone)

The following special features will testify to the richness, representativeness, diversity, naturalness and effectiveness of the area as complicated and interested biogeographic and unique conservation zone.

* The State including part of adjacent Himalayas from one of the richest biogeographic region of the Indian sub-continent.
* The region includes some unspoilt natural vegetation.
* This region forms Biogeographers most interesting and complicated subject in the study on endemism, migration, zonation and adaptability of species.
* The region represents some unique floral and faunal species of tropical, sub-tropical and alpine forests, has wide variety of biota of various altitudes, has eco-system diversity, naturalness and effectiveness as a conservation unit. The vast snowclad areas and glaciers offer shelter to numerous species of insect and some animal species.
* The region has many rare and endangered species.
* The region is one of the richest avifaunal area of the world where more than 500 species (approx.) covering wide range of group thrive within a small space. A large member of birds are local migrants, others are passage migrants.
* The insect population is extremely rich and greatly diversified. It has vast population of Coleoptera, Diptera, Isoptera, Lepidoptera, Hymenoptera orders of insects. The fauna in general exhibits affinities with Burma, Thailand, South China and Japan.
* The fish fauna is zoo-geographically rich and connects Indo-China, Indo-Malayan and Indian sub-region of oriental region.
* The region is very rich in orchid, rhododendron, lichen, moss, liverwort and alpine flora and fauna.
* The flora primarily exhibits Sino-Japanese, Indo-Chinese and Malayan elements. Tibetan, Central Asian and some European

elements are also evident at high altitude flora.

* The flora exhibits a high degree of endemism.
* The region has some primitive plants and animals which are still in evolutionary stage.
* The region is still yielding more Taxa showing that the flora has not been explored completely.
* The region has several species of mammals including some rare ones which show tridimentional zonations. This region lies at the confluence of Ethiopian, Palaerctic, Mediterranean, Indo-China and Indo-Malayan flora and fauna and possess a number of interesting components.
* The region has more than 70 species of snakes and more than 200 species of amphibians.
* The region is expected to have more than 500 species of Orchids, 300 species of Ferns besides other plants which are rare and endangered.
* The area is connected with a complex mountain chain rising to tremendous heights interspersed with deep transverse river gorges and wide longitudinal valleys.
* The region is extremely rich in medicinal plants and also rich ethnobotanically.
* Owing to various topographical, altitudinal and ecological gradients the region has been the theatre of evolutionary development of several angiosperms.
* This region has wild relatives of cultivated plants.
* This region has immense genetic resources for agrihorticultural research and development.
* Richness of timber plants and profusion of flowering plants of assorted colour are treasure of the region.
* Array of colours of more than 57 species of Rhododendron flora which offer a Kaleidoscopic view are special features of scenic beauty; so are the dwarf and stunded growth of plants which are features of high altitude plants.
* Epiphytic flora especially the ferns and lichens are natures masterpieces of creation in the locality. Physiological, morphological and biological importance of epiphytes are also the feature of the locality.
* Diversity of species in both vertical and lateral planes, variety of habitats, high density of species are also special features.
* The area commands a panoramic view of series of mountain ranges culminating in some of very high peaks.
* High altitude insect population and their adaptations are also special features of the locality.

* Alpine pastures with grass-sedge association, bright colouration of flowers, richness of variety of herbaceous flora are objects of scientists study and beauty.
* Other special features of the zone are the juxtaposition of many biotopes or life zones from temperate cold to alpine snow-contiguity of forests from plains to 7000 m. and above has made the tract floristically and faunistically a rare gift of nature.
* The floral diversity parallel the mammalian and other micro and macro-faunal diversity. Phenophase, phenodynamic and phenometry testify to the diversity and richness of the flora.
* Plant-animal interaction (intimate and reciprocal relation) is widely evident in the zone.
* The rich varieties of flora and fauna and their inter-relationship provide inexhaustible opportunity for study and research; the forests are nature's richest laboratories.
* This region forms the gateway of migration of flora and fauna from the east.

All these qualities enumerated characterise and quantify the area for a comprehensive biogeographic study.

6

Forest Cover

(From Satellite imagery interpretation)

The author as Director, Forest Survey of India, a Department under the then Ministry of Agriculture and Co-operation, made an analysis (visual) of Landsat images on band 5 black and white of 1980-82 and the result was published in December, 1982. Prior that, the then Preinvestment Survey of Forest resources in the same Ministry carried out similar studies for northeast India and Orissa using band 5 and 7 black and white and false colour composites on band 4, 5 and 7 for the year 1975-76 and imageries in band 5 (black and white) of the year 1980-82 (on 1:1 million scale). (Refer maps 4, 5, & 6)

The National Remote Sensing Agency (N.R.S.A.), an organisation under the Department of Space, made similar studies for the years 1972-75 and 1980-82 using false colour composite diapositives on 1:1 million scale. The report was published in December, 1983.

The Forest Survey of India (F.S.I.) found that the forests of Assam hill, Manipur, Nagaland, Tripura, Meghalaya, Mizoram and Arunachal Province have suffered a denudation (recession) of 10,770 sq. km. in 5 years (i.e., from 1,57,040 sq. km. in 1975-76 to 1,48,270 sq. km. in 1980-82); the N.R.S.A. found that in 7 years (i.e., between 1972-75 and 1980-82) denudation was 7132 sq. km. (1,30,310 sq. km. in 1972-75 to 1,23,179 sq. km. in 1980-82).

A comparative study on the State of Arunachal may be seen as under :-

(Area in sq. km.)

Description	*Finding of Forest Survey of India from 1980-82 Imageries*	*Findings of National Remote Sensing Agency for*	
		1972-75	*1980-82 imageries*
1	2	3	4
1. Closed forest	36,880	50,460	49,486

1	2	3	4
2. Closed forest with shifting cultivation	10,694	-	-
3. Open forest	19,543	978	2,618
4. Open forest with shifting cultivation	1,585	-	-
5. Non-Forest	11,884	19,628	9,681
6. Gap	2,994	-	-
	68,702	51,438	52,104

There seems to be little difference in the observations. The definition used by F.S.I. of open forest and open forest with shifting cultivation made the areas under items 3 and 4 much higher than those of N.R.S.A. who, it seems, considered some such forests as closed forests. So there are same discrepancies in area assessment.

The vegetation of Arunachal, therefore, covers more than 60% of the geographical area of the State. There is some fallacy in the interpretation of N.R.S.A. who found an increase in forest cover in 7 years. This cannot be accepted.

The findings of F.S.I. and N.R.S.A. put the forest cover of India to 4,73,000 sq. km. (not 7,48,000 sq. km. or 22.8% of geographical area under Forests) which is about 14% of the geographical area of the country. But these observations have been reconciled by both the organizations to arrive at a common finding us dealt in Part I.

Land under shifting cultivation

F.A.O./U.N.F.P.A. Project (RAS/77/POQ) reported (1980) that 36 blocks were affected by shifting cultivation mostly in the middle hills. High altitude dwellers like Monpas of Tawong, Memba and Khamba of Sabarsiri and Siang do not practice shifting cultivation. Khamptis and Singphos of Tirap do not practice shifting cultivation. Apatanis do not practice shifting cultivation. Adis are the largest community among shifting cultivators. Total area under shifting cultivation in Arunachal Pradesh is 703 sq. km. (practiced by 2,70,000 persons).

This practice in Manipur, Nagaland, Mizoram, Meghalaya and to some extent in Tripura has been responsible for complete eradication of vegetation over vast areas. The character of vegetation has completely changed and entire animal community disturbed and probably eradicated from their original home.

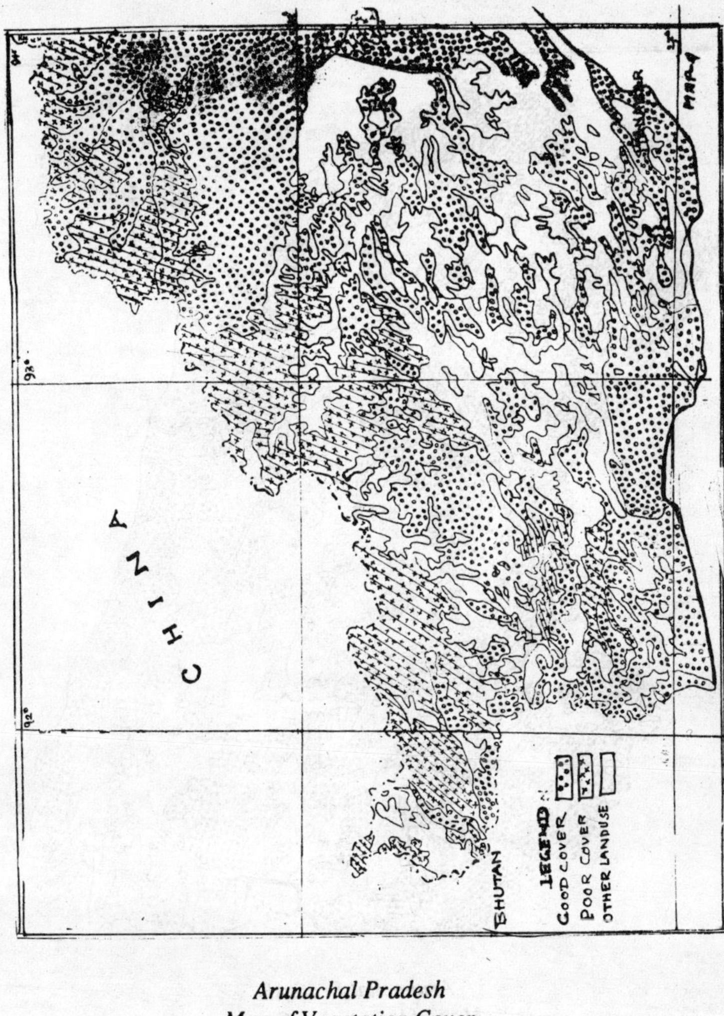

Arunachal Pradesh
Map of Vegetation Cover

Arunachal Pradesh
Map of Vegetation Cover

Arunachal Pradesh
Map of Vegetation Cover

7

BIOGEOGRAPHY

Various Biogeographers have defined biogeography in their own language. Pierre Banserean defines "Biogeography covers origin, distribution, adoption and association of plant and animals, It never the less deals with interaction between plants and animals, .heir evolution in changing environment and also involves recent and ancient geological history of the area." Others think that man's impact on the landscape, landuse pattern, urbanization, Industry, browsing and grazing pattern, fire cycle, lumbering auteecology and population structure, structure of vegetation, dynamics of vegetation, analytical phytosociology also cover the definition." Udvardy M.D. puts, ethology, ecology, animal ecology, systematics, synecology, auteecology, pedology, evolution, geology, phytogeography and even embryology and areography under the definitions of zoogeography.

Biogeographical Background of Arunachal Pradesh

Ecologically and biogeograhically the Indian sub-continent is one of the most fascinating region of the World. A great variety of ecological conditions prevail and a great diversity of wildlife exists.

Eastern Himalayas from Sikkim to Arunachal Pradesh lie close to the confluence of Ethiopian, Palaearctic, Mediterranean and Indo-Malayan Flora and fauna and possesses a number of interesting components from each of these realms as well as several peculiarly indigenous genera. The Ethiopian element is represented by the Chinkara, the Hyena and the Ratel the palaearctic by Lynx, Wolf, Markhor, Urial and Hangal, the Chinese by Red Panda and Musk deer and the Indo-Malayan by the Hoolock Gibbon, the Elephant, the Goat antilope and the Mouse deer. (Diagram 1 and Part I)

Wild mammalian fauna of the Himalayas shows tridimensional zonations viz., longitudinal and altitudinal. In each zone, a number of sub-zones can be recongnised.

The eastern Himalayan Takin, Tahr, Flying squirrel, Limbless lizard *(Ophisauras)*, the Skink *(Leiolopism)*, the toad lizard *(Phrynocephalus)* are all rare animals today. Affinities with Burma, Thailand and South China are predominant.

Abundance of fruits, assorted colourful flowers, abundance of nectars and pollens, abundance of insects, horizontal and vertical contiguity of forest corridor are essential for sustenance and survival of avifauna and several other animals. Accumulated unburnt leaf litter on the forest floor is essential for pheasant, grouse, partridge, fowl and ground dwelling birds. The Arunachal forests provide most of these environmental conditions.

Zoogeographical Classification

P.L. Sclater (1858) classified zoogeography based on bird distribution. But later Illiger (1815), Swainsen (1835) made attempts to catalogue faunas. Woodward (1815) used the Phylum moluca and based 27 zoogeographical regions on them. Schmarda (1853) characterised his 21 land 'realms' by various leading species or higher Taxa of mammals, birds and insects largely corresponding to the sub-regions of Wallace (1876). Sclater's zoogeographical system was widely recognised but not influenced by the Darwin-Wallace theory. However, detailed discussion on this system is not being done here. In 1876, Alfred Russel Wallace published 'The Geographical Distribution of Animals' in which he supported Sclater's regional system as preferable as to that of Allen, a view with which most of his contemporaries agreed. Wallace emphasised historic and evolutionary factors and also climates, vegetation, geological history etc. into consideration.

Wallace in 1876 divided oriental region into Indian sub-region, Himalayan or Indo-Chinese sub-region and Indo-Malayan sub-region; the Arunachal region falls under the region.

In 1888, Blanford proposed the divisions-Tibetan, Himalayan, Indian, Malabar or Ceylonese, Burmese and South Tenasserim. Sharp, however, proposed divisions of India into Indian peninsular sub-region, Indo-Malayan sub-region, Indo-Chinese sub-region, Himalo-Malayan sub-region and Himalo-Chinese sub-region.

Talbat (1939) recognised eight major divisions of India (on the basis of distribution of Lepidoptera mainly Rhopalocera) viz., Ceylon, Peninsular India, Northwest India, West Himalayas, Northeast India, Burma, Andaman & Nicobar Islands. Talbot's Northeast India shows affinity to the Palaearctic and partly to the Oriental, with infiltration from the South.

Alcock (1910) recognised six territories in the distributional pattern of fresh water Crustacea, viz. the western territory, the western himalayan territory, the northeastern frontier or territories, Burma-Malay territory, peninsular territory and Indo-gangetic plain territory.

Annandale (1911), based his division on the distribution of fresh water sponges Polyzoa (this confirmed Alcock's division).

Christopher (1933) made his division on the distribution pattern of

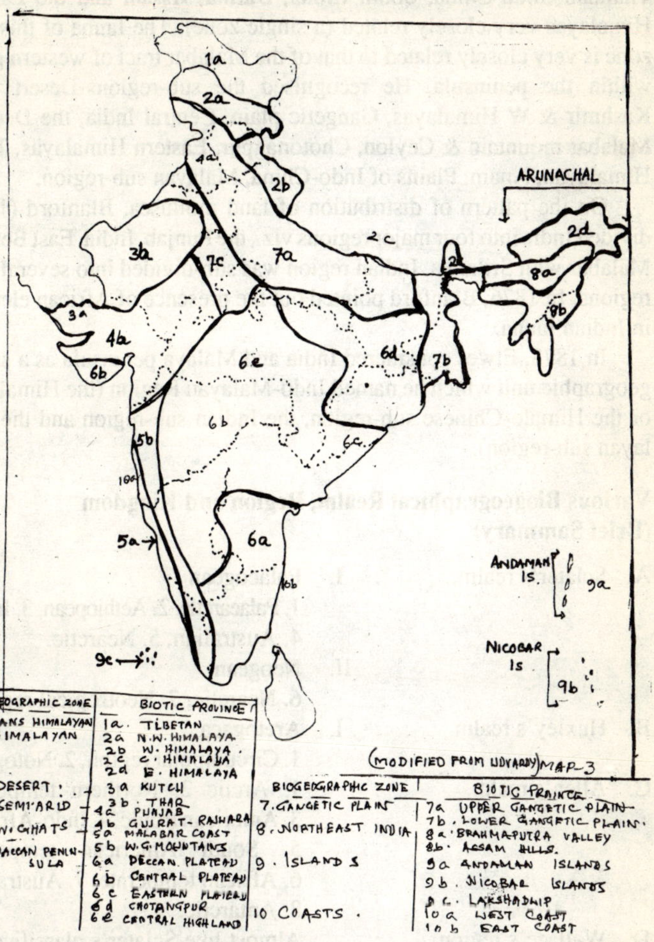

Map : Biogeographical Classification

Culicidae viz., Trano-Indus area, the Indo-gangetic area, the peninsural area, Malabar-Ceylon area, Assam-Burma area and Himalayan area.

Malcolm A. Smith (1931, 1935, 1941) recognised that the reptiles of Thailand, Indo-China, South China, Burma, Assam and the Eastern Himalayas very closely related (a single zone). The fauna of this vast zone is very closely related to that of the Malabar tract of western ghats within the peninsula. He recognised the sub-regions-Desert area, Kashmir & W Himalayas, Gangetic plain, Central India, the Deccan, Malabar mountain & Ceylon, Chotonagpur, Eastern Himalayas, Trans Himalayan, Anam, Plains of Indo-China, Malayan sub-region.

On the pattern of distribution of land mollusca, Blanford (1870) divided India into four major regions viz., the Punjab, India, East Bengal, Malabar with Srilanka. Indian region was sub-divided into several sub-regions. In 1876, Blanford pointed out the presence of African element in Indian fauna.

In 1873, Elwes considered India and Malaya peninsula as a single geographic unit which he named Indo-Malayan Region (the Himalayan or the Himalo-Chinese sub-region, the Indian sub-region and the Malayan sub-region).

Various Biogeographical Realm, Region and Kingdom (Brief Sammary)

A.	Sclater's realm.	:	I.	Palaeogean 1. Palaearctic, 2. Aethiopean, 3. Indian 4. Australian, 5. Nearctic.
			II.	Neogean 6. Nearctic, 7. Neotropical regions.
B.	Huxley's realm.	:	I.	Aretogaea 1. Circumpolar region, 2. Notogaea.
C.	Allen's realm.	:		1. Arctic 2. Northern temperate, 3. American tropic, 4. Indo-African, 5. South American temperate, 6. African temperate, 7. Australian, 8. Antarctic.
D.	Wallace's region	:		Almost like Sclater's classification except 3 which is Oriental.
E.	Engler's kingdom	:		1. Northern extra tropical, 2. Paleo Tropical, 3. South America, 4. Old oceanic.
F.	Blandford's realm	:		1. Aretogaea, 2. Notogaea, 3. Neogaea.

There are other conceptions. Mention must be made of SPATE, O.H.K. who in his work "India and Pakistan:- "A general and regional

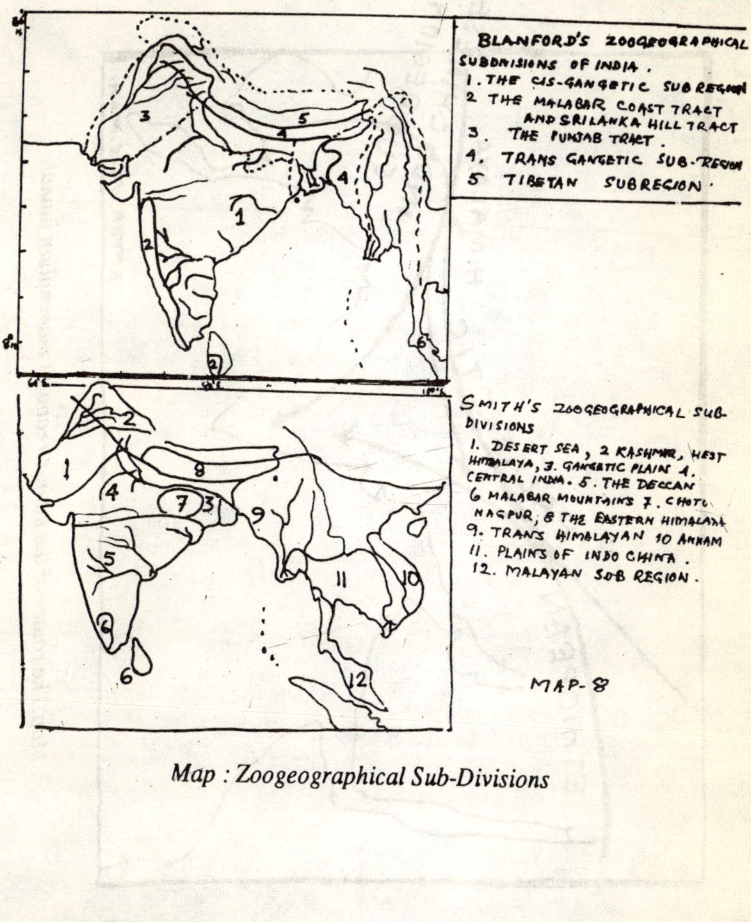

Map : Zoogeographical Sub-Divisions

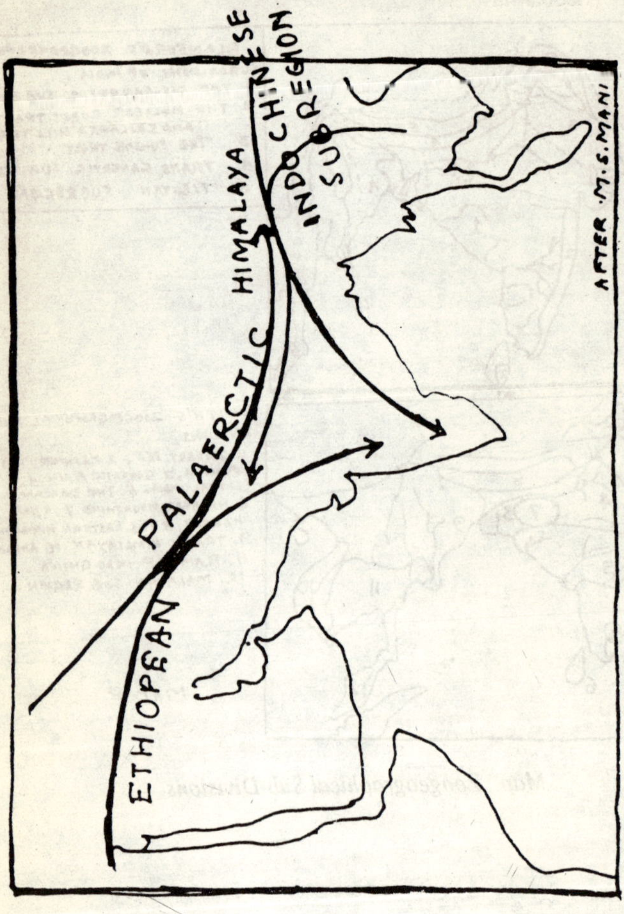

Map:The routes of the main-streams of mammalian faunal

geography" 1957 has summarized various classifications and put India under four broad groups viz., The mountain rim, The Indo-gangetic plains, The Peninsula, The Islands. He puts Arunachal Pradesh under 'The Mountain rim', Eastern Himalayas tagging Bhutan with it; Andaman & Nicobar has been put under the group 'The Islands'.

Himalayan Uplift and the Biogeographical Dynamics

The Himalayan uplift and the biogeographical composition and distribution patterns are interlinked. Original Gondwana flora and fauna in both composition and general character were influenced and modified by the uplift of the Himalayas. It was primarily through the formation of biogeographical routes of interchange and barriers with Eurasia, and to a limited extent with parts of Africa. The Himalayan uplift established a direct connection between the ancient Peninsular Block of India and Asia in the north-east, whereas, the Peninsular block was rammed against the young Tethyan sediments of Asia forcing these sediments to flow around in the form of the Himalayan and Burmese Arcs. (Refer Maps 7, 8 & 13)

With the disappearance of the Tethys sea, the north-east and the older peninsular block came into contact with the young rock giving rise to the north Burma-National Indian (obviously also Arunachal Pradesh) region. This region is the confluence of flora and fauna of the south-east Asia which migrated west and south-west India. This also helped migration of South and South-east Chinese, Indo-Chinese and Malayan elements in India. Arunachal Pradesh, Nagaland and Manipur regions may be called floristic and faunistic gateway of India. The area holds a pivotal place in the historic process of progressive evolution of the present day flora and fauna.

The faunal and floral movement in the Himalayas forest had been from the east to the west. In Peninsular region movements had been from the east and also from the west. The north-western route opened only where the entire Tethys had been completely obliterated; then migration took place from Europe, Africa and other places.

This Arunachal-Assam-Nagaland region is the western-most boundary of the range of many Indo-Chinese mammals, like certain squirrel and the eastern-most limit of the distribution of many peninsular species such as the spotted deer.

The mammals of the study area including Assam, Nagaland and other states are primarily Indo-Chinese, though there is admixture of the Peninsular and Ethiopean elements; the palaearctic montane elements also intrude in it. Several relict species of peninsular India (in western ghats) are also distributed in the project area.

Lacuna in Biogeographical Classification

In Biogeographical Divisions, the Zoologists and the Botanists have not jointly worked to collate various theories. Even the Zoologists have postulated their own theories (Malaclost, Herpetologist, Dipterist, Lepidopterodogist, Ichthyologist, Ornithologist and others). The proposed theories of phytogeographical (based on vegetation and climatic relationship) and zoogeographical divisions of India depict only partial picture of the present day distributional patterns of isolated and often taxonomically and ecologically highly specialised and frequently geographically greatly localised groups of flora and fauna, with over-emphasis on the climatic condition and geographical history. So the biogeographical sub-divisions do not have any relationship to the centres of differentiation, radiation, evolution and affinity. Some of these are artificial, meterological or actually monsoon rainfall divisions rather than biogeographical. Even the close relationship between biogeographical and geomorphological evolution of a region has not been taken into account.

Vegetation structure is changing and so the faunal structure and composition; they interact with each other and are indivisible components of nature and constitute an indivisible whole. So it may be summed up—"The ecology and biogeography of India are thus essentially part of a complex mosaic and the events in India are an integral phase of the larger biogeographical and geomorphological evolutions and involves, therefore, a number of problems of great fundamental importance, based on ideas of Continental Drift which climaxed in the uplift of the Himalayas and cannot be studied in isolation."

Statistical Methods in Regional Zoogeography

A statistic treatment of the Sclater Wallacean zoogeographical region largely verified their strength as well as their weaknesses. Anton Handlirsch, the indigenous Viennese palaeontologist (1913) catalogues over 16,000 insect genera of the world and established the degree of insect endemicity for each of the regions.

Regions	*Number of genera*	*Endemic genera*	*Endemic/Non-endemic ratios*
1. Neotropical	5617	3437	1:0.63
2. Nearctic	3467	797	1:3.35
3. Palaearctic	4956	1859	1:1.67
4. Aethiopean	3968	2249	1:0.76
5. Oriental	4137	1641	1:1.52
6. Australian	3101	1400	1:1.21

The degree of endemicity expresses the distinctiveness of the region; the Neotropical, Aethiopean and Australian regions lead in this respect. From the treatment of genera common to several regions, Handlirsch conducted that the strongest ties are between the Palaearctic and Nearctic, the Nearctic and Neotropical and less strongly between the Palaearctic and Oriental and between the Oriental and Australian regions. These rations serve to confirm the validity of Wallace's region, as qualified by later amendments; an exception is the strong affinity between the Nearctic and the Neotropical regions shown by Handlirsch's data. This can be explained by the fact that the insects are older and radiated earlier, than mammals.

8

Phytogeography

The flora of Northeastern India (Arunachal Pradesh obviously) may be ranked as richest in India, and probably in the whole of Asia. This region abounds in forests, meadows, marshes and swamps each with its characteristic plants and special ecologies and also forms a rich reservoir of genetic variability of great agricultural value. The presence of a large number of primitive plants makes this region the "Cradle of flowering plants".

This region is natural home for many species of Orchids and Rhododendrons. The forests are economically highly valued. Bamboos grow in abundance.

The northeastern region is a region of Tertiary mountain, is characterised by humid tropical climate and remarkable for the wealth and diversity of vegetation. Phytogeographically this region is a highly transitional zone with intermingling of species of very many regions.

Griffith (1847), Clarke (1889), Bor (1938, 1942), Biswas (1941, 1943), Kingdonward (1960), Das (1942), Burkill (1925) have described various aspects of flora of the northeastern region. Biswas, Bor, Panigarhi and Joseph, Roa and Rabha have worked on the vegetation of Arunachal Pradesh. Yet the exploration of this vast forested territory has not been done over even one half of the area. This area is most humid and extends to tropical, temperate and alpine regions. Monsoonic rainfall is a distinct feature. Conifer occurs over comparatively limited areas; Oaks have 20 species, Rhododendron 57, Orchids about 600 and Bamboos over 40.

Most useful observation has been made by H.T. Burkill in his article, "The Botany of the Abor Expedition" published in `Records of Botanical Survey of India', 1924. His observations are imperative in the study of endemism, richness of flora, migration of species and their distribution. Some salient points Burkill's observations are summarised below:-

The flora of Aborlands has 635 species common with Sikkim, 631 with Khasi and Jaintia hills, 540 species with Northern Burma, 340 species with Shan hills and North Thailand, 298 with Java, 272 species with Southeastern China, 257 species with Bhutan, 220 species with Sri Lanka, 190 species with Borneo, 154 species with Eastern Malaya, 108 species with Australia, 103 species with Africa (tropical), 84 species with

Japan, 83 species with Pacific Island, 52 with American Continent, 30 species with Eurupe and Mediterranean.

The Aborland flora has strong Chinese affinities (529 species). Burkill expected more affinities, but further study was necessary. The Malaya peninsula from Java to New Guinea has 366 species of Aborland plants, 108 species occur in Australia and 83 species in Pacific Island. Even species are common with Deccean, Chotonagpur, Malabar and Sri Lanka which are mostly herbs.

There are several theories regarding this 'Discontinuous distribution' of forest species in India, Arunachal and elsewhere which are:-

a) *Himalayan glaciation theory*-Due to glaciation in Pleistocene many species migrated.
b) *Southern route across Indian Ocean theory*-This theory postulates land connection with Australia to South India through the Assam-Arunachal hills.
c) *Deccan Trap theory* - This theory postulates land connection with Malaya peninsula.
d) *Satpura hypthesis* - This envisages that Satpura-Vindyan ranges had an altitude of 1650-1950 m. during pleistocene forming a continuous range with the Himalayas.

Burkill's Views on 'Discontinuous Distribution' of Plants in India

He observed poverty of rain forest endemic species in the Himalayas which is due to climate.

Burkill's views have cleared the phytogeographical problems to two different types of floras. The Malayan element in the hills of South India is much older and had migrated through the Bay of Bengal of Sri Lanka and Deccan route. The Himalayan element is glacial in age and its spread to South-Indian hills was brought by glaciation. This very phenomenon destroyed the rain forests Malayan element from the Himalayan ranges to the west and further checked its westward migration. The glacial theory of distribution of plants to South-Indian hills thus gains further ground in the light of Burkill's studies.

Burkill (1924) worked on Aborhill flora and established a distinct relationship to that of the hills of Southeastern Assam, Khasi hills and with Sikkim. He also found endemic species at all altitudes from 330 m. to 1950 m. (96 species). The lower altitude having more of endemic flora. Aborhill flora has many species common to Africa, Malayasia, America and Europe.

The Himalayan region has the richest endemic flora of the country. The upheavals of the Himalayas in geological areas resulted in great changes in topography climate etc. of these mountains and the consequent separation of floras. The climatic fluctuations resulted in large scale migration and exchange of floristic element leading to the congregation of diverse elements from all parts of the world. This is responsible for the present day endemics and spaciation in the Himalayas, especially at higher altitudes.

Akahills

N.L. Bor in his exploration of Akahill vegetation recorded preponderance of Lauraceous species. Predominent formation is of *Phoebe-Beilschmiedia-Engelhardtia*. In the outer slope of Piri, this formation is in a heavy rainfall pocket. Most of the associated species are similar to Sikkim and Darjeeling Himalayas. Species in Dinup Ko valley and Pankim La valley near Shergya also are similar to Sikkim and Darjeeling Himalayas.

The area is extensive (83,580 sq. km.) and have a wide ranges of altitude upto 7000 m. The width is about 200 km. The districts of Jalpaiguri and Darjeeling in North Bengal, Exhibits about 500 tree species and about 4,000 species of flowering plants in an area of about 8,000 sq. km; As such, being situated in the trijunction of a rich floral and faunal zone of China, Burma and in close vicinity of Indo-China and having an area about 10 times. The aforesaid two districts, the State of Arunachal is likely to have much more plant and animal species.

Vast foot hill areas of Arunachal Pradesh is covered with tropical evergreen and semi-evergreen forests of which some genera and species viz., *Dipterocarpus, Artocarpus, Duabanga, Mesua, Shorea assamica, Terminalia, myriocarpa, Michelia champaca* which are of economic importance. The sub-tropical forests have *Schima, Castonopsis, Engelhardtia, Michelia* and *Quercus* of economic importance. In the temperate hills (1500-2500 m.), economically important species are *Betula, Acer, Juglans, Quercus, Magnolia, Rhododendron, Castanopsis, Pinus walliciana, Abies spectabilis, Tsuga dumosa* and *Picea spinulosa.* Alpine areas have myriads of species of herbs, dwarf, rhododendrons and junipers.

Areas botanically unexplored are Northern region of Subansiri, major portion of Tirap and Lohit, eastern region of Kameng and western region of Siang districts.

In the Sub-alpine vegetation at 3000-4500 m. in the Akahills of Kameng district dominant genera and species are *Abies, Tsuga, Pieca,* bushy *Rhododendron, Juniperus recurva, Berberis, Salix, Cotoneaster, Agapetes, Vaccinium, Sorbus* and *Rubus.* There is profusion of herbaceous species of the families Ranunculacece, Rosaceae, Polygonaceae, Gentianaceae etc.

The vegetation of Alpine areas at 4500-5500 m. is stunted, gnarled, dwarf, shrubs and *Rhododendron anthopogon, R. nivale* frequented with *Rheum, Saussurea, Sedum* and *Saxifraga.*

Chinese Element in Burmese and Arunachal Flora

The Chinese elements in Indian and Burmese flora are conspicious. The genera and families are *Rhododendron, Primula,* Celastraceae, Gesneraceae, Polygonaceae etc.

The relationship between the flora of the Himalayas and the Western China has been explained by Kingdom Ward (1921) on the presence of North East Frontier but in the area. This belt has acted as barrier as well a bridge between the Himalayan ranges to the west and Chinese ranges to the east, which have got closely related flora. He states, "the three areas, Himalaya, Northeast Frontier and Western China form a sort of letter Y, the three limbs each composed of parallel ranges and are separated from one another, but connected independently with desert plateaux of Central Asia.

To sum up, the Northeast Frontier belt, and more specially the Mekong-Salween divide, is primarily a barrier, botanical and zoological marking the eastern unit of the Indo-Malayan, or Oriental region, for at least 1200 km.

Other Foreign Elements in Arunachal Flora

Podocarpus neriifolia, Pinus merkusii are Marlayan. *Juniperus communis, Taxus baccata* are European. Two species of *Juniper* and one each of *Picea, Taxus, Tsuga* and *Pinus* roxburgii are common both in the eastern and western ranges of India.

There are Malayan, North African, Tropical, Subtropical, Mediterranean and European elements in Arunachal flora. Cruciferse and Caryophyllaceae which have several species in Alpine Himalayas are of Mediterranean element.

Endemism in Flora

Although the study area has land connection with a number of countries with distinct floral elements, it has a high proportion of endemic flora. Among the Dicots, India has 65.5% of endemic flora; the Himalayas have highest number of endemic Dicot (Endemism is 30% in Ceylon, 72% in Newzealand, 80% in Australia, 82% in Hawai islands).

Wides have 38.5% distribution in India among the Dicots. These have a lot of species that migrated from neighbouring regions. These are Burmese, Malaya, Mediterranean and European, the African, the Australian the American, the Tibetan and Siberian in alpine belt, and a number of Chinese and Japanese species occurring within temperate regions of India. The Malaya element dominate in Indian flora.

Burmese forests upto 1624 m. are made up of Indo-Malayan forms such as *Dipterocarpus, Elaeocarpus, Engelhardtia, Garcinia, Caryota, Calamus, Ficus, Aroids, Raphidophora, Pothos,* Tree ferns and undergrowth of Zingiberaceae, Acanthaceae and Gesneraceae.

Temperate forests between 1624 and 2500 m are characterised by *Schima, Gordonia, Exbucklandia, Quercus, Magnolia, Acer, Rhododendron,* Araliaceae, *Ficus,* epiphytic ferns, *Aschynathes* and mosses. There is preponderence of Urticaceae, *Impatiens* and *Strobilanthes.* Above 2500 m. Conifers comprise *Abies, Tsuga, Pinus smythiana,* Shrubs of *Rhododendron, Euonymus, Ribes, Rubus, Pieris Rosa, Deutzia, Hydrangea, Buddleia, Schijandra* and *Smilax.*

Bamboos (more than 40 species) occur from foothills to upper temperate region. Between 3200 m. and 3800 m. in Alpine and subalpine areas, there are species of *Allium, Lilium, Meconopsis, Thalictrum, Smilax,* Umbelliferae, *Potentilla, Spiraea, Pyrus, Berberis* etc.

Some endangered and endemic plants

Family	*Species*	*Endemic*	*Status*
Cyperaceae	*Carex rara*	Endemic	V. rare
	C. remota		V. rare
Palmae	*Livistona jenkinsiana*		Rare
	Phoenix rupicola		V. rare
	Wallichia triandra		Rare
	W. distichia		Rare
Araceae	*Lagenandra undulata*	Endemic	Rare
Dioscoreaceae	*Dioscorea laurifolia*		Rare
	D. orbiculata		Rare
Zingiberaceae	*Hedycium longipedunculatum*	Endemic	
Urticaceae	*Baehemeria tirapensis*	Endemic	
Reflesiaceae	*Spira himalayana*		Extremely rare
Gesneraceae	*Rhynchoglossum lazulinum*	Endemic	Extremely rare
Solanaceae	*Pania belladona*	Endemic	Rare
Styraceae	*Huodendron biaristatum*		Very rare
	Ainiphullum fortunei		Very rare
Ericaceae	*Rhododendron concinnoides*		Endangered
	R. nuttalli		Rare
	R. sartapaui		Endangered
	R. subasiriense		Endangered
Rubiaceae	*Nertera slnensis*		Very rare
Araliaceae	*Leptodermis scabrida* (Mishmi hills)		Rare
	Merrilliopanax cordifolia		Rare
Caesalpiniacae	*Phanera khansiana*		Very rare
Aquifoliaceae	*Ilex venulosa*		Threatened
Saurariaceae	*Saurauia griffithii*		Rare

Orchidaceae

Dr. S.N. Hegde listed 170 species of Orchids from Kameng district: some from Subansiri and Tirap. He feels at least 500 species will be available in Arunachal State. Panigrahi and Joseph listed 50 species of Tirap. Sahani (1953) considered 6 species of Orchids of Kameng to be endangered and rare. Hegde collected mostly Coelogyne (13 species), Dendrobium (29 species), Bulbophyllum (17 species), Era (14 species.).

Tropical Evergreen forest Orchid (170 m. to 800 m.)-*Aerides fieldingii, Bulbophyllum* species, *Coelogyne flavida, Dendrobium moschatum, D. aphyllum, Eria flava, E. pannea, E. stricta, Arudina graminifalia* and other species.

Sub-tropical Forest Orchid (801 m to 1800 m)-*Bulbophyllum gutterlatum, B. helenae, Dendrobium densiflorum, D. crysemthum, Eria amica.*

Some rarely distributed species are *Arachnis clarkei, Cymbidium devonianum, Dendrobium falconeri, D. sulcatum,*. In damp forests-*Cymbidium macrorhizon, Eulophia zollingeri.* Temperate forests (1801 m to 2500m)-*Bulbophyllum prostratum, Coelogyne, elata, C. occultata, Eria spicate.*

Pleione humilis, Acanthe phippium sythetense (rare), *Balleyia yunanensis* (very rare), *Esmeralda clarkei* (very rare), *Geleola falconeri* (rare), *G. lindleyane, Anoectochilus sikkimensis, Arachnis clarkei, A. cathcartii, Arandina graminifolia, Cymbidium macrorhizon, C. munronianum, Dendrobium densifloram, D. sulcatum, Diptomeres hirsata, Neolaemiopsis barbata, Epipogon roseum, Paphiopedilum fairrieanum, P. insigne, P. venustum, Vanda coerulea, Vanda punita, Malaxis acuminata, Haberania goodyeroides, Calanthe densiflora, Galeola falconeri, Goodyera procera.*

Other rare species (other than orchids) are-*Abies delavayi, Angiopteris evecta, Balanophora dioica, Camellia caduca, Coptis teeta, Cyathea gigantea, Phyllostachys bambusoides, Picea prachytyla, Rhododendron arizelum, R. edgewrothii, R. nuttallii, R. santapaui, R. stenaulum.*

Some Characteristic Plant Species

Zeylandium (Podostemonaceae), *Alniphyllum fortunii, Rhododendron biaristatum,* recorded from Kameng and Subansiri were also recorded from China and Yunan. *Munronia pinnata* has discontinuous distribution. *Coptis teeta* is endemic. The occurrence of *Magnolia griffithii, M. gustavi* and the genera *Mangletia, Euptelea, Tetracentron, Pycnorrhena, Haematocarpus, Aspidocarya, Decaesnea, Holboellia, Stauntonia, Paravalia, Exbucklandia, Altingia, Houttuynia, Merica, Alnus* and *Betula* are also of significance from plant evolution from plant evolution and migration point of view.

A study of vegetation of Apalanama valley in Dafla hills (Subansiri) revealed the presence of very rich flora especially of orchids, legumes and grasses besides other species. N.L. Bor recorded rich floral elements from Akahills (between Bhorali and Dhansiri rivers) in Tropical to high temperate belt. Several such species are-

Sapria himalayana (Tirap), *Balanophora dioica, Boschiniaepea homalaica, Aeginitia indica, Rhododendron santapaui* (Subansiri), *Lysimachia sàntapaui* (Begi & Amjee), *Merrilliopanax cordifolia* (Begi & Amjee), *Panax belladona* (Tirap), *Rhynchoglossum lazulinum* (Bhalukpong), *Boehmeria tirapensis, Oberonia sulcata, Hedychium longipedunculatun* (Amjee), *Pentasacme uallichii, Diplomeris pulchella* (Orchid), *Rhopalocnemis phalloides* etc.

New Taxa from Arunachal Pradesh

Asraoa triandra, Arisaema setosum, Arthromeris jarrettii, Dalbergia bhutanica, D. auarensis, D. psenudo (Robusta), *Hedychium radiatum, H. robustum, Legenandra undulata, Lasimachia, santapaui, Molineria praineana, Rhychoglossum larulinum, Rubus ghanakantae, Sadiria subsessiliflora, Pholidota corvallariae, Var. breviscapa, Alniphyllum fortunei, Carex eleusinoides, Houdendron biaristatum, Illicium cambodianum,* and several others.

Fern Flora

An investigation in 1980 in Kamang District revealed the presence of 214 species of fern in 71 genera and 21 families. A thorough exploration of the State will probably record the presence of more than 350 species of ferns which is unique for the State. A Botanical excursion of Tirap revealed the presence of 128 species of ferns.

A Few New Taxa from Namdapha

72 new taxa have been recorded from this national park alone. Without going into much detail mention may be made of a few general which are *Agapetes, Aneilama, Anthoxanthum, Arisaema, Bulbophyllum, Chirita, Corybus, Eulophia, Gaultheria, Hedycium, Knoxia, Maesa, Rubus, Smilax, Tacca, Trias, Buellia. Polygonum, Panax* etc.

9

Mammals : Mammal-Geography

The Arunachal region has served as a faunal gateway through which Indo-Chinese element of Oriental faunas and also that of Palaearctic could spread to India and colonize the country.

This region is the westernmost boundary of the range of many Indo-Chinese mammals, like certain squirrels, and the easternmost limit of the distribution of many peninsular species, such as spotted deer. Arunachal mammals are Indo-Chinese though there is variable admixture of the Peninsular and Ethiopean elements, the Palaearctic montane elements. (Refer Map 10).

Richness of fauna in Arunachal and adjoining Assam, Nagaland, Manipur, Meghalaya areas may be marked to some extent as follows:-

Order/Family	*Genera*	*Species*	*Sub-species*
Insectivora	6	9	15
Chiroptera	15	41	48
Primates	4	7	10
Pholidota	1	1	1
Carnivora	23	31	34
Proboscida	1	1	1
Perissodactyla	1	1	1
Artiodactyla	10	12	12
Lagomorpha	3	3	3
Rodentia	21	41	59
Cetacea	1	1	1

Total: 11 order, 28 families, 86 genera, 148 species and 186 sub-species.

Of the total 135 genera of mammals of India, 85 (or 63%) are represented in Northeast India. Of the 11 orders, the Carnivora are the richest in genera by Rodentia and Chiroptera. In the number of species and sub-species Rodentia rank the highest.

In addition to Indo-Chinese faunal invasion, invasion has also been of Ethiopean and Palaearctic elements, the former coming from Western gateway and the latter north-eastern gateway, north-western gateway

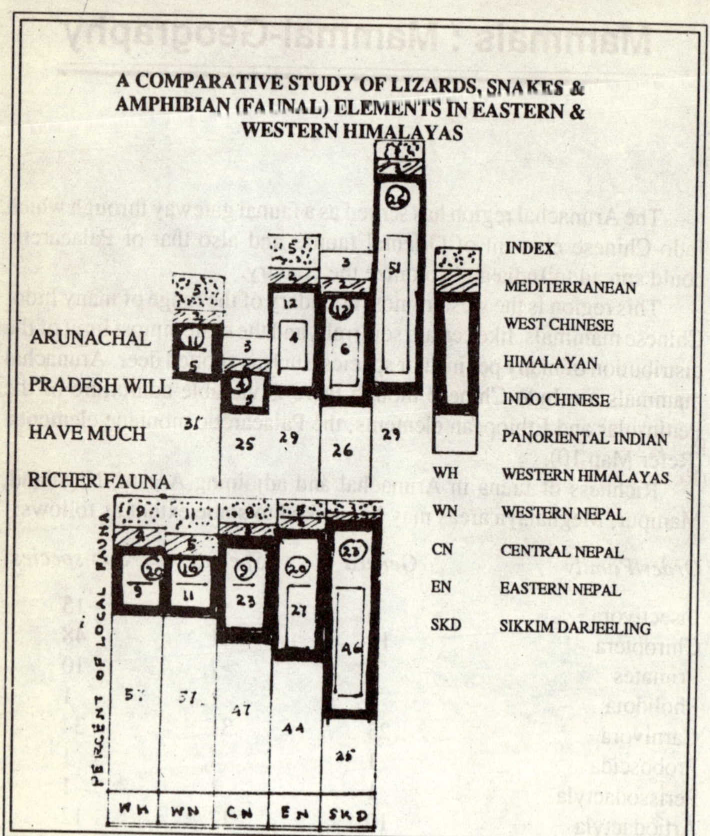
A COMPARATIVE STUDY OF LIZARDS, SNAKES & AMPHIBIAN (FAUNAL) ELEMENTS IN EASTERN & WESTERN HIMALAYAS
ARUNACHAL
PRADESH WILL
HAVE MUCH
RICHER FAUNA
INDEX
MEDITERRANEAN
WEST CHINESE
HIMALAYAN
INDO CHINESE
PANORIENTAL INDIAN
WH WESTERN HIMALAYAS
WN WESTERN NEPAL
CN CENTRAL NEPAL
EN EASTERN NEPAL
SKD SIKKIM DARJEELING
PERCENT OF LOCAL FAUNA
WH
WN
CN
EN
SKD

DIGRAM 1

and from the temperate montane fauna of the larger holarctic region. Western elements form 21% of Assam fauna, the remaining 69% are eastern. Western elements form 21% of Assam fauna, the remaining 69% are eastern. Western genera are *Suncus, Crocidura, Myotis, Eptesicus, Canis, Vulpes, Felis, Herpestes, Panthera, Elephant, Rhinoceros, Anteetope, Lepus, Hystrix, Mus* etc. There are faunal resemblances between westernghat, easternghat, Western India, Northeastern India which are not being discussed here.

Mammals

Some of the mammals are—

Common tree shrew (Lohit valley, Diagun), Anderson's tree shrew (Mishmi hills), Mishmi tree shrew (Siang, Mishmi hills), Sikkim large clawed shrew (Mishmi hills), Brown toothed shrew (Mishmi), Grew shrew (Mishmi hills), Burrowing shrew (Mishmi hills), Indian flying fox, Pearson's horse show bat (Mishmi hills), Sculleji tubenosed bat (Mishmi hills), Rhesus macaque (Mishmi hills), Assamese macaque (Mishmi hills), Kamrup capped langur (Abor hills), Hoolock gibbon (Mishmi hills), Yellow throated marten (Mishmi hills), Hogsons spotted linsang (Mishmi hills), Common palm civet, Jungle Cat (Mishmi hills), Marked palm civet, Golden cat (Mishmi hills), Leopard (Mishmi hills), Indian elephant, Wild bear, Musk deer, Asiatic two horned rhinoceros, Clouded leopard, Barking deer, Nog deer, Gaur, Mishmi Wild Buffalo, Kalimpong serow, Yunnan giant flying squirrel, Pasticoloured flying squirrel, Anderson's Squirrel, Gray's squirrel, Stevens's squirrel, Himalayan striped squirrel, Orange billed himalayan squirrel, Malayan giant squirrel, Tailed porcupine, Swinhoe's crestles himalayan porcupine, Long tailed tree mouse, Long tailed field mouse, Millards rat, Himalayan rat, Hinton's Himalayan rat, Thomas white-bellied rat, Edward's rat, Little Indian field mouse, Mrs. Jackson's jungle mouse.

Mammals (Protected)

Some protected mammals are—

Hoolock, Capped langur, Golden langur, Slow loris, Tiger, Panther, Snow leopard, Clouded leopard, Golden cat, Leopard cat, Marble cat, Fishing cat, Binturong, Spotted linsang, Sloth bear, Red panda, Flying squirrel, Hispid hare, Indian elephant, One horned rhinoceros, Wild buffalo, Himalayan tahr, Serow, Mishmi Takin, Musk deer, Pygmy hog, Indian pangolin, Chinese pangolin, Dolphin.

Reptiles and Amphibians (Protected)

Monitor lizard, Water lizard, Freshwater crocodile, Gharial, Python.

Birds (Protected)

Some protected birds are—

Assam bamboo partridge, Bhyth's Baza or lizard Hawk, Black crested Baza or Lizard Hawk, Bengal Florican, Black necked crane, Blood pheasant, Great Indian hornbill, Rufous necked hornbill, Wreathed hornbill, Indian pied hornbill, Bearded vulture or Lammergeier, Shaheen falcon, Monal pheasant, Sclater's Monal pheasant, Peacock pheasant, Tibetan snowcock, Satyr's tragopan, Greybellied tragopan, Temminck's tragopan, White eared tragopan, White winged wood duck.

Endangered animals (Ref. Maps 11 and 12)

Some endangered animals are—

Marsh crocodile, Indian egg eating snake, Large Whistling Teal, Pink headed duck, White winged wood duck, Bengal florican, Red Parda or Cat bear, Golden cat, Indian clouded leopard, Indian tiger, Snow leopard, the smaller one-horned or the Javan Rhinoceros, the Asiatic two-horned or Sumatran Rhinoceros, Pigmi hog, Himalayan musk deer, Swamp deer, Wild yak, The Asiatic wild buffalo, Hispid hare.

Some vanishing birds that once frequented all over Arunachal Pradesh and have now of doubtful existence are *Rhodonessa caryophyllacea, Cairina scutulata.*

Some vanishing mammals that were distributed in Arunachal Pradesh in past, but have doubtful existence now or are extinct in the locality:-

Rhinoceros unicornis (one horned rhinoceros), *Rhinoceros sondaicus* (Javan rhinoceros), *Rhinoceros sumatrensis* (Sumatran rhinoceros), *Cervus duvauceli* (Swamp deer), *Panthera unica* (Snow leopard), *Felis temmencki, Caprolagus hispidus* (Hispid hare).

Avifauna (Birds)

The study area including adjoining areas are one of the richest avifaunal zone of the World. About 550 species of avifauna have been recorded from Sikkim himalayas. Of these, more than 50% are passerine birds, the largest being the Timaliidae (Laughing thrushes, Babblars etc.) with 61 sp.; and Turdidae (Chats, Thrushes etc.) with 56 sp.; the Sylviidae (Warblers) with 60 species and Muscicapidae (Flycatchers) with 27 sp. and the Fringllidae (Finches) with 22 sp. Some of the birds are exceptionally beautiful. Woodpeckers (15 sp.), Cuckoos (16 sp.) and Owis (14 sp.) and Hornbills are special features of the zone. Accipitrine birds number 40 and include five Himalayan Lammergeyer and Hodgsons feathertoed hawk eagle. There are about 12 species of pigeons and dove. This study area is equally rich.

ALTITUDINAL DISTRIBUTION OF AVIFAUNA

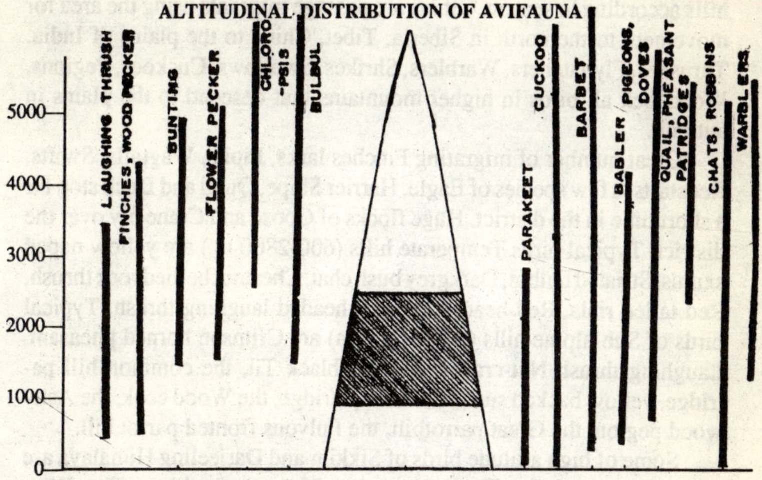

92% BABBLERS, 82% LAUGHING THRUSHES, 100% WOOD PECKERS, 52% WARBLES, 37% REDSTARTS OCCURS (BETWEEN 200 M TO 1400 M ALTITUDE.)

MAXIMUM ZONE OF OCCURRENCE

SNAKES OF EASTERN HIMALAYAS

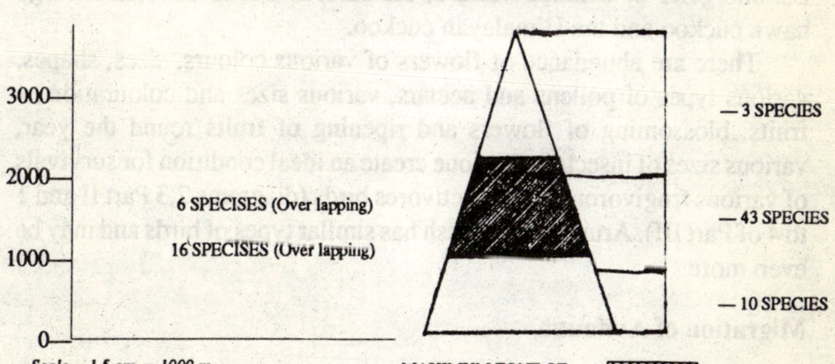

Scale—1.5 cm. = 1000 m

MAXIMUM ZONE OF OCCURRENCE

Diagram 2:

The game birds are of 8 species and consist of the Indian red jungle fowl, the black-backed pegion, the monal, the blood pheasant etc. There are several species of partridges and quails.

A large number of birds are local migrants moving up and down the hills according to season. Others are passage migrants using the area for movement to the north in Siberia, Tibet, China to the plains of India. Thrushes, Flynatchers, Warblers, Shrikes, Swallows, Cuckoos, Pegions, Woodcock all birds in higher mountains that descend to the plains in autumn.

Great number of migrating Finches larks, Pipits, Wagtails, Swifts, Redstarts, a few species of Eagle, Harrier Shipe, Quail and Duck stop far a short time in the district. Huge flocks of Goose and Crane ply over the district. Typical birds Temperate hills (600-2800 m.) are yellow naped sxulus, Striated bulbul, Darkgrey bush chat, Chestnut bellied rock thrush, Red tailed riala, Red headed tit, Red headed laughing thrush. Typical birds of Sub-alpine hills (2800-4100 m) are Crimson horned pheasant, Laughing thrush Nut-craker, Crested black Tit, the common hill patridge, yellow backed sunbird, Snow patridge, the Wood cock, the Ashy wood pegion, the Great parrotbill, the Fulvous fronted parrot bill.

Some of high altitude birds of Sikkim and Darjeeling Himalaya are Finches, Parrot bills, Chats, Pheasants, PIgeons, Sunbirds, Tits, Warblers, Cuckoo and Flower pecker.

Some birds with long altitudiual ranges are:

The Indian sparrow hawk, the Himalayan griffon vulture, the Lammergeier or bearded vulture, the Rufous Turtle dove, the Large hawk cuckoo and the Himalayan cuckoo.

There are abundance of flowers of various colours, sizes, shapes, various types of pollens and nectars, various sizes and colouration of fruits, blossoming of flowers and ripening of fruits round the year, various sizes of insects in the zone create an ideal condition for survivals of various frugivorous and insectivores birds (diagrams 2,3 Part II and 1 to 4 of Part III). Arunachal Pradesh has similar types of birds and may be even more.

Migration of Avifauna

The migration of ducks, geese, cranes in India occurs through Siberia-China over Tibet into Eastern India. On this long march the vast migratory bird population need transit facilities in the Himalayas. The conservation of habitat in the Himalayas, as such, is essential.

Salim Ali has studied the birds in the region and has recorded 379 species in his book entitled "Birds of North-East India! It is not known if the investigation is a thorough one as a small state of Sikkim has about 550 species of birds.

A distribution chart for Sikkim birds is produced in a diagram. In this area altitudiual distribution of 397 species has been worked out as follows:

0	- 1000 m	- 31 species
1001	- 1500 m	- 46 species
1501	- 2000 m	- 58 species
2001	- 2500 m	- 67 species
above	2500 m	- 177 species

It seems the richness of birds species is between 1500-3000 m and the distribution is comparable with Sikkim area.

Some high altitude birds of Arunachal Pradesh are as follows:-

Pink headed duck, Mountain quail, Blood pheasant, Tragopan bluthii, Tragopan satyra, Tragopan temnicks, Sclater's monal pheasant, Humes Bartailed pheasant, Snow partridge, Monal pheasant, Tibetan snow cock and Tibetan partridge.

Plant-Animal Interaction

Colouration, shape size and scent of flowers, scent and size of fruits (Diagram 5) are correlated to various behavioural pattern of insects and avifauna species. Bees are sensitive to red, yellow and blue colours; hive bees prefer blue and yellow flowers: Bats prefer drab green colours; Beetles and moths prefer red, purple, white and pink flowers; Flees prefer scarlet or yellow or red flowers. This zone has all these assorted flowers. All flowers, their colours and various species of insect interact among themselves.

An analysis of colouration of flowers among tree herb and shrub communities in eastern Himalayas is as under:-

Colour	*Tree*	*Shrub*	*Herb*
White	33	33	24
Yellow	23	24	22
Purple	4	18	18
Red	7	10	8
Pink	5	9	13
Blue	8	6	15
	100	100	100

A comprehensive research scheme can be drawn on the subject.

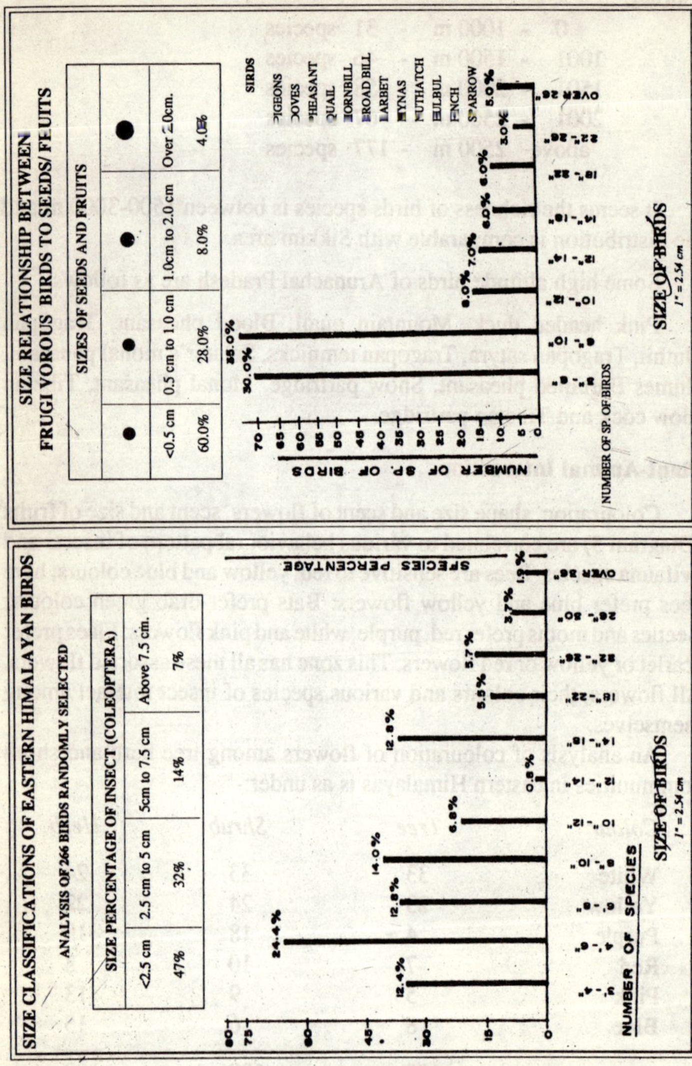

Diagram 3: Size of Birds

Fish, Amphibia and Reptiles (Diagram 1)

Within the Oriental Region, the Indo-Chinese sub-region has much larger numbers of genera and species in nearly all groups, except the Amphibia, than the Indian sub-region. Fifty percent (50%) of the genera of Amphibia are autochthonous in the Indian sub-region (better developed in western ghats). For all the other major elements the Indo-Chinese sub-region is the centre of differentiation and radiation.

In this region the presence of monotypic genera is considerable Snakes (68%), Chelonia (75%) are monotypic. Except the amphibia, other specialised genera occur in Eastern Himalaya, Northeastern Himalaya, Thailand, Cambodia, Malaya and China.

Fish

Fish fauna of India (India has 1700 species under 600 genera and 200 families) is dominated by the Indo-Chinese element: though small amounts of Palaearctic, Etheopean and Malayan elements are also present. None of the peninsular outochthomous genera have penetrated Sri Lanka. Shebbeare and Shaw have identified 131 species of fish in the plains and hills of Darjeeling. Late Hora and Gupta reported 58 species of fish from Kalimpong and Siliguri forest areas.

Among the game fishes notable are the Masheer, Katli, Indian trout and Goonch. There are three varieties of Masheers.

The Indo-Chinese elements occur in swift streams. *Bargarius bagarius, Batasio, Siturus, Gagata* (Sisoridae), *Lepidopygopsis* and *Acrossocheilus hexagonolepis* have been recorded from Arunachal. *Acanchophtalmus pangia* occurs in the Eastern Himalayas and extend through north-east to Burma southwards. *Gagata, Batasio and Silona* occur in north-east and also in Indo-China sub-region. *Balikora brucei* occurs in North-Eastern Himalaya; *Osteobrama cotio* and *Balasio balasio* occur in North-Eastern Himalayas.

Endemic fish fauna of Eastern Himalayas are:-

Psilorhunchus balitor, Balitora brucei, Naemacheilus beavani, N. savona, N. scaturigina, Gagata nangru, Glytothorax rebeiroi, Tor progenius, Schizothorax progatus, and *Lepido cephalichtys.*

Amphibian Fauna (Diagram 4)

Reptile and amphibian fauna in eastern Himalayas have the following number of species of different altitudinal zones are as follows:-

0	-	900 m	-	132 sp.
900	-	1800 m	-	135 sp.
1800	-	2700 m	-	71 sp.
2700	-	3700 m & above	-	38 sp.

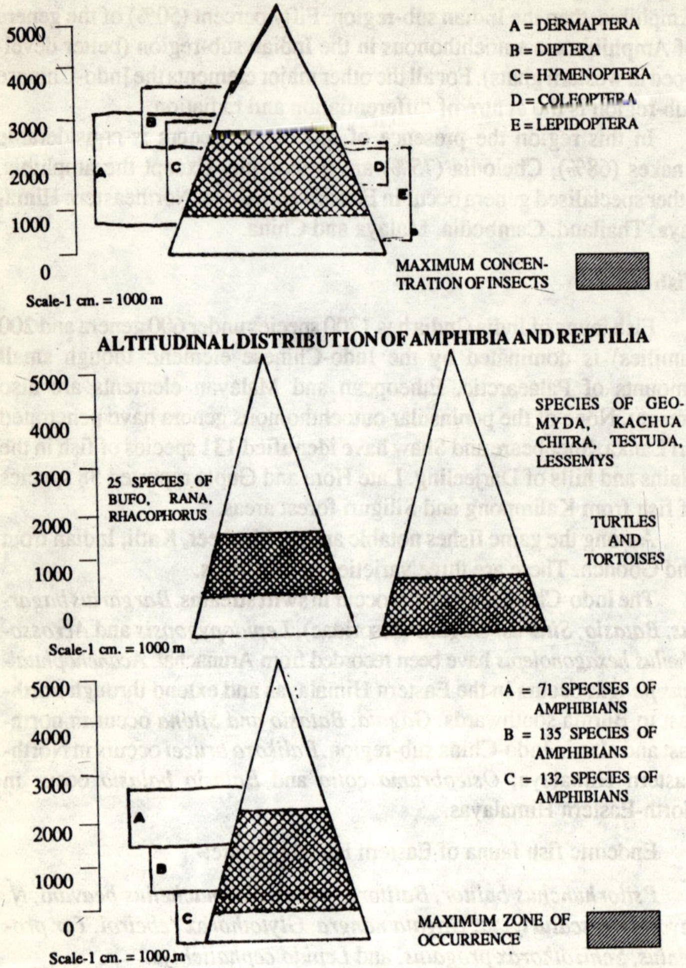

DIAGRAM 4

ALTITUDINAL DISTRIBUTION OF HIMALAYAN INSECTS (MAJOR ZONES OF OCCURRENCE)

So far Cataloguing and Taxanomic descriptions have been made, but no detailed ecological work has been done. The wealth of insects and amphibia and their marvellous adjustment with the special environmental condition of high altitude should be of special interest.

Tylotricton verrucosus (Salamandar) occurs from 1600 m to about 2600 m. *Varanus salvator* occurs from plains upto 2300 m. *Platuyurus* sp. extend from 900-2800 m. *Leiolopisma sikkimensis* occurs from 900-3100 m., *Lagossoma maculatum* occurs from 2300-3100 m and *Calotes versicolor* climbs upto 2600 m.

Of the frogs some species of *Rana, Philautus* (2 sp.), *Rhacophoras* (4 sp.), *Bufo* occur in upper temperate and alpine Himalayas. Typical tree frog upper temperate hill is *Philautus annandalii. Bufo melanostictus* occurs upto 3500 m. Rana sp. *(R. monticola, R. formosa, R. himalayana, R. liebigii,* and *R. limonocharis)* occur from plains to 2300 m.

A diagrammatic representation of the amphibia/reptiles has been presented (Diagram 4).

Geomyda, Kachua, Testudo, Lissemys, Trionyx and *Chitra* occur in the foot hill forests of the district.

Snake Fauna

Darjeeling hills and plains have at least 76 species of snakes. Although maximum concentration of species centres round 800-1570 m., several species predominate in the planes and foot hills, others inhabit 1570-2200 m. though a few species extend in the upper temperate hills and alpine pasteres. Middle hills have (i) Wall's blind snake, (ii) Stoticzka's blind snake, (iii) Jerdons blind snake, (iv) Bermese Diards blind snake, (v) Collared dwarf snake, (vi) Anderson's Keelback, (viii) The buff striped keelback, (viii) Goggle eyed Keelback or hooded tree snake, (ix) Black bellied worm eating snake, (x) The black-bordered rat snake, (xi) The ringtacted Dhaman, (xii) Common wolf snake, (xiii) King Cobra upto 2300 m.), (xiv) The Large spotted viper, (xv) The Himalayan pit viper.

Some species of the genera *Mabuya Lygosoma, Elaphe, Callophis, Naja, Vipera, Trime Ptyas, Zoaays, Liopellies* occur in upper temperate and sub-alpine hills.

Distribution of Sphingidae (Moth)

India and adjoining areas have about 401 species having distribution such as Western Himalayas 62 species, Eastern Himalayas 135 species, South India 75 species, Sri Lanka 58 species, Burma 49 species and Andaman & Nicobar Islands 22 species. Obviously, of the 135 species, in eastern Himalayas Arunachal can claim substantial number of such species.

Butterfly

India and adjoining areas have 1443 species of which the distribution is as-North-Western Himalayas 421 species, Eastern Himalayas 853 species, South India 315 species, Sri Lanka 238 species, Burma 1041 species, Andaman & Nicobar Islands 265 species. Arunachal can claim most of the 853 species that occur in eastern Himalayas.

The greatest bulk of Lepidoptera are Indo Chinese and Malayan elements, though some Manchurian and extremely few Turkmenian forms are found. Most of the species belong to Oriental tropical forest genera. The Indo-China genus of butterfly *Chilasa* is mainly a species of eastern Himalayas. *C. epy cides* extend to Burma, China and Tiwan *C. slateri* is common to Burma, Malaya, Sumatra and Borneo. *Graphium* extends to China and Nepal. *Papilio* has large range of distribution. *Papihomemnon* occurs in South Japan, Borneo, Sunda island, Nicobar island, Burma and Sikkim. *P. agenor* occurs in Sikkim, Assam, Burma and Nicobar islands. The genus *Troids, Polydorus, Aemona, Armendia, Graph* have wide distribution.

Turtle and Tortoises

Geomyda tricarinata, G. indepeninuslares, Kachuga tectum, Testudo elongate, Chitra indica, Trionyx hurum and *Emyda granosa* are same of the forest species.

Insect

The insect life of the Himalayan forests is extremely rich and greatly diversified. The wooded slopes of higher mountains are populated by a greater abundance of mostly all sorts-May flies, Stone flies, Grass hoppers, Crickets, Dragoflies, Caddies flies, Bugs, Moths, Midgets, Ghats, Mosquitos, Fleas etc. The greatest bulk of the species are ground living, wingless and are carnivorous and debris feeder. They live in rock crevices, on moss and on other dwarf plants, on snow edge, in torrential streams of water, ponds and in spring.

From recorded distribution of various species of Coleoptera, Diptera, Hymenoptera, Orthoptera, Lepidoptera, Rhynchota etc. vertical zonation map was prepared to show what zone is rich in various types of insects (No. 4 diagrams).

The area is exceedingly rich in Lepidoptera. Some of the rare and most beautiful moths occurs in this area. Brilliantly coloured beetles also occur in the area.

Size of insects and size of insectivorous birds were mapped and a relationship has been established (Diagram 3).

Abundance of Coleoptera has been recorded between 1200 m and 1900 m, Dermaptera between 1500 m and 2200 m, Diptera between 1500 m and 2000 m and Hymenoptera between 1300 m and 2000 m, though the distribution of the species attain a much wider range.

Insectivorous Birds Vs. Insect Colouration

In this locality the abundance of moths butterflies and Colourful beetles which go parallel with abundance of insectivorous birds. White coloured moth is widely eaten by birds, light yellow orange coloured moths are less accepted, but brilliantly coloured ones are not accepted. Moth concentration is maximum in foot and middle hills though quite a number of species do occur in upper temperate zone.

In a local study each of 61 species of moth was found to be host to one specific plant family and the preservation of the plants of this family (Urticaceae) is imperative. Some important host plant families are Leguminosae, Urticaceae, Ancardiaceae, Euphoribiaceae, etc.

High Altitude Insects

Entomological survey of high altitude (3000-6000 m) in Western Himalayas has been conducted by M.S. Mani and recorded under the proceeding of Zoological Survey Vol. 14, No. 2. These insects are exposed to high light intensity, reduced atmospheric pressure, physical and topographical peculiarities, oxygen deficiency, ultraviolet radiation, high wind velocity, low temperature, atmospheric humidity and aridity.

At 3000 m. 394 species were identified belonging to the Orders Ephemerida, Plecoptera, Odonata, Orthoptera, Dermaptera, Heteroptera, Homoptera, Coleoptera, Hymenoptera, Neuroptera, Trichoptera, Lepidoptera, Diptera, Thysanura and Collembola. These species are either terrestrial, transitional, aquatic (Stagnani water), aquatic (torrential), in vegetation mats, phytophagous, scavengers, predators, non-phytophagous. Insects occur in barren rocks, in alpine meadows or snow, in lakes. Occurrence of various Orders of insects at different altitudes are:-

Ephemerida-5000 m, Plecoptera-5000 m, Orthoptera-4800 m, Dermaptera-4500 m, Heteroptera-5400 m, Homoptera-4000 m, Coleoptera-5600 m, Hymenoptera-4800 m, Trichoptera-4500 m, Lepidoptera-5800 m, Thysanura-5400m, Collembola-6000 m.

There is immense scope for study of insects of glacial and high mountain areas of Arunachal Pradesh. The floral and faunal species diversits of Eastern Himalayas outnumber such species of Western Himalayas.

No through investigation on the occurrence of high altitude May flies, Stone flies, Grasshoppers, Crickets, Dragon flies, Caddis flies,

Bugs, Cicads, Beetles, Ants, Wasps, Bees, Butterflies, Moths, Midges, Mosquitoes, and Flies have been made. The greatest bulk of these high altitude insect species are ground living forms-that typically lack wing and can rarely take to wing. They are carnivorous, carrion feeders, debris feeders. They live under stone, in soil, in rock crevices, on moss, on dwarf plants, snow edge, on snow field, in glaciers, in torrential waters, ponds, lakes and streams. The bulk of invertebrates are terricolous and of endogerous type but many are found in glacial ponds, lakes and streams of rivers. At 6800 m. on the Himalayas (record of presence of animal in highest region) occur a complex of Collembola, Diptera, Acarina and Salticid spiders. Over 70% of the high altitude species are endemic.

Dragon flies (Odonata-36 species, Rhynchota-11 species, Cicads (Cicade)-4 species; Bugs-22 species, Diptera-2 species, Wasps-4 species, Bees -8 species, Orthoptera-8 species.

Biogeography of Namdapha

Dasman (1973) suggested a system of classification of World's Natural Regions for the purpose of conservation.

According to him, Namdapha and adjoining areas are in palaearctic sub-region, the biotic provinces of the Himalayan-Karakoram and Szechwan; and in the Oriental sub-region, the biotic provinces of the Bengal rain forest, Burma monsoon forest and the Burma rain forest containing elements of all these biotic provinces, with an exceptional range of climatic conditions, topography and long duration of uninterrupted biotic evolution.

Mani (1974), however, considers the eastern parts of the Eastern Himalayas, Eastern Tibet, the Yunnan Province of South China, Annam, Hainan, the northern parts of Indo-China, Assam, Burma and the northern part of Thailand constitute together a single natural unit of mountaneous iand, the Indo-Chinese sub-region.

Kurup (1974) considers that this region serve a great faunal gateway through which Indo-Chinese elements of the oriental fauna and also that of the Palaearctic could spread to India. Indo-Chinese mammals predominate in the north-eastern region with an admixture of Palaearctic and Ethiopean elements. This region has utmost significance in the evolution of the present day flora and fauna.

Of the total of 135 genera of land mammals in India 85 (or 63%) are represented here. Carnivora have 23 genera and 41 species and 34 subspecies followed by Rodentia with 21 genera and 41 species of Chiroptera with 15 genera and 41 species comes next. Among the families Muridae, Viverridae and Vespertillionidae are well represented. Viverridae has most genera in this area that other parts of India.

The following table will present the richness of animal population in Namdapha National Park. This will indicate the possibility of identifying immense resources when thorough investigation will be made of the entire State. A number of new species including one in mammal, two in amphibia, four in pisces and fifteen in insects were evident. A total of 33 species of animals has been recorded for the first time from India of which 29 is insect species. The result of biosphere survey added 197 new records of insects, ticks, scorpions, molluscs and vertebrates.

Group	*Family*	*Genera*	*Species*	*New species*	*New to India*	*New record*
Dictyoptera	8	9	13	-	-	-
Orthoptera	10	8	36	-	-	-
Mantodia	2	10	10	-	-	10
Dermaptera	6	12	14	2	1	-
Isoptera	3	11	13	1	-	-
Odonata	4	10	11	1	-	10 *
Neuroptera	1	2	3	-	-	2
Lepidoptera (Papilionidae)	1	5	21	-	-	1
Nymphilidae	1	8	31	-	-	-
Pieridae	-	8	11	-	-	6 *
Danaiodae	-	2	4	-	-	- *
Satyridae	-	4	7	-	-	3 *
Lycinidae	-	10	10	-	-	9 *
Coleoptera (Carabidae/ Carculionidae)	-	16	22	-	17	17
Histeridae	-	12	18	1	3	-
Passalidae/ Lycanidae	-	9	16	2	-	3
Cucujidae	1	8	13	1	-	13
Silvaridae	1	5	8	-	2	4
Chrysomelidae	1	35	56	-	3	37
Cerambycidae	1	11	12	-	-	12

Group	*Family*	*Genera*	*Species*	*New species*	*New to India*	*New record*
Diptera	10	30	41	2	3	13
Staphylinidae	-	12	26	-	-	-
Scaradidae — Coprinae	-	13	55	-	-	-
Hynenoptera	11	21	28	-	-	-
Arachnida						
Ixodidae	1	6	10	-	-	3
Scorpions	1	2	3	-	-	1
Molluscs	10	20	23	-	1	10
Pisces	1	19	28	4	3	-
Amphibia	3	5	14	2	-	6
Reptilia	7	15	15	-	-	15
Birds	12	46	64	-	-	-
Mammals	8	15	20	1	-	-

* The summary based on collection so far made indicates the existence of a rich genepool and detailed analysis all of animal taxa clearly reveal an assemblage of primitive and recently evolved forms, suggestion of *an area under continuous process of evolution*. Wildlife resources (mammalian) have been listed elsewhere.

10

Conservation Need

(Refer Maps 1 and 2)

From the map of altitudinal distribution of sanctuaries and National Parks read with altitudinal contour (diagram ^), it may be seen that these protected areas cover forests from foot-hill upto 3500 m. They cover tropical evergreen, tropical semi-evergreen, sub-tropical and temperate forests to the maximum. A low level secondary grassland is represented by D'Ering sanctuary. Namdapha offers an ideal habitat from many species of birds, bees, wasps and animal to travel long and varied types of altitudinal distances through cover of forests from foothill to high altitudes; but this only may not be adequate to conserve the vast bulk and varied of flora ad fauna of the State. The complex vegetation of mixed bamboo brakes (there are more than 40 species of bamboos), varied types of conifer forests, mixed conifer and broad-leaved forests, Conifer and Rhododendron forests, (more than 57 species of Rhododendron), extensive sub-alpine forests, high level dwarf Rhododendron and other flowering plants, vast alpine grasslands where extensive and varied flowering shrubs occur, vast glacial beds that offer aquatic substrata for hundreds of insect and others species in Summer and last but not the least high latitude rocky substratum which form the habitat of insects, have not been adequately represented. These high altitudes forests offer a rare habitat for some of the world's most beautiful animals and birds viz., Snow leopard, Musk deer, Mishmi takin, Tragopan, Monal phcasant, Partridges, Woodcock and many other species of birds.

Effort has been made in consultation with the Chief Wild Life Warden to identify a few more areas which will give complete protection to most of the birds and animals mentioned in the aforesaid paragraphs.

Tawang and Walang sanctuaries have been proposed in West Kameng and Lohit districts respectively. While Tawang extends from an altitude of 2000 to 4500 m. in western boundary close to Tibet and Bhutan, Walang extends from 2000 to 5000 m. in the border with China; the latter covers alpine, sub-alpine grasslands besides pine forests, conifer forests and broad-leaved forests.

The swamps of Taley valley have not been thoroughly explored. It has conifer and broad-leaved forests in the hilltop. The rare Salamandaer

(Tylotrotrichon verrucosus) is sure to occur in the swamp.

Lado Sanctuary in east Kameng district extend from 2500 m to 5000 m and covers various types of forests including alpine grasslands. Palin sanctuary in Miri hills is primarily a broad-leaved forests extending from 2000 to 5000 m. Dibang Valley sanctuary extends between 2000 m to 4000 m.

Both Tawang and Walang sanctuaries were marked earlier by the State authority for constituting as wildlife sanctuaries and detailed data were compiled. Latest satellite imagery justifies inclusion of this two as they still maintain good forest cover. A small area designed as Namdapha extension was also marked earlier by the State authority and now it has been decided to include in the existing Namdapha sanctuary. All these areas have been selected on the basis of Toposheet analysis, satellite imagery interpretation and in consultation with the forest officials. It will be necessary to visit the areas before submitting final proposal for demarcating and aligning boundary lines.

Diagram 1 show the altitudinal ranges of the existing wildlife sanctuaries and national park and those of a few sanctuaries now proposed. The richness of flora and fauna of the locality and the country as a whole is due to imigration and colonisation of flora and fauna from China, Tibet, Japan, Africa, Europe, Malaya, Siberia and other countries. Unlike the fauna, the flora is characterised by the dominance of relatively young intrusive elements so that, ancient forms are of lesser importance. The fauna however, has unmistakable evidence of the profound influence of the Himalayan uplift. Humid tropical Asiatic elements came through the Assam gateway. Later the migration came from west and north where Boreal, Steppe, European, Mediterranian and Tropical African elements crept in.

As such the rare and vast floral and faunal resources of the Himalayan region have to be protected and resources augmented. Distructive processees like shifting cultivation, fire, felling and other biotic depredation have to be stopped. Not only the State of Arunachal Pradesh needs such protection, all other hill states viz., Nagaland, Manipur, Mizoram, Tripura, Meghalaya and the district of North Cachar and Karbi Anglong need similar protection. These states have several wildlife sanctuaries viz., Intanki, Dampa, Gonti, Trishna, Keibul, Lamzoa and several others.

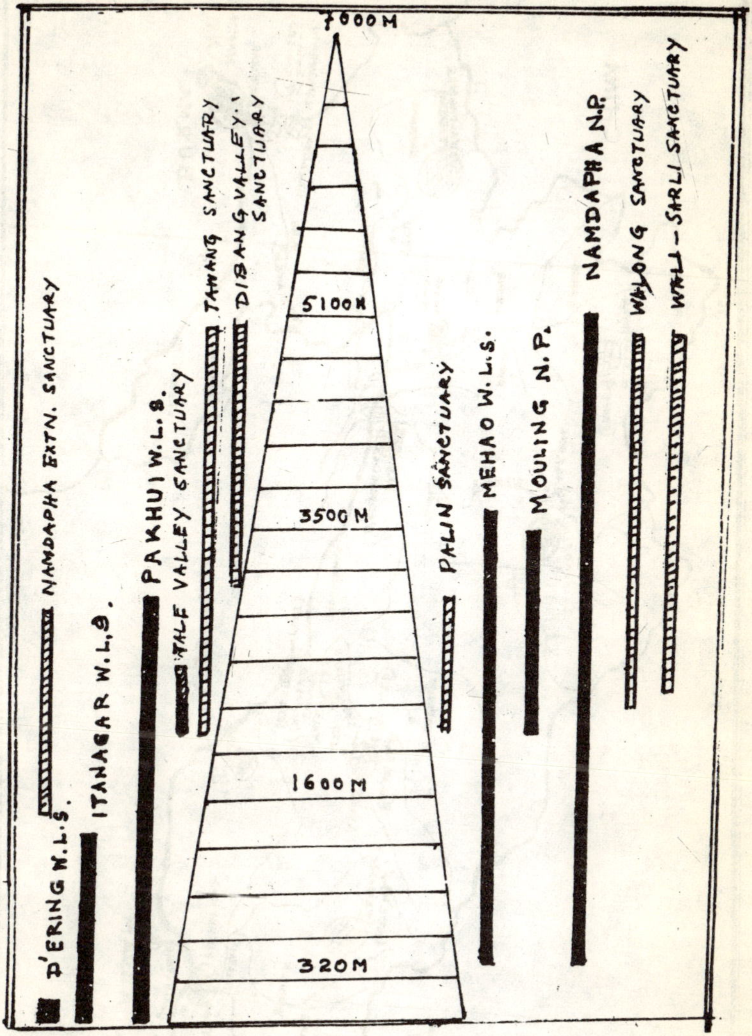

Digram: Altitudinal Spread of National parks and wildlife sanctuares in Arunachal Pradesh

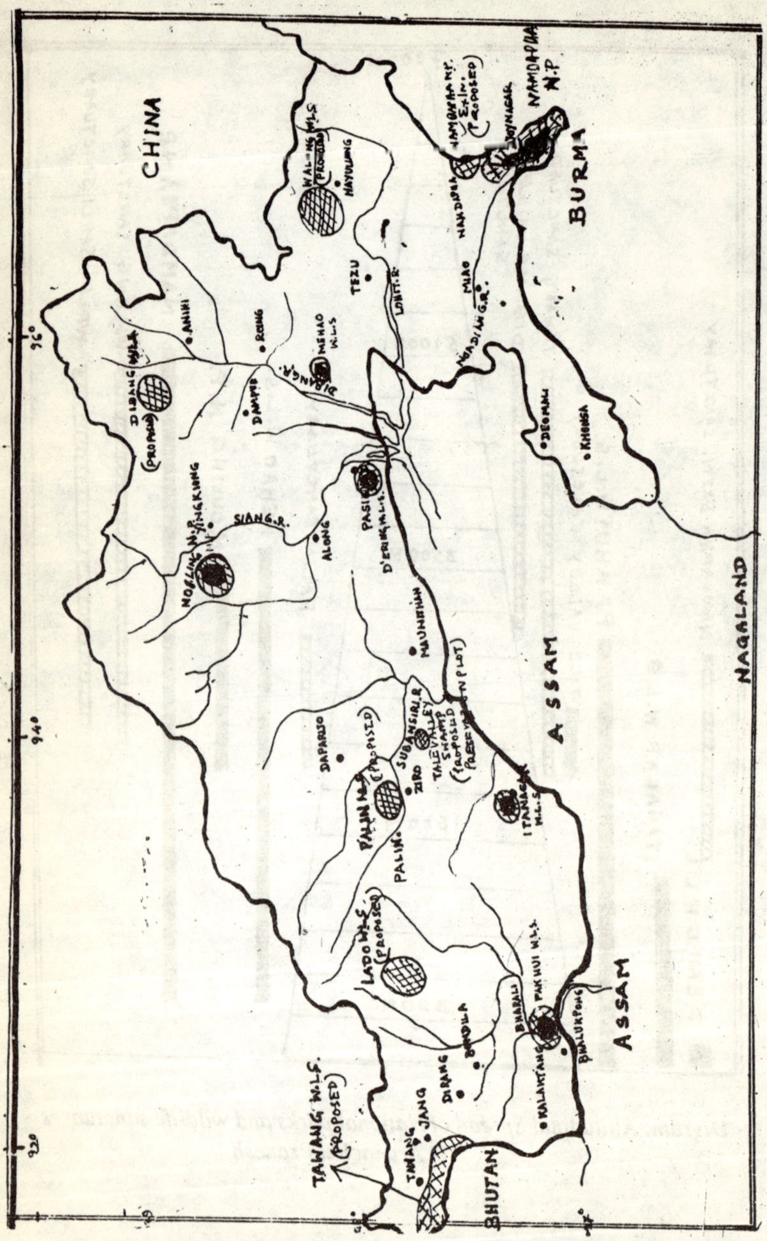

Map 1: Reference Existing Wildlife Sanctuaries, National Parks Proposed Wildlife Sanctuaries

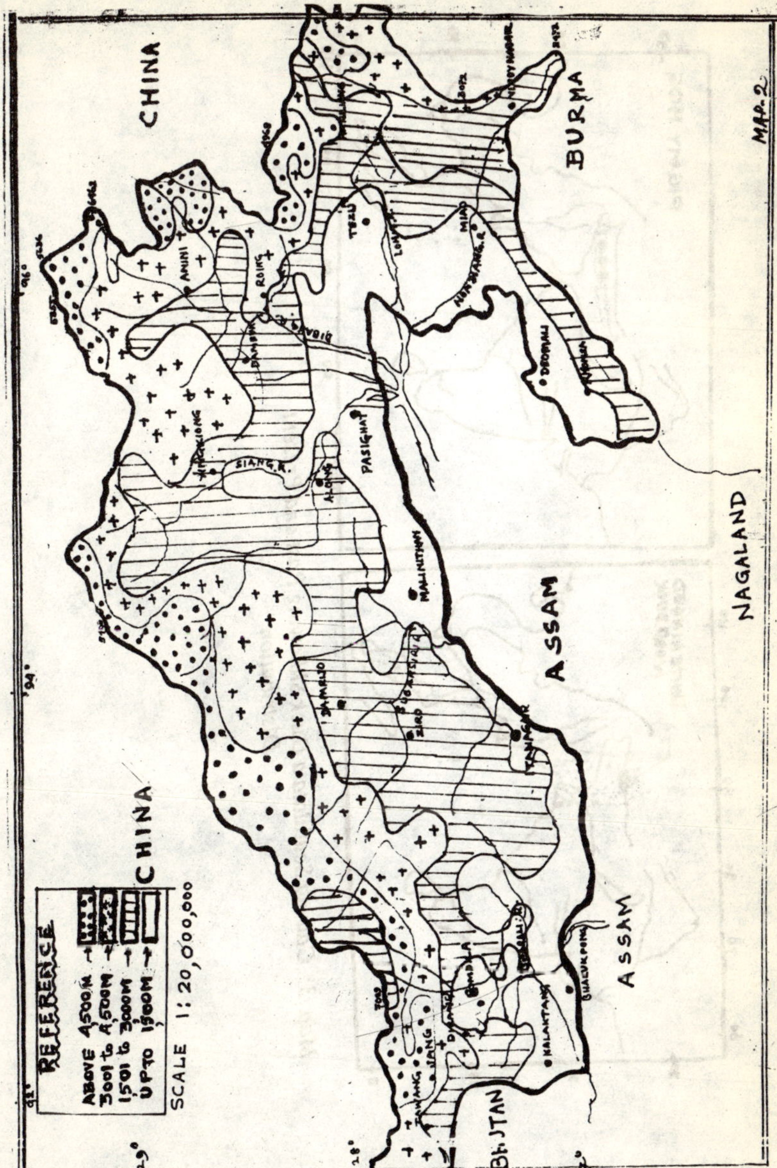

Map 2: Arunachal Pradesh Showing Altitudinal Divisions

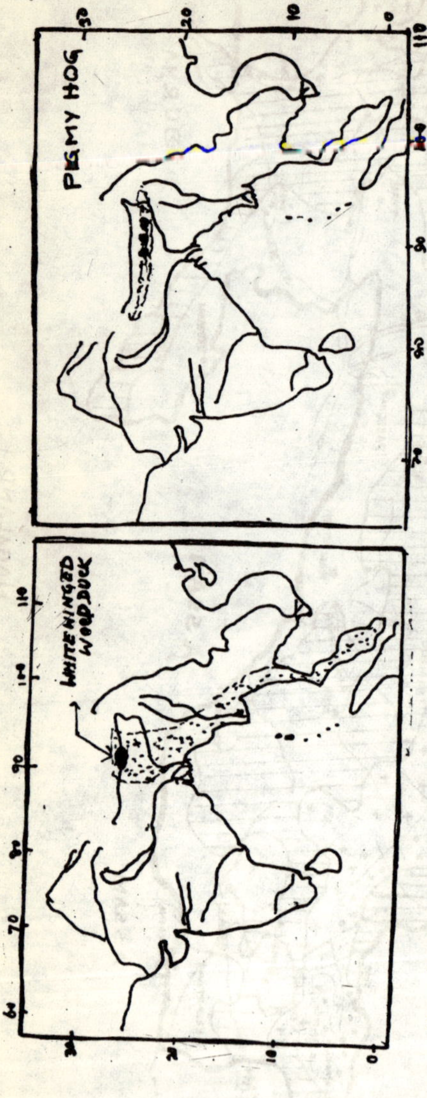

Map 3: Cairina Scutulata a Duck Shhowing Past and Present Distribution

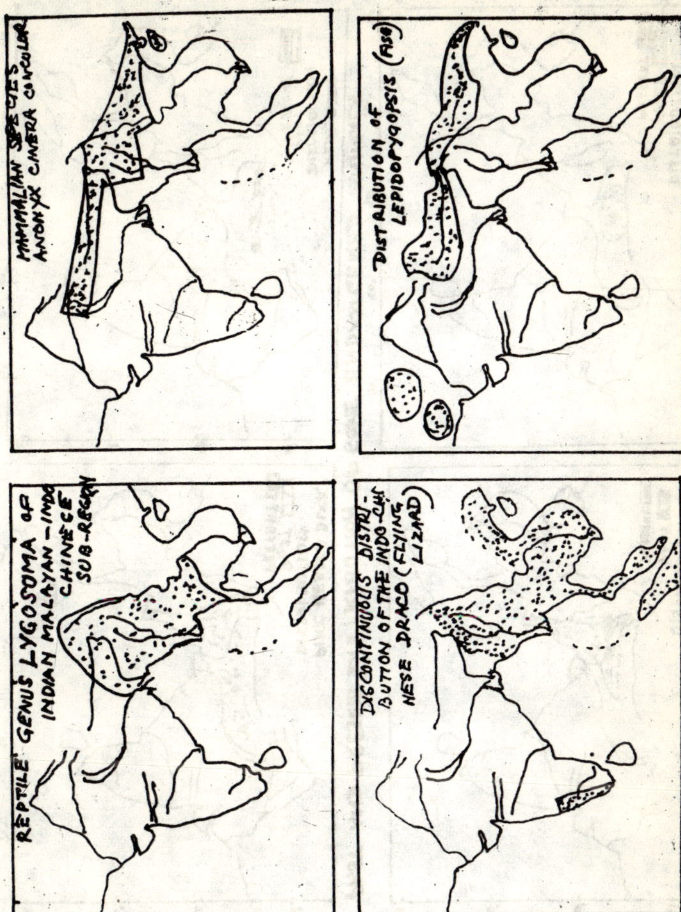

Map 4: Distribution of some Animals in Arunachal and South East Asia.

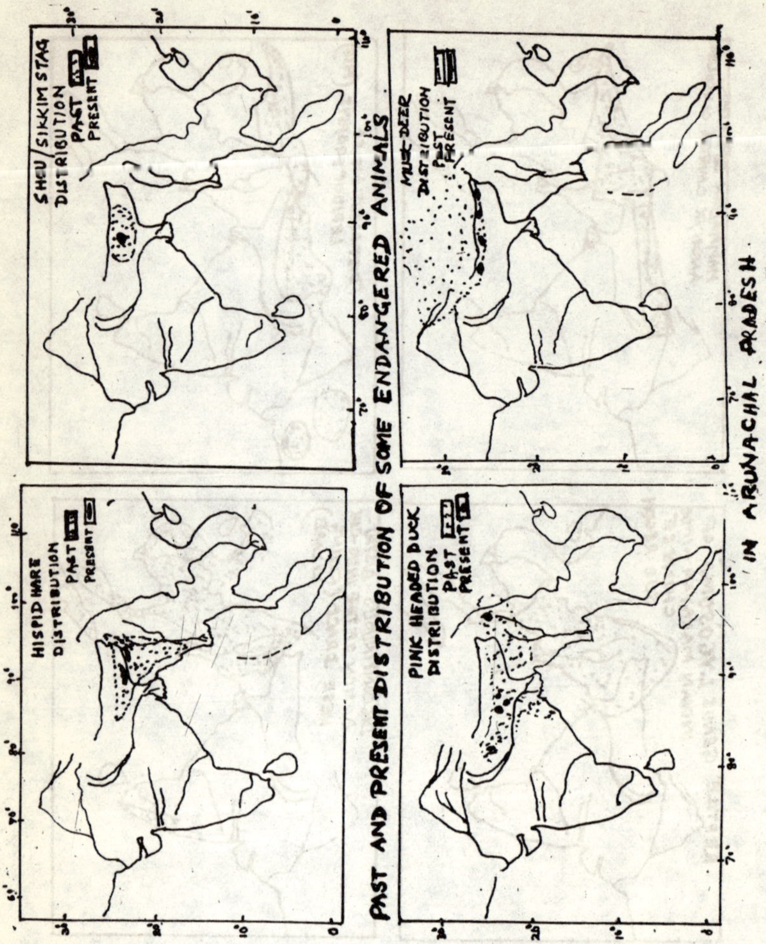

Map 5: Past and Present Distribution of some Endangered Animals in Arunachal Pradesh

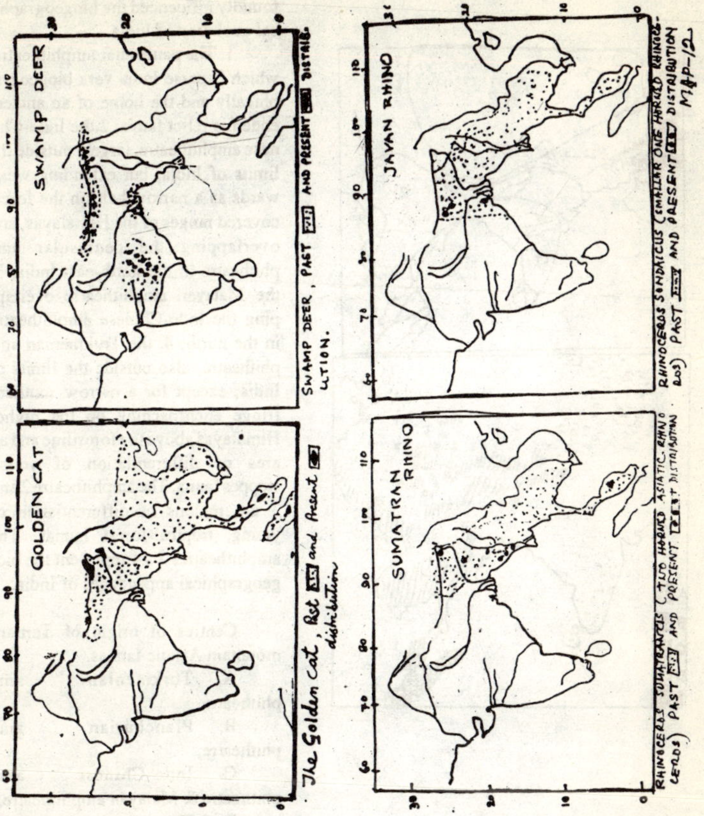

Map 6: Arunachal Pradesh Past and Present Distribution of some Endangered Animals

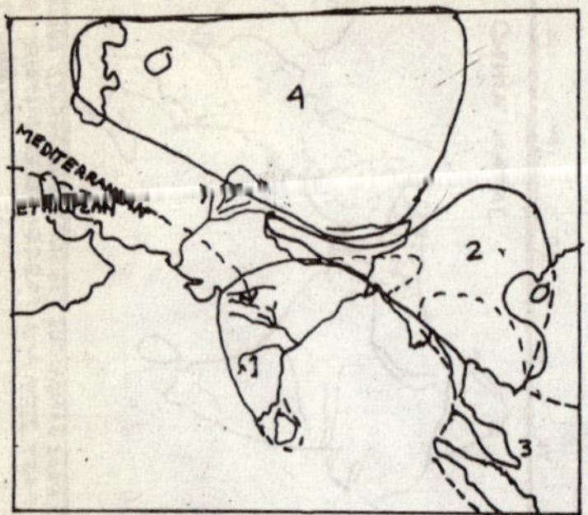

Major amphitheatres of differentiation, evolution and radiation of flora and fauna, which have profoundly influenced the biogeographical evolution of India.

1. The peninsular amphitheatre, which is perse India vera biogeographically and the home of an ancient endemic relict fauna; 2.the Indo-Chinese amphitheatre, largely outside the limits of India, but extending westwards as a narrow belt on the forest covered ranges of the Himalayas, and overlapping the peninsular amphitheatre in the north-east India; 3. the Malayan amphitheatre overlapping the Indo-Chinese amphitheatre in the north; 4. the Trukmenian amphitheatre, also outside the limits of India, except for a narrow southern fringe enchroaching on the higher Himalayas above the forestline and an area of differentiation of young steppes fauna. The amphitheatre 2 and 3 are regions of differentiation of young, tropical-forest faunas. The amphitheatres 2 - 4 represent the biogeographical appendages of India.

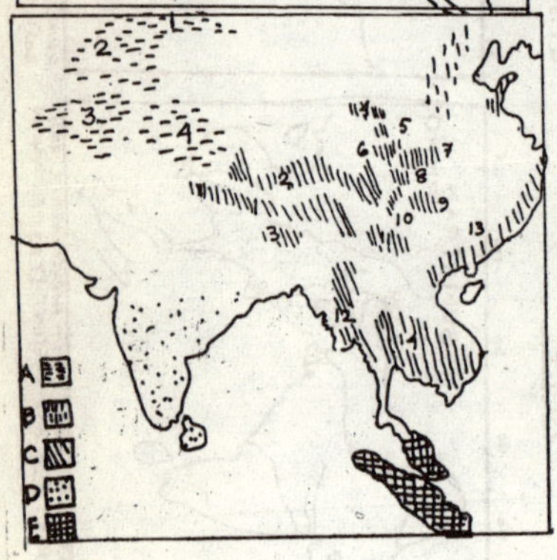

Centres of origin of Tertiary mountain Asiatic faunas.

A. Turkmenian amphitheatre,

B. Planchurian amphitheatre,

C. Indo-Chinese amphitheatre & Malayan amphitheatre,

D. The Peninsular amphitheatre lies outside the region of Tertiary mountains and is a much more ancient centre of evolution of flora and fauna. The amphitheatres A and B belong to the Palaearctic Realm, and C, D and E to the Oriental Realm. The subcentres A are

1. Ala-Tau-Tien-Shan mountains.
2. South Turkestan.
3. Afganistan and
4. Northwest

Himalayas. The subcentres in C are

1. the forest covered Himalaya to Assam, 2. Tibet, 3. Khasi-Jaintia hills, 4.-11. Eastern Tibet and Yunnan, 12. Burma, 13. Indo-China, 14. Thailand.

BIBLIOGRAPHY

A. Proceeding (1978) National Seminar on Resources Development and *Environment in the Himalayan Region*

Borthakur, D.N. et al - Shifting Cultivation in the North-Eastern Region-(1978 National Seminar).

Burrard, S. and Heron A.M.- The Geography and Geology of Himalayan Mountain (1932).

Jain, S.K. et al - Plant resources of the Himalayas (1978 in National Seminar).

Khajura H. Wild Mammals of the Western Himalayas.

Mehra, K.L. et al - Plant genetic resources of the Himalayas (1978 in National Siminar).

Rao, Y.P. The Himalayas-Their Climate (1978 in National Seminar).

Seth, S.K. Forests and Forestry, in the Himalayas (1978 in National Seminar).

Singh, J.S. et al - The grass cover of the Himalayas Region (1978 in National Seminar).

Swaminathan, M.S. Agricultural Systems in the Himalayas (1978 in National Seminar).

Thingran V.G. et al - Aquatic life-Resources etc. (1978 National Seminar).

B. Abdulali H. 1965. The Birds of Andaman & Nicobar Islands. Jan. of Bom. Nat. His. Soc.

Alcock A.W. 1910. A cataelogue of Indian Decapoda etc. Rec. Ind. Museum.

Annandale T.N. 1911. Fresh water sponges etc., Fauna of British India.

Arunachal Forest News, 1980, 1987. Govt. of Arunachal.

Biswas K. 1941. The flora of Akahills. In For. Rec.

Blattee 1986. A note on the Organisation and activities of Andaman Forest Deptt.

Blandford W.T. 1970. Some notes on Reptila and Amphibia from Central India Jan. Asia. Soc. Bengal.

Bor N.L. 1938. A sketch of vegetation of Akahills. Ind. For. Rec.

Burkill H. 1924. The Botany of Abor hill expedition. Rec. Bot. Sur. India.

Champion H. & Seth S.K. 1968. Revised Classification of Forest types in India.

Chatterjee D. 1939. Studies on Endemic Flora of India and Burma. Journal Asia. Soc. Bengal.

C.M.F.R.I. Bull. 34 Sept. 1983. Mericulture potential of Andaman & Nicobar Islands.

Christopher S.R. 1933. Fauna of British India.

Day F. 1870. Of the Fishes of Andaman Island.

Elwes H.J. 1873. On the Orinithographical distribution of Asiatic Birds. Proc. Zoo. Soc. London.

Good R. 1947. Geography of Flowering plants.

Jayakar K.C. 1974. Ecology and distribution of Fresh water fishes, amphibia & reptilia.

Kurup G.U. Mammals of India and Mammalian Geography of India.

Kurup G.U. 1966. Mammals of Assam and Adjoining Areas.

Kurz S. 1875. Flora of Burma.

M & B Committee. Deptt. of Science & Tech. Preliminary inventory on potential area for Biosphe Reserve.

M & B Committee Deptt. of Environment. 1981. The Namdapha Biosphe Reserve.

Mani M.S. 1968. Ecology and biogeography of high altitude insects.

Mani M.S. 1974. Ecology and Biogeography of India.

Mc. Vean D.N. 1976. Reports on Landuse in Andaman & Nicobar Islands.

Mukherjee A.K. 1982. Endangered animals of India.

Nair S.C. 1983. Natural Resource Conservation Dev. Report.

Prasad B. 1942. The Zoogeography of India.

Puri G.S. 1960. Indian Forest Ecology.

Rao G.C. et al 1985. The fauna of the Bay. Journal of And. Sanc. Society.

Rao V.M.K. 1983. Interim Report of Biosphe Reserve on North Andaman & Nicobar Is.

Sahni V.C. 1983. Bot Expedition to Great Nicobar.

Sclater P.L. 1891. On the Geographical distribution of birds-Ibis.

Smith M.H.1934, 1938, 1943. The Geography of mammals.

Spate UHK 1951. India & Pakistan-A General & Regional Geography.

Talbol G. 1939. Fauna of British India.

Wallace A.R. 1876. Geographical distribution of Animals.

Ward K. 1921. The distribution of Flora in S.E. Asia. Jouranal Ind. Bot. Soc.

PART III

QUALITATIVE AND QUANTITATIVE ECOLOGY AND ENVIRONMENT OF THE VEGETATION OF EASTERN HIMALAYAS

This part is dedicated to the memory of those foresters of the districts of Darjeeling and Jalpaiguri who laid down their lives for the protection of forests and wildlife and their conservation.

Contents

Synopsis of Charts, Maps Diagrams and Histograms etc.

Preface

Very reliable quantitative assessment on density and volume per hectare on various commercial timber trees of all the Himalayan States viz., Jammu and Kashmir, Himachal Pradesh, Uttar Pradesh, Western Nepal, Bhutan, Arunachal Pradesh (and also of all the hill states of eastern Himalayas) have been compiled by Forest Survey of India. Qualitative information of the types of flora, dominant, co-dominant and conspicuous vegetation including ground flora is available in the works of Champion and Seth (1908) and regional working plan authorities and publications in The Indian Forester and Bulletin of Botanical Survey of India. But comprehensive quantitative assessment an all the synusia of vegetation such as ground cover, herbs, shrubs, climbers, epiphytes and trees has not been done on the flora of eastern Himalayas over a large area to represent the entire eastern and Northeastern Himalayan flora.

Himalayan flora is so rich and varied and covers such a wide area that a team of botanists and ecologists will only be able to undertake such type of work for the entire Himalayan ranges. For the present quantitative estimation of vegetation two districts of Darjeeling hills viz. Jalpaiguri and Darjeeling covering about 8300 sq. km was selected.

Very careful preliminary study was made in selecting the area of study so as to cover tridimensional natural features such as latitude, longitude and altitudinal wide ranges. It was also considered which region would cover maximum altitudinal ranges yet accessible, represent (i) all the forests from plains to at least 3500 m uninterrupted, (ii) rich foot hill vegetation, (iii) areas free from shifting cultivation, (iv) wide range of species of moist deciduous, (v) areas free from fire hazards and (vi) at least 50-60 percent of the species of entire eastern and north-eastern India (Himalayas and the adjoining offshore of Himalayan mountains).

No region was found as productive and interesting as representative as the aforesaid two distrtics which satisfy all the major qualities described in the previous paragraphs. Within the small expanse of 8300 sq. km. floral resources of alpine, upper, middle and lower temperate regions, montane region and undulating plans were available for study.

The flora of this region had attracted the attention of Sir J.D. Hooker (1854), J.S. Camble (1895), Sir D. Prain (1903), Dr. D. Brandis (1906), I.H. Burkill (1916), A.M. and J.M. Cowan (1929) and others (Kurz, King, Clarke, Anderson and several others). Most of the works of these explorers were confined to listing the vegetation and no information on density, frequency, cover, abundance, etc. was available. None employed modern methodology and techniques in such studies.

The study area is characterised by Bhabar, Terai, Duars and alluvial flats in the plains and Himalayan mountains upto an altitude of 3500 metres and covers a wide variety of forests. The floral elements are diverse and very large in number and included herbs, shrubs, climbers, epiphytes, grasses, sedges, other monocots and ferns. Although plants and animals are considered integral parts of the forests very little attention has been paid about the role played by flowers (nectar, pollen etc.) fruits (size, colour, edibility etc.) fodder plants, insect-host plants and layers (vertical and spatial) of vegetation in providing food and shelter to the wildlife. The present study takes some such factors into consideration.

The present work, which is the result of the author's cumulative efforts spread over several years in the field and laboratory is expected to meet the needs of a long-felt want in the matter of quantitative information relating to the diverse types of vegetation of the districts of Jalpaiguri and Darjeeling. Extensive data have been collected on various layers of vegetation and these have been analysed by employing standard methods. It is also expected that the data and their analyses will form the basis of future work and stimulate further research on the plants of this area. This eco-botanical and environmental study is also likely to prove extremely useful in experiments on plants-genetics, plant-breeding, plant-geography and to scientific forestry management, agricultur, horticulture, etc; and as the study area is also extremely rich in avifaunal and animal population in having about 600 species of avifauna, 100 species of mammals and a large numer of species of insects (Coleoptera, Diptera, Isoptera, Orthoptera, Lepidoptera, Odonata etc.), reptiles, lizards and amphibia, this should attract greater attention of the zoologists, biologists and ecologists as well.

The study presented in this work is likely to furnish sufficient data to assess the forest wealth of this area, such as commercially important timber resources, availability of medicinal plants, wildlife food resources and plywood, box-wood and furniture-wood resources. Our country is passing through a phase of industrial and agricultural revolution. In order, therefore, to meet the challenge of the time in the context of our population explosion, we should not only endeavour to preserve our forests, we should also make a proper assessment of the existing forest resources.

The plants collected by the author were indentified in the herbaria at Shillong, Lloyds Botanic Garden, Darjeeling, the Botanic Garden at Sibpur and in the Forest Research Institute, Dehradun. The specimens were finally kept in the herbarium of the Lloyds Botanic Garden Darjeeling and at Forest School at Kurseong.

It is hoped that the methodology of study, analysis, findings will indicate a new line of approach for similar studies and provide voluminous data for ecological, botanical, environmental, biological, genetical and commercial studies of the flora of this region,. The foresters, botanists, agriculturists, agronomists, ornithologists, ecologists and environmentalists will find the work useful in the pursuit of their scientific studies.

11

Inventory Design - Methodology

The broad objectives of the study are:-

i. to determine the density and percentage of occurrence of species.
ii. to determine the frequency of occurrence and dispers of species.
iii. to bring out salient ecological features of herb, shrub tree and other layers of vegetation.
iv. to identify various composition of crop and to group them into forest types.
v. to list the conspicuous and representative vegetation of the forests.
vi. to enumerate qualitative and quantitative characters of vegetation.

Pilot Study:

No adequate and reliable field data was available on which a realistic and workable design could be made. As such a pilot project was made to determine optimum plot size and the number of plots to be inventoried in different types of forests.

Stratification:

The basic approach in respect of stratification was to divide a heterogeneous forest composition into homogeneous sub-population to facilitate the course of investigation. Geographical contiguity, density of crop, condition of terrain and working in the forest etc. were some of the basis for stratification.

On close examination of forest composition of the plains it was found that *Shorea robusta* (Sal) was the dominant crop over extensive areas and two plant families formed bulk of the associates. These two families were dominant where Sal was of low occurrence. In the hills no such classification was possible; but the attitude formed the basis of classification.

Forests in Plains:

In the plains an analysis of the percentage of occurrence of three major families viz. *Dipterocarpaceae* (Shorea), *Lauraceae* and *Meliaceae* has been found to indicate the following percentage distribution in a sample of 50 observations at various places:

Type	*Shorea robusta*	*Lauraceae*	*Meliaceae*	*Others*
Pure Sal	50	10	8	32
Scattered Sal	25	13	12	50
Wet mixed	6	15	20	59
Dry mixed	3	13	14	70

Based on the above observation, four strata namely Sal, Scattered Sal, Wet mixed, and Dry mixed were considered for forests in the plain.

In case of hill forests, it was observed that there was a variation in the vegetative characters with the changes in altitude. As such, the altitudinal variation was taken into account to stratify the hill forests. Thus 10 strata were considered as detailed below:-

Strata	*Altitudinal Range*	*Major genera*
1	300 m.	*Shorea - Terminalia - Stereospermum - Syzygium - Phoeba - Beilschmeidia*
2	900 m.	*Terminalia - Schima - Bauhinia - Castanopsis*
3	1400 m.	*Castanopsis - Engelhardtia-Beilschmiedia*
4	1600 m.	*Castanopsis - Schima - Nyssa - Acer*
5	1800 m.	*Michelia - Castanopsis - Enhelhardtia.*
6	2000 m.	*Castanopsis - Schima - Michelia-Acer - Engelhardtia - Nyssa - Garuga*
7	2200 m.	*Acer - Quercus - Machilus - Michelia*
8	2400 m.	*Machilus - Quercus*
9	2600 m.	*Machilus - Quercus - Michelia-Acer*
10	2800 m.	*Tsuga - Abies - Rhododendron*

Plot Size and Layout:

Plot size (i.e. size of quadrat) is an important characteristic in assessment of the vegetation. Larger units may involve bias and is also inconvenient from operation point view while a plot of small size is easy to lay out.

Plot of various sizes were laid out at point 'O' representing one corner of the quadrat. 3 strings ab, cd, ef were stretched at an angle of 45^0. Diagonal distance was calculated and marked on the string ab whereas on strings cd and ef the side lengths of the squares were marked m - m", n - n" & p - p" and the right angle intersection of these points were marked on the diagonal as m', n' & p' respectively. At 'O' a permanent post was fixed. The recorder himself stood at the corner diagonally opposite to 'O' a permanent post was fixed. The recorder himself stood at the corner diagonally opposite to 'O'. One person each with a ranging pole indicated the position along the strings cd and ef. The fourth man enumerated the species and individuals and marked the same to avoid double counting. The positions are shifted accordingly for each expanding quadrat.

Based on the above method quadrats of 2m x 2m were initially laid out. The size was gradually increased to 4m x 4m and upto a limit of 10m x 10m according to requirement.

Graphs were drawn co-relating the number of species and the size of the quadrat. It was observed that after a certain interval, an increase in the quadrat size did not result in the number of species and thus there was no significant gain. The curve depicted a tendency for stabilization. From the. above method it become evident that for ground flora including shrubs and herbs, the optimum size of the quadrat was 4 metre x 4 metre and for trees was 10 metres x 10 metres square. The picture is rather different in hills where the size for trees was 10 metres x 10 metres square; for herbs and shrubs it was 2 metres x 2 metres square and 4 metres x 4 metres square respectively.

Allocation of Number of Units (sample size):

The primary requisite is the choice of a suitable sample size (No. of units). A larger sample may be necessary from each stratum so that the total variance of the estimate gets reduced.

The analytical procedure for obtaining the number of plots is given as below:-

Stratum	*Dry/Mixed*	
Character under study	Total	trees
	Y	Y^2
	20	400
	10	100
	12	144
	5	25
	7	49
	4	16
	6	36
	7	49
	21	441
	23	529
	115	1789

$$(Y) = \frac{y}{n} = 11.5$$

$$S^2 = \frac{\sum y^2 - (\sum y)^2/n}{n-1}$$

$$= \frac{17890 - 1322.5}{10 - 1} = 51.8$$

Where S = standard deviation

y^2 = Sum of square of individual observations

n = number of samples considered

Similarly, variance for all the strata were calculated as shown above. The number of sampling units needed to achieve within desired precision for infinite population is given as below:-

$$n = t^2 \frac{\left(\sum_{J=1}^{M} P_j S_J\right)^2}{E^2}$$

where n = total no. of sampling
PJ = areas proportion
SJ = variance of strata
t = 1.96
E = Allowable Error.

Thus the total number of sample plots obtained as above were 856. Allocation of the total sample units proportional to the size of the strata was as follows:-

Strata	*No. of plots*
Pure sal	194
Scattered Sal	237
Wet Mixed	201
Dry mixed	224
Total	856

Hills Forest:

Similar analysis was carried out for all the strata in the hill forest. Optimum number of plots obtained for ten strata were as below:-

Sl No.	*Strata metres*	*Approximate Area of each Strata sq. km*	*No. of plots*
1.	300	180-01	80
2.	900	160-22	65
3.	1400	145-53	55
4.	1600	135-02	50
5.	1800	120-00	45

Sl	*Strata No.*	*Approximate metres*	*No. of plots Area of each Strata s. km*
	6.	2000	110-32	40
	7.	2200	110-32	30
	8.	2400	105-38	35
	9.	2600	92-15	30
	10.	2800	70-02	16
			1228-65	446

LAYING OUT OF QUADRAT

Figure

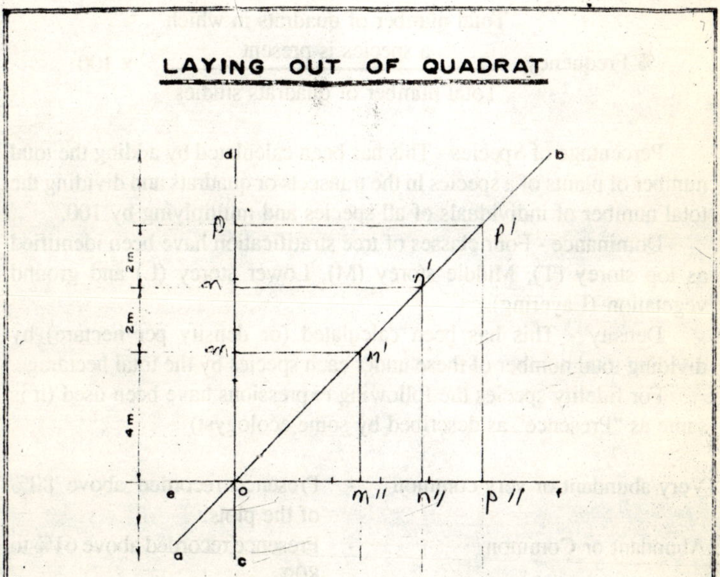

Orientation of sample grids in map:

An area of 1,684 km^2 in plains was divided into 1/2 km. x 1/2 km square grid. This generated about 6,000 grids for the area. Each sample plot was located at the centre of the grid. Each of these square plots was numbered and with random number table of Fisher & Yates, a total of 856 plots were randomly selected for the plain forest area. The quadrats were laid out at the centre of each grid i.e. at the inter section point of diagonals.

Similar procedure was adopted for selection of 446 number of plots in the hill forest area.

Presentation of data:

The data collected in the field were analysed and calculated as follows:

Frequency - It is calculated as the percentage of total number of quadrat or transects in which a particular species occurs.

$$\%\ \text{Frequency} = \frac{\text{Total number of quadrats in which a species is present}}{\text{Total number of quadrats studies}} \times 100$$

Percentage of Species - This has been calculated by adding the total number of plants of a species in the transects or quadrats and dividing the total number of individuals of all species and multiplying by 100.

Dominance - Four classes of tree stratification have been identified as top storey (T), Middle storey (M), Lower storey (L) and ground vegetation (Layering).

Density - This has been calculated (or density per hectare) by dividing total number of these under each species by the total hectarage.

For fidelity species the following expressions have been used (it is same as "Presence" as described by some ecologyst)

Very abundant or very common	-	Presence recorded above 80% of the plots.
Abundant or Common	-	Presence recorded above 61% to 80%.
Frequent	-	Presence recorded above 41% to 60%.
Sporadic	-	Presence recorded - 21% to 40%
Rare	-	Presence recorded below 20%

Apart from the occurrence of plants inside the quadrats, a full list of vegetation falling within 10 metres of the plot foundary have been made as enumerated at various chapters.

In analysis of crop frequency, Raunkiaer's findings were comparied. Raunkiaer (1934) while comparing different plant communities determined frequency percentages of all species present and ground them into frequency classes, a procedure called valence analysis.

Frequency	Class	A	species occurring with frequencies of	1 - 20 %
"	"	B	- - - - -	21 - 40 %
"	"	C	- - - - -	41 - 60 %
"	"	D	- - - - -	61 - 80 %
"	"	E	- - - - -	81 - 100%

In his analysis Raunkiaer found that the Classes A, B, C, D and E included 53%, 14% 9%, 8% and 16% of species respectively. He found that it was generally true for all communities that the percentage of species in different frequency classes gave a frequency polygon. This is summarized in Raunkiaers "Law of Frequency" which state that A>B> C $\lesseqgtr$ D > E, Class A includes the numerous scattered rare species with low frequencies and class E will always tend to be high as it includes all the dominant species in plant community.

Form his "Law of Frequency" Raunkiaer suggested "frequency diagrams" as an expression of the homogeneity of a stand. According to him floristic uniformity varies with the values for classes A to E. When classes B to D are relatively high, the stand is homogeneous. In general, the higher the class E, the greater is the homogeneity of the stand. The study area is biotically affected and therefore the pattern of diagram is varied. Species that has high density and frequency is the most important species. High frequency has been considered very useful in plant classification and has been adopted in the present classification forest types.

Shape - In the study a square quadrat has been fixed because it is easy to lay such plots. Circular plots are difficult to lay out in the hills. Rectangular plots, of course, have advantage since it include larger linear variations.

Size and Number - This has been arrived at by species - area curve and quadrat number - Quadrat size curve.

Cover (area or space) - is the area occupied by all the individuals of a species and is expressed in percentage of the quadrat. Cover is of two types viz., Canopy cover and Basal cover, the former being applied for trees and the latter for herbs and shrubs.

In his analysis Raunkiaer found that the Classes A, B, C, D and E included 53%, 14%, 9%, 8% and 16% of species respectively. He found that it was generally true for all communities that the percentage of species in different frequency classes gave a frequency polygon. This is summarized in Raunkiaer's "Law of Frequency" which state that A>B> C<≥D<E. Class A includes the numerous scattered rare species with low frequencies and class E will always tend to be high as it includes all the dominant species of plant community.

12

Factors of Locality

Climatic

The Climate of the region is of the monsoon type. In the plains there are three distinct seasons, the hot, rainy and cold; in the hills it is humid and cold. Southwest monsoon causes heavy rainfall from May to September; maximum rainfall occurs in the months of July and August. There is very little precipitation during the winter months.

The cold is quite severe in the Darjeeling district. Dews are heavy in the hills and plains. Frost is not common in the region except in higher hills.

Reliable meteorological data on the intensity of light in the forest is not available.

Snow occurs occasionally at Tonglu, Sandakphu and Phalut in winter but does not persist long.

Regional climate is the result of large meteorological phenomena over a greater part of the country, but micro-climate in the forest is the result of minor changes in temperature, relative humidity, evaporation etc. and this has given rise to many freak distribution of species.

The rainfall chart I gives full detail of rainfall in the observed field. Chart II shows the average monthly minimum and maximum temperature in the hills and plains.

The temperature governs the growth and development of the vegetation. The temperature decreases with increase of altitude and there is a negative correlation between temperature and altitude. The lowest minimum temperature occurs in the month of April, May and June. Chart II shows annual temperature range.

Temperature

The plains forests lie in the moist tropical zone. The average day temperature varies from 15° C to 21° C from November to February, between 27° C to 32° C from May to September, and between 24° C to 27° C during the rest of the year. In the forest tract there is an appreciable variation between day and night temperature throughout the year and sometimes winter nights are quite severe. During the months from June to September moist heat during the day and early evening are oppressive but the nights are cooler.

In the Tista Valley and foot hills the climate is sub-tropical while in the hills it is temperate. The bulk of the area of forests is in the temperate climate. The winter is extremely cold and extends from December to March. The summer spreads from April to Mid June and during this period Darjeeling is extremely cool. The rainy months are from June to September. The mean maximum and minimum temperatures of Darjeeling are 47⁰ F to 35⁰ F (8.25⁰ C to 1.67⁰ C) during January and the hottest seasons they are 66⁰ F and 58⁰ F (18.70⁰ C and 14.30⁰ C) respectively. The hottest time of the year is May. The lowest minimum temperature is reached in February but January on the whole is the coldest month. In Darjeeling temperature fall below the frezing point. Above 2,200 metres there is snowfall once or twice a year. In Singalila Range snow persists for number of days.

Chart - I.
Distribution of average monthly rainfall
(In milimetres)

Month	*Jalpaiguri*	*Darjeeling*	*Kalimpong*
January	32.2	22.1	19.6
February	17.9	12.2	6.3
March	75.5	55.54	35.8
April	129.3	94.4	52.7
May	254.2	202.6	78.4
June	669.2	752.1	340.0
July	955.2	863.5	520.1
August	776.3	715.5	396.1
September	540.5	520.4	263.3
October	101.0	97.4	57.5
November	17.0	12.9	7.5
December	4.8	2.7	1.1
Total :	3573.1	3351.2	1778.4

Average annual rainfall in mm. of a few area is as follows:-

Buxaduar	5650	Tista bridge	2080	Rangirum	3230
Bholka forest	3900	Sukiapokri	3090	Takdah	2370
Alipur duar	4320	Kurseong	4800	Samdakphu	3300
Kalchini	4320	Samsing	7500	Batasi	2450
Siligiri	3450	Ramam forest	1800		

Chart - II
Average monthly Maximum and Minimum Temperature at some stations in the region (in °C)

Month	*Darjeeling*		*Kalimpong*		*Jalpaiguri*	
	Max.	*Min.*	*Max.*	*Min.*	*Max.*	*Min.*
January	9.68	2.25	15.78	8.03	23.82	10.84
February	10.01	2.97	16.83	8.86	25.13	12.37
March	13.36	6.22	20.08	11.72	28.82	16.39
April	16.44	9.40	27.72	14.58	30.09	20.24
May	17.87	11.83	23.65	16.88	30.31	22.72
June	19.03	13.75	24.26	18.92	30.31	24.26
July	18.98	14.47	24.26	19.31	30.69	24.81
August	19.20	14.19	24.20	19.03	30.86	24.81
September	18.65	13.37	23.76	18.54	30.47	24.09
October	17.38	10.34	22.77	15.57	31.13	21.07
November	13.86	6.22	20.02	10.78	27.94	15.79
December	11.28	3.14	17.00	7.92	25.19	11.61

Edaphic

(a) Geology

(1) Jalpaiguri district

The rock formations in the district occur in the following order of antiquity downwards.

Recent and Sub-Recent	-	Alluvium and older alluvium consisting of boulders, pebbles and sand beds.
Miocene (Siwaliks)	-	Boulder bed, pebble bed and Nahan Sandstone.
Permo-carboniferous (Demuda series of Lower Gondwanas)	-	Fine to medium grained calcareous sand stones interbedded with thin slaty beds. Carbonaceous shale and thin seams of flaky coal. Pebble and conglomerate beds at and near the base.
Precambrian (Buxia Series)	-	Schist, dolomitic limestone and quartzite.
Archsean (Daling series)	-	Slates, Phyllites, banded haematite schists and Darjeeling gneiss.

Archsean System:

The Darjeeling gneiss very frequently passes into mica schists, or into a variety intermediate between the two, e.g. a felspathic micaschist or gneissese schist. The actual gneise is composed of grey quartz, white felspar and dark brown and silvery mica. It varies in texture. Banded haematite quartzites found on the hills north-east of Raimatang and elsewhere.

Buxa Series:

Rocks of this series are observed in the Titi river in the vicinity of Lankapara and is well developed in Buxa and Jainty hills. The lower most rock type here is a dark grey dolomite which forms a band ranging in thickness between 30 and 90 metres.

Gondwanas:

There are a few isolated exposures of Gondwana rocks in the area. Sandstone, shale and flaky coal are exposed below a thick cover of boulder and pebble beds, on the right bank of Daina river, close to the India-Bhutan border. The sandstone is greyish white, soft, medium grained and is visible for a distance of 50 metres along the right bank. The coal is much weathered and its presence in situ can only be inferred by the occurrence of black clayey soil with lumps and flakes of coal.

Another exposure of Gondwanas occurs on the left bank of Chamurchi river close to the India-Bhutan boundary and in Jainty.

Siwalicks:

The siwaliks comprise micaceous and arkose sandstones, bluish and greyish siltstones, conglomerates and pebble beds.

Older Alluvium:

After the uplift of the Himalayas, a fore-deep extended from east to west fringing the Sub-Himalayas of Western Duars. This fore-deep was gradually filled up by the quarternary terraces in areas east of Tista river. Evidences of upheaval of the fore-deep are noticed in the vicinity of Chalsa Railway Station, at Matiali and Nagrakata where such terraces form elevated plateau a few hundred metres above the present level of the flood plains.

Economic Minerals:

Coal:

Exposures of flaky graphitic coal are exposed in the Chamurchi and Diana river. Flaky graphitic coal an shale are noticed in Turturi stream.

Limestone:

Limestone of aggregate thickness of more than 80 metres is exposed along minor slips and scarp faces in some localities. Intercalations of chlorite-schists are noticed also. On a five metre thick portion of the band appears to be free from schists.

The results of the analyses show that the limestone of this area siliceous as well as high in magnesia. Limestone also occur at Buxaduar along the path leading to Chunabhati.

(2) Darjeeling district:

(a) Rock :

The geological formation on the southern hills consists of unaltered sedimentary rocksand over the rest of the area different grades of metamorphic rocks. The southern areas have a characteristic feature where the older formations rest on the younger, displaying a complete reversal of the original order of superposition.

From an ancient sea that accumulated sediments this great range was elevated in the tertiary period. The mountains are made of folded rocks. The formations have granite intrusions at many places.

The present configuration of high peaks and deep valleys have been carved by wind, water and snow. The products of disintegration of the mountains have been swept over the sub-mountain tract as the rivers debouch into the plains.

Darjeeling gneiss (Daling series) which is the most important rock rests under variety of foliated and banded metamorphic rocks, partly sedimentary and partly igneous in origin. The foliated types are usually mica schists. The banded rocks are gneisses and have been formed by injections of granitic fluid along the micaceous layers of the schists. Where soaking has been thorough, gneisses approach granites in composition and are made of biotite, muscovite, quartz and feldspars.

The sedimentary varieties of Darjeeling gneiss contain garnet, sillimanite, Kyanite and staurolite. It also carries bands of quartzite. The tertiary beds crop out between Sukna and Chunabhati, the coal bearing Gondwanas below Tindharia, the daling rocks between Tindharia and north of Gavabari and the Darjeeling gneiss over the rest of the distance.

The Daling series also occur between Kalijhora in the Tista-Valley and Rangpu and extends into Sikkim. It is present in the Rangit Valley below Darjeeling and Ghum. Elsewhere in the Darjeeling hills Darjeeling gneiss is common. From the deposition of low grade metamorphic rocks underlying highly metamorphosed ones, some geologists consider the Daling Darjeeling gneiss as two distinct series and maintain the latter have been pushed over the former and separated from it by a metamorphosed under part of a great sedimentary succession of which the Daling represent the lower part. No final decision has yet been reached in the matter and the age and relations of the Darjeeling gneiss are uncertain.

(b) Soil :

Soil of Forest land all over the plains and hills is slightly acidic. Soil has high percentage of organic matter and nitrogen.

Cryptomeria soils are acidic and the calcium status is is higher than other forests in the hills. Nitrogen content is not higher than other forests (natural). Such soils need sulphate and phosphate fertilizer.

Shorea robusta grows well in a heavy but will drained soil derived from acid rocks. Iron is beneficial as the purest Shorea crops are found in areas where iron ore predominates. Acidic, feruginous, base deficient soil seems to be ideally suited for the luxuriant growth of *Shorea*.

Although very rough survey on the forest souls have been made, no survey on scientific and organized basis has been done. This is a problem of considerable importance to the ecologists. In this region climate plays a fundamental role in limiting climaxes, but there are significant patch of vegetation which are effects of edaphic condition. The geological features of the region are responsible for great diversity of rocks and subsequent formation of soils. There are areas of immature soils.

Sandy leam predominates in the Tista and Rangit valleys having been derived from Daling rocks. Clay or light brown soils is derived from gneiss.

The soil of the plains is allvial. In Bamanpokri the soil is micaceous and sandy. In the Pankhabari region conglomerates and sandstones are predominant. Higher up coal seams, Sandy clays, Sandstones, grey silky slates and schists are common.

Biotic :

The biotic factors in the region are of prime importance because of men and other activities which include destruction of forests for extension of cultivation, rehabilitation of displaced persons, shifting cultivation, fire and grazing; others are the influence of Wild Life and Flora.

(a) Animal damage :

Animals are integral parts of vegetation. They derive food from plants and seek shelter. they also react in shaping development, succession and migration of vegetation. All types of deer cause damage to young shoots and leaves of plants. The leaves of *Chikrasia* and *Toons* are mostly browsed among the plantation species. Damage by elephant is appreciable; they trample seedling, uproot young plants and break others. Nilgai, Gaur, Monkeys and Rhinos also browse in the plantations and trample seedlings. Of all these animals the damage by elephants is very conspicuous. In recent years damage by elephants have considerably increased owing to large scale destructions of habitat. Bears strip off barks of Cryptomeria in the hills.

(b) Plant damage :

Massive lianas such as the species of *Butea, Bauhinia, Millettia, Mucuna, Combretum, Vitis, Entada* etc. and various other scrambling plants grow their coils over and through the canopy of the forest embracing in a firm not work. The weight of these massive climbers is considerable and they strangulate, break and deform the host plant. In moist areas masses of *Ipomea, Atylosia, Dioscorea* etc. from an impediment to the free growth of plants. *Mikania cordata* has recently been found to have caused large scale damage to the plantations. *Eupatorium odoratum* and *Lantana camara* damage have reduced in recent years and *Mikania* has probably ousted them. *Loranthus* species are the chief plant parasites. They cause damage to the plantations of *Michelia champaca, Shorea robusta, Gmelina arborea, Biscofia javanica* and many other species. In the hills troublesome weeds in the plantations are *Rubus* species, *Viburnum erubescans, Girardiana palmata, Arundinaria* and *Strobilanthes* species. *Viscum album* a parasite has been found to damage *Quercus lineata* plantation in the hills. Lichens and mosses in the hills cause danger in case of fire.

(c) Fire damage :

(1) Prior to the reservation of the forests made in 1885 the forests got burnt annually. Each piece of shifting cultivation was a potential focus from which fire originated. Only the damp sites escaped. Strict fire protection was enforced when the management of forests was taken over by the Forest Department. Since then natural regeneration of Sal *(Shorea robusta)* has ceased to establish. Mention must be made of 125 sq. Kilometers of Baikunthapur Sal forest where fire and grazing control was not strictly enforced. Coarse grasses have invaded the area, humus has been comparatively burnt away, trees are crooked and deformed, and the regeneration of Sal in plenty.

(2) In the Sal forest of Jalpaiguri district no natural regeneration can be seen and most of the crops are middle aged to mature. The effect of these fires and subsequent protection on plant succession exert a great influence on the ecological balance of the flora. Annual fire transformed some forests from a moisture to a drier type and ultimately to grassland. Young growths are all killed by fires and the light that are admitted brings in grass which again suppresses tree regeneration. At one stage controlled burning was introduced in the forests, but it has been found that fire of any kind is inimical to evergreen forest. Evergreen species are fire tender but some deciduous species including *Shorea robusta* are fire resistant. Periodic burning helps Shorea to regenerate. Strict fire protection renders a forest towards moist climax type.

(3) In the hills, the risk of fire damage is great due to very dense growth of bamboos *(Arundinaria* species). In the year 1878, 1879, 1882 and 1909 fire swept the Conifer - Rhododendron forests of the upper hills of Darjeeling Division and caused great damage to the dense conifer forests. Even now one finds the charred and blackened stems of trees standing in grassy blanks. The grasses of such blanks offer a great scope for ecological studies. Extensive grassland occurs in the upper hills which are results of fire followed by grazing. There is no fire climax type of grassland in the tract. Senchal pasture was artificially created in 1879.

(4) Fire repeatedly kills young growth and gradually transform forests to grassland. Fire of any intensity is inimical to evergreen forest. Some species of bulbous and rhizomatous herbs which flower when there is no cover and some such species throw their leaves viz. *Curcuma zedoaria, Pancratium verecundum, Hypoxis aurea, Grewia species, Ochna pumila, Careya herbacea* etc.

(5) In forest management and silviculture, burning has been practiced as one of the methods for obtaining regeneration. Burning destroys organic acids and liberates bases. Burning destroys fungi and promotes multiplication of bacteria and reduces Protozoa. Burning changes chemical composition of Soil and their permeability.

(d) Grazing :

The plain forests are being grazed illicitly by about 80,000 cattle daily which is sure to result in eradication of tender species and survival of tough and limited number of species upsetting the pattern of composition. In the hills selected fodder species are being collected heavily and trees are being lopped. This direct impact on several selected species is definitely going to cause obstruction to their function of flowering and fruiting and propagation.

Urination and dropping of grazing animals and wild-life species have considerable effect on the various species in grazing ground. Different animals have their own pattern of grazing and the vegetations suffer charges accordingly. Herb layer of forest floor is mixed with tree regeneration, and therefore, study on herb flora cannot always be done in isolation. Soon the tree regeneration grows taller surpassing the height of the herbs, and the cycle continues. Such mixed layer of vegetation suffers from grazing, browsing scratching and pawing of domestic and wild animals. A 200 kgs. cow or buffalo will void about 0.050 m^2 of ground each day. It is estimated that the forests of Jalpaiguri district (Plains area) 1,800 km^2 is grazed by about 50,000 cattle; they add 10x 50,000 kgs. or 5,00,000 kgs. or 5,000 quintals of dung everyday which cover about 12,500 m^2 of forests everyday. Large dung dropping brings disaster for the plants beneath and around it. The smothering effect of dung has species-specific effects. Close to dung, herbs grow profusely. Animals dislike to graze close to their dropping and avoids to graze on plants growing near the dropping. Even the animals are unwilling to eat a herb fouled by member of its won species. Detailed study on this line has not been done. Besides that a grazing cow or a buffalo walks about 5 - 6 kms. a day. Hoof pressure is great (1,000 - 1,200 gms. 1 cm^2) for cow or buffalo. Such pressure causes damage to herbs, shrubs and soil. Vast elephant population in North East India cause a great change to vegetation. On an average a Buffalo/Heifer urinates about 51 lit. a day. This leads to 5 x 50,000 or 2,50,000 litre a day on forest soil. But its effect on vegetation has not been studied. Plants with epigeal germination are most sensitive. Dicot plants have mostly epigeal germination. Grasses have hypogeal germination and they survive and multiply.

(e) Shifting Cultivation :

Selected foot hill areas have suffered the depredation of shifting cultivation. Such burning and felling of forest for temporary cultivation has set back the successional trends in the forests. In the foot-hills of Kalimpong forests invasion of bamboo and climbers are results of shifting cultivation and burning. This was destroyed natural vegetation and fertility of soil and extensive erosion is the result. After abandonment of cultivation scrubby vegetation that succeeds may not always hold the soil against erosion. May such devastation can be seen outside the reserved forest all over the district of Darjeeling.

(f) Population :

The incidence of population per square kilometre is about 600 at present and there has been steady increase by 10 percent per square kilometre. Accordingly the demand for forest produce has increased tremendously and there has been constant pressure for more and more forest produce for which forests have suffered a lot.

(g) Fungal damage :

Trees get attacked by a variety of fungi. Fungal epidemics are not uncommon. Most menacing Shorea robusta fungi are *Fomes albomarginatus, Fomes lamaoensis* and *polyporus shoreae*. Many other fungi belonging to genera *Deedalea, Ganoderma, Hexagonia, Lenzites, Polystictus, Stereum* and *Tremetes* damage Sal.

Many species of Fungi also damage *Acacia catechu, Albizzia* sp., *Bambusa* sp., *Toona* species, *Eugenia* and *Syzyzium* species, *Teactona grandis, Terminalia tomentosa, Abies densa, Taxus baccata* and *Tsuga dumosa.*

(h) Lopping :

This is local problem in some parts of the hills where many tree species are damaged. The trees become crooked, get fungus attacked and ultimately die. Removal of leaves from the trees for the purpose of fodder minimise the formation of litter and forest soil is impoverished in organic matters. Innumerable plant species are lopped of which following suffer most:-

(1) Hills :

All *Ficus* species, *Saurauia* species, *Pilea* species, *Elatostema* species, *Eriobotrya* species, many grasses, species of the families Urticaceae and Leguminosae. Some ferns are also used as fodder.

(2) Plains:

All species of *Ficus*, *Premna* species, *Bombax ceiba*, *Tinospora cordifolia*, *Trema orientalis*, *Callicarpa arborea*, *Litsaea* species, *Garuga pinnata*, *Artocarpus chaplasa*, *A. lakoocha*, *Bauhinia* species, *Streblus asper*, *Trevesia palmata* and very many species of grasses.

(i) Insect damage :

Insect oe considerable damage to the trees in the plains. *Hoplocerambyx spinicornis* is a borer of *Shorea* and does extensive damage to trees. *Lymantria mathurai* is a leaf defoliator of *Shorea* and causes considerable damage. *Hypsipyla robusta* and *Pagiophleeus longiclavis* cause damage to the *Meliaceous* species *(Toona ciliata, Chikrasia tabularis* and *Swietenia macrophylla)*. *Tonica nivifera* the shoot borer cause considerable damage to *Bombax ceiba*. *Hapalia machaeralis* causes defoliation and *Hyblea puera* skeletonises, *Tectona grandis* over considerable areas. *Michelia champaca* is damaged by *Urostylis punctigera*. Young *Tectora grandis* plantations are attacked by the cankergrub *Dihamus cervinus*. *Calopepla leyana* attacks *Gmelina arborea*. In the hills *Bactocera horsfieldii* attacks twings of *Cryptomeria* and *Juglans regia*. Another grub of *Cryptomeria*, *Alnus*, *Michelia* and *Eucalyptus* is *Phassus hepialidor*.

Insects are parts of plant communities; some destroys plants, others live in forests, decompose dead matter of forest litter and help luxuriant growth of vegetation.

(j) Microfauna :

The action of micro fauna and flora has not yet received much attention as they deserve. Their effect is very rapid and is evident by the quick rot, deformation of trees and decomposition of litters.

(k) Drought :

Drought may be very serious for vegetation even in areas of heavy rainfall. The drought period occurs either in premonsoon or post monsoon periods. Tops of the trees dry up and plants are killed outright; sometimes top become stagheaded. Smaller trees are killed either sporadically or in patches.

Physiographic :

The tract dealt with in this paper comprised the northern two districts of Darjeeling and Jalpaiguri covering an area of 8318.97 sq.km. The area lies between 88°18 E longitude and 26°20' N and 27°13' N latitude. The bounderies are briefly:-

North	-	The Sikkim and Bhutan States
East	-	Assam State
South	-	Coochbehar District
West	-	Nepal State and Purnea district of Bihar.

Major portions of the Darjeeling district are mountineous and is intersected by numerous torrents, deep narrow valleys and precipitous hill slope are special features of the district.

The summits and upper slopes of the important ridges are occupied by forests. The Tista and Ranjit valleys have forests in the slopes. The hilly areas are confused labyrinths of ridges and narrow valleys. The highest ridge is at Phalut in the Singalila Range. The ridge is 3,600 metres high at Phalut and descends to Manebhanjan (2,000 metres) and it is on the upper slopes of the ridges that the forests of Tonglu and Singalila Ranges are situated. The ridge undulated from Manebhanjan east-wards upto Ghum and then rises steeply to Tiger Hills. The Takdah - Pesoke ridge which originates from Tiger Hills descends to the junction of Rangit with Tista and also covered with forests. Another ridge on which Darjeeling is situated bifurcates and both descends down to Rangit via Takvar and Lebong. Most of the forests are situated between 2,000 and 3,6000 m. In the Kalimpong hills the southern portion of Chel, Neora and Jaldacca forests adjoining the Jalpaiguri district have moderate slopes. The tract then rises abruptly and form the outer ranges of the Himalayas. The altitude varies from 100 metres to 3,400 metres. The hills consist of numerous valleys - the slopes of which vary considerably from moderate to precipitous. the main rivers flow from north to south directions.

The forests of Jalpaiguri are in the plains and the rivers flow from north to south. There are innumerable rivers and streams. Some originate in the hills and some in the plains. During cold weather small rivers dry up, the water having gone underground.

13

Salient Features of Vegetation

General Observation:

One of the outstanding features of the physical geography of the study area is the diversity of plant cover which varies enormously in density both in vertical and lateral layers. There is an extraordinary luxuriance of vegetation of diverse forms. The physiognomy of the plant cover is so complex that only a trained specialist is able to interpret it and find out the relationship that exists between the plant cover and the surrounding vegetation.

The past geological history is rather obscure and it is difficult to find out the phylogenetic relation and distribution of species. Some advocated that there was a land bridge between the northern and southern hemispheres and maintain that there had been a free infiltration of the flora of the southern hemisphere into the eastern Himalayas.

Many areas of this region have not been fully explored though there were geographically quite well known. In the adjoining Sikkim hills the great Himalayas from the innermost line of very high ranges. They rise above the limit of perpetual snow having an average altitude of more than 6000 metres. On their summit are situated the world's highest peaks like the Mount Everest, Kanchenjunga etc. The width of 60 kilometres includes deep ravines and groges. The outer Himalayas is known as *Siwalika* having an average height of 1300 metres. The intermontane zone between the ice covered main ranges of the north and Siwaliks of the south is characterised by the presence of huge gorges. The altitudinal effects on the varying elements of climate, namely, temperature and rainfall are classically represented by the hills between Sukna and Darjeeling. At higher elevations a sudden contrast of climate results from small differences of aspects, slopes and configuration of ground.

The study of microclimate and its influence on the plant species of plains and hills forests throws some light on the ecological aspects of vegetation. The extremes of moisture and temperature are vital for distribution of vegetation types.

In the hills moisture-laden strong monsoon wind influences the character of the vegetation. The ramifying outer spurs have a very heavy rainfall and are densely clad by moist forests The moisture laden breeze of upper layers of the temperate zone passes over the outer spurs only to be arrested by high ridges where they deposit their moisture. As a result a moist temperate flora develops. This is made up of moss clad lichen draped trees and a huge assemblage of cryptogamic species.

The temperate forests extend from 1300 m. to about 3100 m. In the upper limits there are conifer forests. Down below the forests are characterised by broad leaved evergreen species. There is no clear demarcation line between the two zones as one passes into the other either very gradually or abruptly. Conifer forests are represented by *Abies*, *Tsuga* and *Taxus*. Alpine forests do not occur in the limits of West Bengal but there are patches of grasslands and vegetation showing influence of extreme climates. *Pinus patula* and *Cupressus cachmeriana* have been extensively planted. But principal vegetation consists of mixed broad trees with no species growing gregarious over large areas.

Many plants of the region have world wide distribution and many others have discontinuous distribution, occurring only in a few but sometimes widely separated areas. Some species adapt to a very wide range of habitat and easily cross the climatic barriers.

The monsoonic rains as well as the key position among several sub-floral regions viz. Tibetan and Chinese massifs to the north and northeast, Indo-Malayan plateau to the south east, Gangetic plain and the plateau of the Peninsular India to the south and south west make the vegetation of this area very rich and varied.

The Shorea robusta Gaertn, and the mixed forests are the most gregarious and luxuriant but with no regeneration. The relative humidity and rainfall are high. The total light intensity in the case of forest type mentioned above should generally be less then temperate forests, but in the latter type fog reduces the intensity appreciably. There is no periodocity of light during the year. There is abundance of carbondioxide in the lower stand layer, wind action is much less and micro-climatic effect is equable. Canopy is some-what irregular owing to felling of some mature trees and presence of wet-mixed forest type pockets. The stem from is also irregular and the bark form various; crowns have moderate branching; the undergrowths are in profusion; epiphytes and climbers are plentiful.

The best forests of North Bengal plains are moist deciduous with *Shorea* as principal crop with fair admixture of evergreen species hygrophilous in character, atleast 30m. high rich in thick stemmed lianas and in woody as well as herbaceous epiphytes. There are several types of *Shorea* forests where Lauraceae and Melianeae species predominate. Extensive grassland of Savannah type occurs in the plains *Shorea* forests; several rivers cut up such *Shorea* forests and the sandy banks of the rivers bear extensive *Dalbergia, Sissoo, Acacia, Catechu* forests. Birds and animal population in the study area is very high. The riverain grasslands have a maximum concentration of animals. In pure *Shorea* forests top canopy or stratum is somewhat continuous though in the Dry and Wet-mixed the strata it is discontinuous. Death of Scattered trees and felling sometimes make the canopy discontinuous (Profile diagram 1,2,3, and 4).

Overwhelming majority of trees are woody including some epiphytes. The undergrowth largely consists of woody plants, seedlings and saplings of trees, Shrubs and herbs and woody climbers. Some plant families which in temperate hills are represented exclusively by low growing herbs assume woodiness of trees in this area. Several plant families are represented by tree species. Undergrowths, canes, climbers and bamboos form thickets which even elephants can not penetrate; stout woody climbers ascend to the top of the highest trees thus ornamenting them with festoons of fragrant flowers. There are heteromorphic plants such as saprophytes and parasites, though their number is relatively small.

Buttresses are not very common but they are quite conspicuous in *Tetrameles nudiflora, Bombax ceiba, Quercus* species *Exbucklandia populnea, Elaeocarpus* species, *Sloanea* species, *Meliosma wallichii* and in several other species. The barks are mostly smooth except in those of *Shorea robusta, Schima wallichii, Terminalia tomentosa, Castanopsis* species and few other species; lenticels and papery bark in many are quite conspicuous. The forms of a few members of flora are widely divergent from normal type, notably the palms,*Pandanas* etc; some have stilt roots.

Many species have clean boles, branching beginning at the top with many angles of divergence in different species. Crowns are of different shape and size, leaves of some trees are crowded at the end of the branches or shoots. (*Mastixia* sp., *Acronychia, Talauma* etc.) Many trees produce coppice shoots (*Shorea, Tectona, Toona* etc.).

Several trees bear flowers directly on the trunk (Cauliflory). There are climbers and climbing shrubs having characteristic tendrils, many having thorns, prickles and spines, various sized and shapes of leaves and types of barks.

There is an abundance of climbers in the lower hills and Drymixed forests. The majority of them are woody, fluted and have great length and thickness. They twist, hang, twine and form an impenetrable net work.

The epiphytic vegetation; besides lichens, some algae, mosses, liverworts and ferns, consists of large number of Orchids and other flowering plants. They have many morphological forms. There are epiphytes on tree trunks, branches and on tree tops; some are semiparasitic and stranglers.

The vegetation of the hills is characterised by the presence of about 250 species of ferns and tree ferns; innumerable species of liverworts and mosses form the ground cover of the temperate forests. About 140 species of thalose and foliose species of lichens occur in the area. *Trentepohlia aurea-a* brick red algae, is very common. Woody fungal flora consists of more than 75 species. There are numerous tree species of low crown. *Betula, Meliosma, Acer, Exbucklandia, Abies, Tsuga* and few other trees have clean boles of appreciable length. Some trees have crowns in different tiers (*Machilus* and *Litsaea* species).

Status of Germination, Regeneration and Age of Principal Crop:

(a) Germination and distribution of seedlings:

It has been observed that there is a significant difference in the distribution of diameter and number of trees of different species which appeared to be partly due to specific variation or characteristics and partly to habitat and biotic factors. The specific and local differences in the relative abundance of the seedlings prove to be still greater. Relative abundance of various species in different storeys differs significantly. Examples of this type are common in various forests. In *Shorea,* seeds germinate but the seedlings do not pass beyond seedlings stage. Similarly, seeldings of many germinate in full light; in many, the seedlings do not advance beyond the unfolding of first pair of leaves. It seems the effect of biotic factors like human interference, fire and subsequent change in soil condition have been primarily responsible for ruthless elimination of seedlings and saplings of several associates of *Shorea.* One fact which is very much prominent is that seedlings and saplings of the species of *Litsaea, Machilus, Bridelia, Dillenia, Dysoxylum, Terminalia, Aphanamixis, Premna* and a few other species can be seen all over the forests. Representatives of many other top and middle storey species are not found in as much in abundance as they should have been found. This indicates large scale mortality of seedlings in the early stages. Seedlings of many survive under the shade of closed canopy. Seeds of many tree species have been tried in the nursery and forest plantations and they have been found to germinate; even in the moist forest floor the condition for germination is ideal. There are also many species whose seeds are eaten up or destroyed on the forest floor before they germinate. Viability of the seeds of Shorea and some other species is of short duration, and within a few days the seeds are either attacked by insects or microorganisms. For this reason seeds of several economically important plants are either directly sown in the field or in the nursery.

Although most of the species fruit every year, there is periodicity in heavy flowering and fruitings. The regeneration from the trees, if at all are patchy all over and it is never very pure and uniform under any crop. Most of the species of top and middle storey can germinate in shade and are somewhat shade tolerant, but there are some strong light demanders like *Shorea* which generally do not survive long.

(b) Age class representation:

Regeneration of forests requires the presence of a sufficient number of seedlings and poles to ensure replacement. Age determination is a difficult study, but size-class shows that there are variations in the abundance of trees in different size classes. *Shorea,* which forms principal crop in the top storey, does not have any saplings and poles in the lower storey as it is not regenerating. There is however, representatives of other associates especially of other common species in various states of development. Flowering by no means results in ripening of seeds. Sometimes no fruit results, owing to unfavourable weather conditions; this happens in species having short flowering seasons. Birds damage many fruits. The seeds in general have high percentage of germination. Moreover, seedlings of many species that are found on the forest floor do not develop into saplings and pole stage.

The difference in size (and age) class representation between associates of Shorea depends on reproductive behaviour of species and on the ecological tolerance in its young stages. No single associate species is gregarious and as such most conspicuous. Regeneration of most of the economically important species is sporadic

The trees differ widely in frequency and abundance of seed production. In many species seed production is seasonal, in many other flowers, fruits and seeds are borne continually in small quantities or at irregular but frequent intervals. *Shorea* flowers and seeds profusely almost every year; *Schima wallichii* also seeds profusely but *Michelia, Chikrasia, Gmelina* have slack seasons once in three years. Abundance of frugivorous birds in different layers of vegetation indicate profuse flowering and fruiting in different species (GRAPH 1,2,3,4; DIAGRAM 1,2,3, and 4).

(c) Diameter class representation:

The following graphs represent diameter class distribution of some important tree species. Most of the species show a normal size-class distribution. This is to say that the number of individuals decreased logarithmically as the diameter increased. To express mathematically, diameter bears a negative correlation with the logarithm of the number of individuals. This type of curve is typical of mixed forests where there are some regeneration of shade tolerant species. Shorea shows a different trend as there is no rageneration and most of the trees are of higher diameter and age classes.

Graph A. Shows the distribution of diameter classes in *Shorea* forests (low-level); while most of the miscellaneous associates (a few in number) have smaller diameters and absence of mature crop. *Shorea* has most of trees of higher diameter classes (see also Graph F.)

Graph B. Shows the distribution of diameter classes of trees in mature (pure) *Shorea* forests. *Shorea* is well distributed with middle to higher diameter classes and there is more of miscellaneous species in each diamter class than in the former type.

Graph C. Represents a few species of dry-mixed forests where Shorea is absent. In this type there are several species of higher diameter classes (more than in other types); this shows better development and regeneration of miscellaneous species in the dry-mixed forests.

Graph D. Represents the distribution of diamter classes and number of species in each class in wet mixed forests. There is absence of *Shorea* to show that this species is not representative of wet-mixed forests. Other common species are not so well represented as in dry-mixed forests.

Graph E. Shows the distribution of diamter classes in Scattered *Shorea* is considered poor (quality III as per standard site quality classification). The associates have higher diamter classes; the number of trees in each class is also high.

It is obvious that commercially important species other than *Shorea* is poorly represented in the forest and calls for stricter investigation and replacement by planting. It also shows that *Shorea* is not regenerating and with the felling of mature trees planting of *Shorea* is necessary.

(d) Diameter increment in tree crop:

There is practically no quick growing species in the forests so as to work economically on a very short rotation. *Shorea* is slow growing and attains a diameter of 60 cm. in 80 years. *Michelia champaca* and *Chikrasia tabularis* attain this girth in 50 years. Some of the species like *Anthocephalus Chinensis* and *Dalbergia sisso* have been found to grow quickly in the first few years. Thereafter they stagnate. *Cryptomeria japonica* (an exotic, now naturalised) grow to a diameter of 33 cm. in 32 years and 38 cm. in about 50 years.

In the plantations *Tectona grandis* (local exotic) have been found to attain a diameter 40 cm. in 50 years and 50 cm. in 60 years in about 70 - 80 years.

Other economically important species have growth rates of girth/ diamter increments far less than the above species.

(e) Regeneration of tree crop:

The process of natural regeneration in the moist deciduous and monsoon temperate forests is rather complex because of large number of species, comlexity of structure and floristic composition of the forests. There are middle aged to mature trees to produce large quantity of viable seeds.

In the plains, dense undergrowth and unfavourable soil conditions due to excessive dampness interfere with the natural regeneration of

Shorea robusta. Seedlings cannot make their headway through the dense weed. The *Shorea* forests of Jalpaiguri furnish a striking example of the effects of continued fire-protection in altering the character of the forests and bringing it into such a condition as entirely to prevent the natural regeneration of *Shorea*. These *Shorea* forests had their origin in burn savannah lands. Such Savannah land are not present in the western part of Jalpaiguri district, but they are present in the eastern part.

From the study of vegetation in transects and quadrats it has been seen that economically important tree species are not regenerating; even if the seeds germinate, the weeds suppress them. Some of the species of Lauraceae and Meliaceae regenerate profusely. In the clear felled areas *Anthocephalous Chinesis* have been found to regenerate profusely. In the riverain tracts *Dalbergia* and *Acacia* regenerate profusely followed by *Albizzia* and *Bombax*.

In the hills there is no broad leaved crop of gregarious nature and the regeneration status of economically important tree species is discouraging. In the valley forests some regeneration of *Duabanga gradiflora* and *Ailanthus grandis* can be noticed. In the clear felled areas regeneration of *Alnus nepalensis, Betula alnoides, Pruns nepalensis* and *Acer compbellii* is found on the screen and open places. Regeneration of *Symplocos* and *Viburnum* is very common all over the temperate forest areas; a sprikling of Lauraceous tree species is found in the ground vegetation every where.

The Physiognomy of plants:

I Buttress : The trees of the tract dealt with do not have conspicuous buttress. A few that possess are as follows:- *Boobax ceiba, Tetrameles nudiflora, Acrocarpus fraxinifolius, Sterculia alata, Firminana colorata, Ficus* species, *Engelhardtia spicata, Lophopetalum fimbriatum* and *Terminalia belerica.*

II Roots : The species that can be identified from the nature of roots (stilt) are as follows:-
Pandanus species, *Carallia integerrima, Elaeocarpus* species, *Jambosa formosa, Leea robusta and Macaranga denticulata,*

III Cauliflory : Trees that bear flowers and fruits in the main body of stem and branches are as follows:
Gynocardia odorata, Baccaurea sapida, Ficus species, *Anonaceae* in general, *Dillenia pentagyna* and *Saurauia* species.

IV Tendril : Some of the genera or even families have characteristic origin of tendrils. Combined with other characters they can guide identification of families, genera and species. A few examples are as follows:-

		Leaf opposed tendril	- Vitaceae (Vistis)
		Extra axillary tendril	- Cucurbitaceae
		Axillary tendril	- Passiflorae
		Petiole with two tendril like outgrowth	- *Smilax* (Liliaceae)
		Petiole tendril bearing	- *Naravelia* (Ranunculaceae)
		Leaf with coiled tendrillar apex	- *Gloriosa superbu* (Liliaceae)
V	Spines, prickles and thorns	: These an be useful sometime in guiding identifical : tion of species and genera as follows:-	
		: Thick based squat spines	- *Bombax, Fagara*, and *Erythrina* species.
		Various types of spines all over the plant body	- *Calamus* species. In pole stage the plant that bears spines in *Bridelia retusa.*
		Stipular thorns or spines	- *Capparis* species, *Zizyphus* species and *Celastrus stylosus.*
		Stem with recurved and ordinary prickles	- *Toddalia, Rosa* and *Rubus* species.
		Leaves prickly	- Ilex species *Caesalpinia crista, Mezoneuron cuculatum, Mahonia* and *Berberis* species (has also spines).
		Stem thorny	- *Cudranea* species, *Randia longispina* *Premna mucronata.*
		Large recurved bifid stipular spines like buffalo horn	- *Uncaria* (Rubiaceae)
		Maroon coloured fruit 6cm. with long spines	- *Sloanea sterculiacea*
		Spiny fruit, straight spine	- *Castanopsis indica, Castanea* and *Caesalpenia crista.*
		Spiny fruit, spines flattened and forked	-*Castanopsis tribuloides* var *longispina.*
		Large fruit with concial long sharp spines	- Buettneria species - (Climber).

VI	Leaf sheath surrounding the stem	:	Species of Commelinaceae, Polygonaceae, and Xyridaceae have generally such sheath.	
VII	Leaves with glandular dots	:	Species of Rutaceae, Myrtaceae, Hypericaceae, Samydaceae, Lauraceae, Piperaceae and Myristicaceae.	
VIII	Leaves	:	Stipules leaves circular scar on stem	- Magnoliaceae
		:	Stipules interpetiolar	- Species of Rubiaceae and Rizophoraceae.
			Petiole peltately attached to the leaf blades having number of glands	- *Mallotus* species, *Macaranga* species.
			Petiole connected right angles to leaf blades	- *Mallotus philippinensis*
			Heart shaped leaves red stalk	- *Adina cordifolia.*
			Large with prominent nerves	- Species of Magnoliaceae *Dillenia* species, *Saurauia* species and Myristicaceae.
IX	Intra-marginal	:	Such veins are prominent in Myrataceae and Rhizophoraceae, in some *Rubiaceous* species, *Ficus* species, *Drimycarpus* species, *Psychotria calocarpa* (Rubiaceae) and *Goniothalamous* (Anonaceae).	
X	Prominent stipular line	:	Many species of Loganiaceae can be identified from the prominent stipular lines connecting the two opposite leaves.	
XI	Swollen at node	:	*Impatiens* and *Strobilanthes* species have prominent swollen nodes.	
XII	The Bark	:	The barks of yound and old trees often are distinct to guide to the identification of genera and species. Some such trees are as follows:-	
			Fissured bark	- *Shorea robusta, Erythrina* species, *Terminalia tomentosa* and *Schima walliachii*
			Bark exfoliating in patches	- *Lagerstroemia parviflora,*

		Gmelina arborea and *Albizzia stipulata.*
Papery bark	-	*Dillenia indica, Betula* species, *Prunus cerasoides* and *Syzygium praecox*
Powdery bark	-	*Bischofia javanica, Aporosa* species.
Prominent gall on bark	-	*Machilus gamblei.*
Bark grey, handsome owing to wrinkled and reticulate pattern	-	*Sapium beccatum* and *Trichillia connarioides.*
Bark bitter in taste	-	Simarubaceae and some Gentiananceae.
Bark corky, wrinkled, pale buff, pair of bract prominent in cold weather	-	*Hymenodictyon excelsum.*

XIII The Blaze : Some of the trees have distinct colour and smell in its blaze that gives clear indication of the genera or species as follows:-

Blaze giving white latex	-	Apocynaceae, Asclepidaceae, Caricaceae, Euphorbiaceae, *Amoora wallichii* (Meliaceae), Moraceae, Convolvulaceae and Sapotaceae.
Blaze giving yellow latex	-	Guttiferae and Papavaraceae.
Blaze giving red juice	-	*Bischofia javanica,* Myristicaceae, *Butea monosperma.*
Blaze giving red gum	-	*Macaranga* species, *Ougenia oojenioides.*
Blaze smelling resinous	-	*Canarium* species.
Blaze giving colourless gum	-	*Spondius* species, *Lannea coromendelica* and *Dillenia* species.
Blaze smelling garlic	-	*Dysoxylum* species.
Blaze reddish, with	-	*Drimycarpus* species

		scanty beads of sticky juice turning black like tar	
		Blaze and crushed leaves smelling aromatic	- Lauraceae
		Blaze changing from green to burnt siena	- *Carallia brachiata.*
		Blaze white	- *Machilus* species.
		with darker specks, soft mucilaginous	- *Pygeum acuminatum* and *Pygeum montanum.*
		Blaze and crushed leaves smelling bitter almond	
		Blaze pink with white streak	- *Toona ciliata.*
		Blaze resembling betel nut	- *Celtis tetrandra.*
		Blaze thick and spongy with bright pink to purple inner layer and bark with grey horizontal wrinkles	- *Ilex godajam.*
XIV	Branches:	Whorled branches	- *Bombay ceiba* and *Anthocephalous cadamba.*
		Branches right angle to the bole with tips drooping	- *Duabanga grandiflora*, *Myristica* species, *Pol yalthia* species and *Mastixia pantendra.*
		Zig-zig branches	- *Marlea bvegonifolia.*

Flowers and Fruits:

The reporductive organs of plants are less susceptible to the external influence of the environment than the vegetative organs. Floral biology of the forests of the districts has little been investigated. Many birds as well as insects play an important role in pollination and dispersal. The plants in the plains are mostly insect and bird pollinated; in the hills there are several wind pollinated species.

In the present study flower colouration of conspicuous plants (herbs, shrubs and trees) has been grouped according to colour of petals with details of months of fruiting and their place of occurrence. Graphical representation of the flowering sequence in different plant families have been made. It is possible to correlate the abundance of avifauna species in the study area with the abundance of flowering and fruiting trees. There are more than 580 species of birds in the study area which from about 33% of the whole of India. The fruits are found all the year round all over the area.

The fruits and seeds have highly adapted mechanism for dispersal and many are succulent and arillate and form food for birds and animals.

In general there are six prominent shades of colours of flowering plants in the forests viz. white 28.5%, yellow 24.0%, pink 10.0%, Red 12.0%. A list of species may be seen in the statement given overleaf.

Fruiting:

Diagrams 1 and 2 represent the fruiting sequence and patterns of 57 families of flowering plants; it is obvious that these families bear fruits throughout the year.

Flowering:

Diagrams 3 and 4 give the sequence of flowering in several monocotyledon (6 families) and dicotylendons (46 families). These show that the flowering season is distributed all over the year in the study area.

Graphs 1 and 2 show the peak period of flowering in 7 widely distributed families of flowering plants. They flower throughtout the year with peak period in May to October which coincides with the nesting and breeding period of many birds.

Graph 3 includes many monocotyledon families, which also shows the peak period in July and August. Graph 4 represents 57 families of flowering plants and shows two peak periods of flowering.

Fruits, seeds and dispersal :	Pods 1 seeded	- *Derrus, Butea* and *Dalbergia.*
	Pods flat with more than one seed	- *Albizzia* and *Mezoneuron.*
	Pods break into four quadrangular pieces	- *Dosmodium.*
	Pods curved and spirally twisted	- *Pithecolobium*
	Fruits big and roundish	- *Crataeva unilocularis, Careya arborea, Dillenia indica*

		Gynocardia odorata, Pterygota alata, Randia species, *Prunus Hodgsonia, Trichosanthes, Citrus* and *Limonia* species.
Seeds with floss	-	Bombacaceae and Malvaceae.
Drupes	-	Lauraceae, Elaeocarparceae and Rosaceae.
Follicles	-	Apocynaceae, Asclepiadaceae Bignoniaceae and Sterculiaceae
Winged fruit	-	*Shorea, Acer, Fraxinus, Terminalia tomentosa, Combretum, Aspidopterya, Ailanthus, Compositea, Pinus, Picea, Abies, Rumex, Dioscoria, Gouania* (several winged), *Ventilago* (one winged), *Engelhardtia* (winged bract), *Spatholobus* (recurved Samara).
Wind dispersed seeds, spores or zygotes because of minuteness	-	Orchidaceae, Filicales (ferns) Lycopodiaceae, Equisetaceae Musci (Mosses), Hepaticae, Lichens, Pungi and Algae.
Explosive fruits	-	Acanthaceae, Balsaminaceae Euphorbiaceae and Leguminosae.
Seeds and fruits either eaten or disseminated by birds frugivorous	:	Anonaceae, Dilleniaceae, Magnoliaceae, Menispermaceae, Guttiferae, Flacourtiaceae, Ampelidae, Illicineae, Myrsinae, Piperaceae, Moraceae, Araceae, Zingiberaceae, Musaceae, Liliaceae. Lauraceae, Myrtaceae, Elaeocarpaceae, *Berberis, Sterculia, Grewia, Rhus, Prunus, Pyrus, Crataegus, Rubus, Rosa, Ribes, Nyssa, Sambucus, Viburnum, Lonicera, Vaccinium, Symplocos* and *Cissus.*
Viscid fruits	:	*Duabanga grandiflora, Tetramales nudiflora, Coniferae, Oroxylum indicum, Dioscoria* species (has circular wings),

		Ptereospermum, Hymenodictyon, Toona species, *Schima wallichii, Butea* species, *Stereospernum* species and *Lophopetalum fimbriatum.*
Seeds and fruits having adhering mechanism	:	*Setaria* species, *Xanthium strumarium. Centotheca lappacea, Chrysopogon aciculatus, Cenchrus cillaris, Oplismenus* species, *Ageretum conizoides, Bidens pilosa, Taraxacum* species, *Sonchus* species, *Anisomeles indica, Rumex* species, *Urena lobata, Desmodium species* and *Mimosa pudica.*
Seeds and fruits	:	Species of *Potamogeton, Nymphaea, Chara, Nitelia, Blyxa, Ottelia, Lemna, Wolffia, Azolla, Oenanthe, Carex* and *Polygonum.*
Arillate seeds	:	Celastraceae, Euphorbiaceae, Meliaceae, Myrtaceae, Zinziberaceae, Marantaceae. Species of importance are:- *Baccaurea sapida, Dysoxylum* species, *Alpinia* species, *Taxus, Pyrenium* species, *Celastrus* - species, *Casearea* species, *Bischofia* species, *Bridelia, Glochidion* species, *Adenanthera pavonina, Elaeocarpus ganitrus* and *Vitex quinata.*
White fruits	:	*Hedyotis, Morinda, Glycosmis, Callicarpa, Breynea, Chloran thus* species and *Jambosa - formosa.*
Dark blue purple fruits	:	The fruits of Sapium beccatum is purple black, sweet and eaten by Imperial pigeon. *Dichoroa febrifuga* (a large shrub), *Syzygium* species, *Clerodendrum serratum, Berbaris* and *Mahonia.*
Red fruits	:	*Derringea, Taxus, Pyrus,*

		Euonymus, Schizandra, Argyrea species *Stephania, Holboellia, Ilex, Capparis, Rubus, Chisocheton, Sorbus, Ficus benegalensis, Ficus altissima* and some other *Ficus* species and *Daphne cannabina.*
Yellow fruits	:	*Duranta, Gmelina, Casearea vareca, Celtis* sp., *Schefflera* species, *Glochidion, Dillenia pentagyna, Garcinia, Rubus, Mezoneuron, Artocarpus lackoocha, Ficus glaberrima, Ficus retusa, Ficus benjamina* and some other *Ficus* species
Fruits edible by animals	:	*Dilleniaceae, Terminalias, Gmelina, Zizyphus, Spondius, Phoebe, Rumex, Stellaria, Scoparia, Anthocephalus, Ficus, Michelia excelsa, Symplocos, Elaeocarpus, Cassia fistula* and *Myristica.*
Seeds horsehoe shaped	:	Menispermaceae and Resedaceae.
Plumed fruits	:	*Imperata arundinacea, Pogonatherum crinitum, Saccharum* species, *Oplismenus compositus, Phragmites, Imperata* and *Miscanthus nepalenses. Vernonia cinera* and *Aerua scandens* bear hairs. The *Typha* has hairs at the base of fruits.
Plumed seeds	:	*Epilobium, Tamarix* and *Bombax.*

Distribution of Sizes of Fruits:

Fruits of innumerable species were measured and they have been classified into the following sizes along with their percentage distribution in the forests:-

1. Sizes upto 0.5 cm. (diameter) - 60%
2. Size from 0.51 to 1.0 cm. (diameter) - 28%
3. Size from 1.1. to 2.0 cm. (diameter) - 8%
4. Size above 2.0 cm. (diameter) - 4%

14

Quantitative and Qualitative Ecological Observations in Quadrats and Transects

A. GROUND VEGETATION IN THE PLAINS:

This study has been made in quadrats laid out in various grids as described under the design. In each of these places plots of 10 x 10 sq. metres (quadrats) were found to reflect 46 species; in the riverain forests, however, the quadrats of same sizes reflected 38 species. Graphical representation of number of species as ordinate and number of quadrat as abscissa show that 3 - 5 quadrats count give stable and reliable data.

In the tables (Appendix - 1), each vertical column represents a number of such quadrats and as such the F% and D are expected to give a very correct picture of the crop. For economy of space more data could not be presented. These were not necessary either, as data already analysed have been found reliable and adequate Observations were made during and after the rains.

The results of analysis have been tabulated under the following groups:—

(i) Shorea forests.
(ii) Riverain forests.
(iii) Dry mixed forests.
(iv) Plantation (artificial).

(i) Shorea forests (Histogram-8)

Although there are differences in the composition of forests in the top layers of vegetation, the ground vegetation does not differ very much in general structure in the wet mixed, scattered *Shorea* and pure *Shorea* forests.

In this type 112 species have been reflected on the ground vegetation; of the grasses, *Oplismenus* and *Paspalum* have good incidence of occurrence and they are very much shade loving. In shade, most of the grasses are soft and shade tolerant, but as soon as the forests get burnt and the floor is exposed to sun coarse grasses like *Imperata* invade the area.

Of the ferns, *Cyclosorus* and Thelepteris are widespread and indicate well drained nature of soil.

The regeneration of tree species form 47% of the entire ground flora. Of this, *Litsaea, Machilus, Aphanamixis, Sterculia, Terminalia* and *Syzygium* are common. Among the herbs, shrubs and climbers, *Clerodenrum Coffea, Dioscorea, Smilax*, are very common. *Indigofera, Desmodium* and *Crotalaria* have high degree of occurrence, but are not conspicuous.

(a) Wet mixed Forest:

At Moraghat Forest - Analysis of ground flora including ferns, grasses, canes, has been made and described in various degrees in several chapters. Of the ferns *Cyclosorus molliusculus, Abacopteris multiliniata, Stenoclaena palustris and Lygodium flexuosum* are common. Of the grasses *Setaria palmaefolia, Imperata cylindrica, Paspalum scrobiculatum. Digitaria* species, *Axonopus cimicinus* and *Panicum* sp. are common. *Smilax ferox, S. macrophylla, S. prolifera Dioscorea angunia, D. pentaphylla, Curcuma aromatica, Zingiber Globba bulbifera. Alpinia allughus* and other *scitaminaceous* species are conspicuous among the monocotyledons. *Scleria elata* is the only common sedge recorded. Regeneration of *Syzygium cuminii, Terminalia belerica, Meliosma simplicifolia, Bridelia* species, *Litsaea* species, *Sterculia villosa, Dillenia pentagyna, Sloanea sterculacea, Calamus* species are common among the seedling regeneration. Among herbs and shrubs, *Borreria hispida, Morinda citrifolia, Leea* species, *Coffea bengalensis, Laportea crenulata, Forrestia glabrata, Calamus latifolia* and *Daemonorops jenkinsianus* are very conspicuous and common (over 60% frequency).

At Damanpur forests - The ground is moist and there are plenty of *Dioscorea deltoidea, D. daemona, Asparages racemosus, Zingiber clarkei, Z. chrysanthemum, Z. rubens, Digitalia setigera, D. royleana, D. stricta, Piper longuma and P.peepulioides.* Eighty percent of the species referred to in Moraghat forests also occur here.

At Rajabhatkhawa - The density of crop per hectare is 275, which is very high in comparison to other areas. There is preponderance of *Meliosma pinnata, Dillenia pentagyna, Macaranga denticulata, Bauhinia speciosus, Murdannia spiratum, Forrestia glabrata, Agaretum conyzoides, Calamus* species are common in the shrub and herb layers (above 60% frequency).

The statement of analysis shows the abundance of species and their occurrence.

At Apalachand Forests - The ground vegetation has striking similarity with the vegetation of Moraghat. The following are the common and characteristic flora:-

Grasses - *Cyrtococcum oxyphyllum, Paspalum species, Digitalis* species and *Oplismenus compositus.*

Ferns	-	*Abacopteris* species, *Lygodium flexuosum* and *Cyclosorus molliusculus.*
Monocots	-	*Smilax* speices, *Dioscorea* species, *Zingiber* species, *Globba bulbifera, Chlorophytum* species.
Climber and Climbing shrub.	-	*Leguminous* species, *Dalbergia stipulacea. Calamus* sp., *Acacia pennata, Stephania hernandifolia* and *Thunbergia grandiflora.*
Herbs and shrub	-	*Clerodendrum viscosum, Sida carpinifolia, Leea* species, *Coffea bengalensis, Labiateceous* species, *Urena lobata.*
Regeneration of tree species	-	*Machilus* species, *Syzygium* species, *Lannea coromendelica, Terminalia belerica, Meliosma simplicifolia, Litsaea* species, *Baccaurea* sapida.

At Udlabari - The density of crop per hectare is 200. Principal species are *Dillenia pentagyna, Meliosma simplicifolia, Premna bengalensis Syzygium* species, *Styrax serrulatum, macaranga denticulata, Ilex gadajam Litesaea monopetala* and *Machilus gamblei.*

Common ground flora are:-

Grasses	-	*Setaria palmaefolia, Cyrtococcum oxyphyllum, Passpalum* and *Digitaria* species.
Ferns	-	*Thelepteris* species, *Cyclosorus milliusculus, Lygodium flexuosum.*
Monocots	-	*Similax prolifera, Dioscorea pentaphylla, D. Daemona, Curcuma aromatica, Asparagus racemosus, Globba bulbifera.*
Climbers and climbing shrubs	-	*Ichnocarpus frutescens, Leguminous* species, *Piper* species, *Croton caudatus, Millettia* species, *Beaumontia grandiflora, Calamus* species.
Herbs and Shrubs	-	*Clerodendrum viscosum, Leea* species, *Coffea bengalensis, Urena lobata* and *Ervatamia divericata.*
Regeneration of	-	*Terminalia belerica, Toona ciliata, Bridelia* species, *Litsaea* species, *Careya arbores, Sterculia villosa, Callicarpa arborea* and *mallotus philippinensia.*

(b) Scattered Sal Forests:

At Rajabhatkhawa - Ground flora is very abundant as in the case of Wet mixed type and there is no significant difference. *Laportea crenulata, Phrynium capitatum* and *Schumanianthus dichotoma* are present in the swampy areas in the transects.

At Jainty - The density of crop is 362 per hectare. The percentage of *Shorea robusta* is 31.5%. *Aphanamixis polystachia* has *Terminalia belerica, Bridelia retusa, Dillenia pentagyna, Mallotus albus. Schima walliachii.*

The following are other characteristic vegetation:-

Grasses:

Comon are *Cyrtococcum oxyphyllum, Paspalum scrobiculatum, Oplismenus compositus* and *Pogonetherum sacchariodies.*

Ferns:

Common are *Thelepteris* species and *Cyclosorus molliusculus.*

Sedge:

Common is *Scleria elata.*

Monocot:

Common are *Curcuma aromatica, Dioscorea deltoidea, Alpinia allughus, Asparagas racemosus* and *Smilax ferox* etc.

Climbing and Climbing Shrub:

Common are *Eupatorium odoratum, Ichnocarpus frutescens,* species of *Crotalaria, Flemingia, Indigofera, Acacia pennata, Spatholobus roxburghii, Croton caudatus, Millettia* species etc.

Herbs and Shrubs:

Common are *Clerodendrum viscosum, Borreria articulata. Coffea benegalensis, Urena lobata* etc.

Regeneration of Tree Species:

Common are *Lagerstronemia parviflora, Machilus* and Litsaea species, *Lannea coromendelica, Hymenodictyon excelsum, Premna* species, *Careya arborea, Sterculia villosa, Bridelia* species etc.

At Bengduhi - The density of the crop is 317 per hectares. The percentage of Shorea is 13.8; *Schima wallichii* has a percentage of 18.2. *Talauma hodgsoni, Symplocos spicata, S. ferrugenia, Eurya acuminata, Ilex godajam, Ternstroemia gymnenthera* and *Endospermum chinense* are very prominent species of the line. *Ternstroemia* is not found elsewhere in North Bengal. Homalium zeylanicum is another species that is found only in this locality.

Areas where these trees occur have been described under an edaphic type.

The ground vegetation is as follows:-

Grasses:

Common are *Setaria palamefolia, Isachne albens, Paspalum* species and *Oplismenus compositus.*

Ferns:

Common is *Adiantum* species.

Sedge:

Common is *Scleria elata.*

Monocots:

Scitaminaceous species, *Smilax* species, *Alpinia allughus* etc. are common.

Climber and climbing shrub:

Common are *Paederia foretida, Convolvulaceous* species, *Calamus* species etc.

Herbs and Shrubs:

Common are *Clerodendrum viscosum, Borreria articulata, Morinda citrifolia, Coffea bengalensis* etc. Regeneration of *Aphanamixis acronychia, Schima, Stereospermum, Dillenia, Mallotus, Litsaea, Machilus, Glochidion* sp. etc.

In the ground vegetation Scitaminous species is conspicuous of which *Costos speciosus, Globba bulbifera, Curcuma aromatica, C. zedoria, Zinziber capitatum, Amomum aromaticum, Alpinia allughus, A. malaccensis* are abundant and sometimes occur in pure formations. Among the grasses *Oplismenus compositus, Setaria palmaefolia* and *Digitaria* are common. *Stenoclaena palustris, Lygodium flexuosum* and *Abacoppteris multilineata* are the common ferns. *Morinda citrifolia, Phlogocanthus thyrsiflorus, Eranthemum* species *Leea bracteata, Dalbergia stipulacea* are the common shrubs. There are cane brakes in the moist pockets. Regneration of *Lauraceous* species, *Shorea, Sterculia, Litsaea. Aphanamixis, Bridelia* are common.

(C) Pure Sal.

At Moraghat and Tondu.

Analysis of Ground Flora:

In moist forests in Lataguri 1-2 blocks the following are the species:- *Oplismenus compositus, Lasia spinosa, Syzygium balsmefolia, Acorcus calamus, Piper* species, *Pilea* species, *Calamus* species, *Smilax* species, *Dillema sarmentosa, Alpinia allughus, A. mallaccensis, Croton caudatus, Stenocaeana palustris, Abacopteris multilliniata, Maesa chisea, Rupidophora glauca* and *Forrestia glabrata.*

The regeneration of tree spcies is there, but most of them cannot establish owing to human interference. Common speices are *Machilus* species *Litasaea* species, *Amoora* species, *Aphanamixis* species, *Schima wallichii Lannea coromendelica etc.*

In a *Lagerstromemia speciosa* plantation of 1948 scanty of ground flora consisting of *Dellima sermentosa, Saurauia* species, *Mallotws philippinensis, Terminalia belerica, Schima wallichii, Castanopis tribuloides, Clinnamomum cecedodaphne* and *Alpinia allughus* were found.

At Rajabhatkhawa and Damanpur:

There is preponderance of Calamus species all over the forests. This zone is the best cane area in the whole of North Bengal. *Daemonorops jenkinsianus, Calamus latifolius* are the principal species; other species are *Calamus guruba, C. flagellum* and *C. tenuis. Pinanga gracilis* and *Dracaena ensifolia* form pure associates in dark and moist forest floor in the Damanpur forests. There are many marshy areas where plenty of *Panicum, Paspalum, Eragrostis, Cyperus* and *Scirpus* species grow.

At Terai and Foot hills:

The ground vegetation has more of *Leguminous* species viz. *Desmodium cephalotus, D. tiliafolium, D. polycarpum, D. heterocarpum, D. pulchellum, D. motorium, D. laxum, D. caudatum, D. latifolium, Pueraria sikkimensis, P. wallichii, Flemingia bracteata, F. stricta, F. congesta, Flemingia semialata, Crotaria striata, C. sericea.* Other notable species are as follows:-

Grasses:

Common are *Pogonetherum sacchariodeum, Imperata cylindrica* and *Microstegeum ciliatum.*

Ferns:

Common are *Cyclosorus mollisculus* and *Asplenium nidus.* The former form colony on sunny forest floors.

Sedges:

Common is *Scleria elata.*

Monocots

Common are *Smilax* species, *Dioscorea* species, *Asparagus racemosus, Alpinia allughus, Curcuma* and *Zingiber* species.

Climbers and climbing shrubs:

Common are *Eupatorium odoratum, Ichnocarpus frutescens, Dalbergia stipulacea, Vallaris heynei, Croton caudatus, Millettia* species.

Herbs and Shrubs:

Common are *Agaretum conyzoides, Coffea bengalensis, Morinda citrifolia* and *Urena lobata.*

Regeneration of species:

Common are *Shorea robusta, Bridelia, Litsaea* and *Machilus* species, *Dillenia pentagyna, Carea arborea, Premna* species, *Phoebe* species *Lagerstoremia parviflora* etc.

(ii) Riverain forests (Histogram 10)

Typical *Dalbergia - Accacia* forests have not been included in the study as ground vegetation hardly exists there, *Albizzia - Bombax* and mixed forests of the riverain tract have been included in the study. There are 8 species of grasses in these forests of which *Pogonetherum* and *Eragrostis* are not found in (group (i). These two speceis indicate sandy soil. In the moister pockets *Cyclosorus* and *Diplazium* ferns form large colonies in the heavily grazed tract. The incidence of *Jasminum pubescens, Acacia pennata, Vallaris heynei, Eupatorium ororatum, Ipomoea* species, *Rivea* species are very high. Species common to Group (I) are *Smilax, Dioscorea, Desmodium, Indigofera, Crotalaria*, and regeneration of *Machilus* and *Mallotus.*

(iii) Dry mixed forests (Histogram - 9)

The analysis and histogramic representation show 72 representative species which are exactly same number as riverain forests. In fact the soil condition has some resemblance, but the regeneration pattern is almost same as *Shorea* type. The number of species in the frequency class 21-40 are towards highest.

Dioscorea and *Asparagas* species are very conspicuous. *Imperata* has high incidence of occurrence owing to open nature of forests. *Clerodendrum* and *Coffea* have wide range of distribution. In moist pockets *Shorea* occur.

Among the tree species *Aphanamixis, Terminalia, Litsaea, Dillenia, Wrightia,* and *Bauhinia* have good incidence of regeneration. More and more *Wrightia* invades as the soil condition grows drier.

(iv) Plantation (Histogram - 11)

In these areas original virgin forests have been replaced by a few selected species after clearfelling and burning. The original character of the forest is never regained as the plantations are repeatedly tended and fresh regeneration cannot establish. Total number of species that invade the plantations is only 55, (against 112 and 72 in other groups). Of the five sites. Lataguri and Moraghat are wettest; next is Apalchand, while Sevoke and Chapramari are drier. There are fewer numbers of ferns, monocots and grasses. *Clerodendrum* and *Coffea* have wide occurrence; *Eupatorium* and *Ichnocarpus* are commonest among the shruby climbers.

The elimination of species in different storeys in the plantations is a significant point. The habitat conditions gradually get changed and many shade loving and fruit bearing species fail to regenerate.

There is heavy incidence of climbers in the plantations. In a plantation (5 hectares) near Sukna the followig climbers were identified:-

Stephania harnandifolia, Cyclea peltata, Ipomea linifolia, Ipomea pestigris, Impmea vitifolia, Rivea roxburghii, Thunbergia grandiflora, Thunbergia fragrans, Vitis assimica, Ampellocissus barbata, Tetrastigma serrulatum, Aspidopteris nutuns, Trichosanthes bracteata, Cephalendra indica, Mukhia scabrella, Memordica dioica, Cryptolepis elegans, Marsdenia tinctoria, Deeringia celosioides, Mimosa himalayan, Mucuna imbricata, Mucuna sp., *Millettia* species, *Desmodium triquetrum, Abrus pulchellus, Phaseolus calcaratus. Paederia foetida, Caropegia macrantha, Celastrus multiflorus, Dioscorea daemona, Dioscorea bubifera* var. *sativa, Dioscorea pentaphylla* and *Smilax macrophylla.*

Summary of Ground Vegetation Survey

Group	*(i)*		*(ii)*		*(iii)*		*(iv)*	
Types	*Sal forests*		*Riverain forests*		*Dry mixed forests.*		*Plantations*	
	No. of species	*%*	*No. of species*	*%*	*No. of species*	*%*	*No. of species*	*%*
Grasses	9	8.0	8	11.0	6	8.4	5	9.0
Ferns	5	4.5	2	2.8	1	1.4	2	3.6
Fern allies	-	-	1	1.4	1	1.4	1	1.8
Sedges	1	0.9	2	2.8	2	2.8	1	1.8
Monocots	10	9.0	4	5.7	8	11.2	4	7.3
Climbers and climbing shrubs	23	20.4	21	29.2	11	15.0	8	15.0
Herbs and shrubs	17	15.2	19	26.4	12	16.4	9	16.0
Tree	47	42.0	15	20.7	31	43.4	25	45.5
	112	100%	72	100%	72	100%	55	100%

Special features of the ground vegetation in Shorea Forests.

There are 112 species of plants composing the ground vegetation of which 5 species have 100%, 5 species have 90% 6 species have 80%, 9 species have 70%, 9 species have 60%, 9 species have 50%, 12 species have 40%, 15 species have 30%, 21 species have 20% and 2 species have 10% incidence of occurrence.

Grasses form 1.2%, ferns form 5.0%, Monocots form 14.5%, climbing shrubs and climbers form 17.4%, herbs and shrubs form 14.5% and Tree regenerations form 47.5% of the species.

Frequency has been represented in Histogram Nos. 5,6,7 and 8.

In the ground vegetation analysis the terms dominance, sociability and vitality or vigour are generally used. The dominance, however, is expressed by the terms abundant (a), Common (c), rare (r), and present (+). Sociability is expressed by the terms singly (a), patches (p), colonies (c) and groups (g). Vitality or vigour is used to express (i) ephemeral (ii) plants that reproduce by vegetative means and (iii) plants that produce viable seeds.

Since the present observations have been made in quadrats laid over extensive area the species have not been classified in detail as above. But more intensive informations have been collected on frequency, percentage of species and canopy class classifications. Record, however, has been made on dominance and sociability.

Grasses of *Shorea* forests are special features of the forests studied separately in greater detail.

B. GROUND VEGETATION IN THE HILLS:

The methods of determination of the number and size of the quadrats have been described earlier. For the purpose of analysis, the hills have been divided into various zones under three major zones viz., the lower hills, the middle hills and the upper hills and the upperhills hereunder described as Group (i), Group (ii) and Group (iii) respectively in the Text. Each column under each group represents the results of analysis of very many quadrats laid out at a particular altitude at different grids. Occurrence of species has been represented by plus sign (+).

Analysis shows the occurrence of the following prominent plant families and the number of species under each family:

Urticaceae	12	Poaceae	8	Gentiananceae	5
Commelinanceas	3	Fabaceae (Leguminosae)	11	Labiatae	6
Acanthaceae	4	Orchidaceae	3	Asteraceae (Compositae)	11
Rosaceae	5	Umbelliferae	4	Begoniaceae	3
Filicales	11	Scitaminae	4	Liliaceae	3
Cyperaceae	5				

The families Sccophulariaceae, Polygonaceae, Caprifoliaceae, Caryophylaceae, Thymeliaceae, Curcurbitaceae, Gesneraceae, Ternstroemiaceae and Selaginellaceae have two species each. 36 families of flowering plants have been shown prominance.

Filicales have 11 species almost very common in all the temperate areas. Fabaceae (Leguminosae) is wide spread in the lower hills; Asteraceae (Compositae) and Graminae species increase with decrease in altitude, Urticaceae, however, have many species at all altitude from foot hills to the upper hills. Labiatae, Acanthaceae and Scrophularianceae species increase with increase in altitude.

Group (i) - Lower hill forests (plains to 1000 metres).

This zone does not have any special characteristic feature in the ground vegetation as this is the transition zone (ecotone) of the vegetation of plains and hills. Most of the species of plains and middle hill forests occur here. There are, however, some areas, viz., Tista valley forests, Rangit valley forests, Sevoke forests, Sukna forests and Dalkjhar forests which have distinct flora of their own and they have been described and recorded.

(a) Common ground vegetation of Ghoramara an Andere blocks (Foot hill forests from plains upto 1000 metres):-

Capsella bursa-pastoris, Phyllanthus reticulata, Millettia glaucens, Desmodium polycarpum, Commelina obliqua, Myriactis nepalensis, Hyptianthera stricta, Carex cruciata, Carex stramentita etc.

(b) Conspicuous ground flora of the Rangit and Tista valley:-

Vernonia cinera, V. saligna, Blumea myriocephana, B. balsmifera, B. aromatica, B. densiflora, Lectuca polycepha, Laggera flava, L. alata. Wedelia wallichii, Spilanthes acmella; Pogostamon tuberculesus, Dysophylla quadrifolia, Mosta dianthera, Calamintha umbrosa, Anisomeles indica. Gomphostemma mastersii, Hyptis suaveolens, Stachys melissaefolia; Majus pumillus, Linderbergia hookeri, Vandellia cordifolia, V. mollis, V. scabra, V. erecta, V. sessiliflora; Pratia begonifolia, Lobelia rosea; Rapanea capitallata; Caropegi hookeri, Pergularia pallida; Didymocarpus aurantiaca; Rungea pectinata.

(c) Common ground flora of Terai (Dalkjhar forests) but occurring scattered in the foot hills:-

Blumea hieracifolia, B. laciniata, B. runcinata, Xanthium indicum, Saussurea affinis, Senecio ramosus, Conyza viscidula, Laggera pterodanta, Tridex procumbens, Cynura pseudochina; Mazus rugosus, Lindenburgia urticifolia, Limnophila conferta, Bonnaya veronicaefolia,

Scoparia dulcis, Lindernia crustacea; Cardenthera uliginosa, Peristrophe bicalyculata, Nelsonia tomentosa; Plectranthus species, Pogostemon bengalensis, Dysophylla crassiculalis, D. quadrifolia, Anisomeles indica, Tournefortia hookeri. Streptocaulon slyvestre, Exacum tetragonum, Wahlenbergia gracilis and Melilotus indicus.

Groups (ii) & (iii) - Middle and Upper hill forests (10001 to 2920 metres)

These areas are very rich in the abundance of species and number of individuals. In the temperate hills flowering herbs occur in abundance. Analysis show that there are 8 species consisting of 173 individuals in a quadrat of 1 sq. metre; another area showed as many as 18 species; another area showed 14 species and 350 individuals (*Selaginella* forming thick mat on the floor has not been counted). Quadrats of 2 x 2 sq. or 4 sq. metre gave 24 species; 3 x 3 sq. or 9 sq. metre gave 36 species, 4 x 4 sq. or 16 sq. metre gave 46 species and 5 x 5 sq. or 25 sq. metre gave 52 species. It is impossible to count the number of species beyond 2 x 2 sq. metres quadrats as there are abundance of individuals.

In the temperate hill, *Selaginella*, Mosses and Ferns are rich in species diversity and they, except the ferns, form thick carpet on the forest floor.

Various paragraphs enumerate the number of families, genera and species that are significant in their occurrence in the hills. At 2500 metres most conspicuous during the rains are *Strobilanthes wallichii, Myriactis lutea, Brunella vulgaris, Pilea umbrosa, Begonia josephi, Potentilla mooniana Biswarea tonglensis, Parochetus communis, Epilobium tetragonum, Chitria macrophylla, Impatiens arguata, Scutellaria violacea, Chitria urticaefolia, Lecanthus wightii, Anaphalis subumbellata*, and *Begonia laciniata*.

Gregarious growth of *Aneilema thomsonii, Chitria macrophylla, Cautleya lutea* and *Stellaria sikkimensis* have been observed in *Acer Machilus, Quercus*, and *Cryptomeria japonica* mixed forests.

Common genera of ground flora in some areas of temperate forests:-

At Bhanjang - *Dicroa, Elatostema, Begonia, Impatiens, Oplismenus, Strobilanthes, Rubus, Calamintha, Silaginella* and *Pilea.*

At Durbin - *Cattleya, Urtica, Carex, Hydrocotyle, Vitis, Polygonum Rubus, Rubia, Daphne, Oplismenus* and *Viola.*

At Tiger hills - *Impatiens, Rubus, Pilea, Eragrostis, Scleria, Polygonum, Oplismenus, Viola, Begonia, Smilax, Boeninghasiana, Symplocos* and *Daphne.*

At Rangiroom - *Oplismenus, Elatostoma, Impatiens, Polygonum, Ophiopogon, Aneilema, Plectranthus, Carex, Rubus* and *Melissa.*

At Rongbong - *Swertia, Fagopyrum, Anaphalis, Polygonum, Carex, Geranium, Hydrocotyle, Eragrostis, Selaginella, Oplismenus, Achyranthes* and *Jasminum.*

At Plantation of Cryptomeria - *Vitis, Viburnum, Urtica, Smilax. Selaginella Ilex, Jasminum, Dapnne, Rubus, Achyranthes, Symplocos, Marlea, Oplismenus* and *Cauttleya.*

Detailed study of the ferns grasses, fungi, mosses and hepatics are necessary as they are indispensible components of these forests.

C. FERNS OF FOREST FLOOR IN PLAINS, FOOT HILLS AND TEMPERATE HILLS:

Many ferns that occur on the forest floor belong to the genera *Dryopteris, Pteris, Thelepteris, Cyclosorus, Athyrium* and *Diplazium.*

Cyclosorus molliusculus is common along the road-sides in sal forests. *Abacopteris multilineata* occurs extensively in Sukna forest and Tista road to Gangtok. On the marginal land also one finds *Pteris biaurita, Cyclosorus aridus, C. dentatus C. parasticus , Pteris dactilina, Pensiformis., P. griffithii, Thelepteris uliginosa.* Along Tista-Singtam road *Blechnum orientale* Linn flourishes. *Diplazium esculentum* is exeedingly common in West Bengal plains. It has the habit of stunted tree fern. *Nephrolepis cordifolia* is common in moist situation in Lebong Peshoke-Tista Road, Gorubathan-Lava road. At 1000 metres *Alsophila glauca, A glabra, A. latebrosa* and *A. ornata* are common in the hills.

In the hills between 900 - 3000 metres common are *Acrophorus stipellatus, Microlepia strigosa, M. platyphylla, M. Pilosula, Dennstaedtia scabra, Pteris longipes, Asplenium unilalerale, A. tennefolium, Athyrium setigerum, Dryopteris* sp. and *Colysis* sp. tree ferns such as Cyathea *spinulosa, Hemitelia decipiens,* and *H. brunoniana* are not uncommon in virgin forests. In shady slopes at 2000 metres *Adiantum pedatum* grows luxuriantly.

Pteris wallichiana has large digitately compound fronds and is common from 1700 to 2200 metres. *Diplazium stoliczkae* var. *horsutipes* is common in *Quercus pachyphylla* forests of tiger hills.

In the *Quercus - Rododendron* forests *Athyrium thelypteroides, A setioferum, Ctenitis apiciflora, C. nidus, Cystopteris tenuisecta, Polystichum prescottianum and Dryopteris paleacea* are prominent on forest floor.

In the shades of *Cryptomeria* plantations *Diplazium stoliezkae* var. *hirsutipes* and *Selaginella monospora* are quite common. Scattered in them are *Pteris aspericaule, P. cretica* and *Acrophorus stipellatus.*

In Very moist situation common ferns are *Diplazium polypodioides. D. bellum, D. maximum, D. latifolium,* and *Angiopteris evecta. Angiopteris* is common in moist shady places near Sukna and other places in th duars.

Fern Growing on Rocks and Gravelly Soil:

Comm are *Cheilanthes farinosa, C. albomarginata, C. tenuifolia, onychium siliculosum.* At 2500 metres *Woodsia elongata* is a characteristic species.

On stony walls are found *Adiantum lunulatum, A. incisum* and *Nephrolepis cordifolia* (at 900-1200 metres) at higher altitude. *Athyrium puncticaule Dryopteris chrysocoma. Dannstaedtia appendiculata, Stenoloma chinensis,* and *Lindsaya cultrata* grow gregariously.

Ferns of Tista Valley Shorea Forests:

Very common - *Bolbitis* (Acrosticum) *heterochita, Diplazium austriale, Thelepteris uliginosa, T. aurita ching, Cyclosorus molliusculus, Pteris vittata, P. ensiformis.*

Other less common - *Asplenium falcatum, Adiantum incisum, A. lunulatum, Onychium siliculosum. Cheilanthes tenuifolia C. anceps., C. farinosa, Pteris stenophylla, P. semipinnata, P. subindivisa, Microlepia splenucae, Abacopteris multilineata.*

Epiphytes - *Vittaria elongata*

Common Ferns of Senchal Forests:

Very common. - *Mecodium exertum, Asplenium tennuifolium, A. cheilosorum, Coniogramme procera. Dryopteris chrysocoma, Diplazium stoliczkae* var. *hirsutipes.*

Epiphytic Ferns - *Loxogramme involuta., L. lanceolata., Phymatodes stewartii., Hymenophyllum sinosianum.*

Ferns on Rocks - *Mecodium* (Hymenophyllum) *javonicum, Vandenboschia* (Trichomanes) *pyxidifera, Didymoglossum* (Thichomanes) *insigne.*

Common Ferns of Tiger Hill (Machilus - Quercus Association):

Very common- *Dryoathyrium boryanum, Loxogramme chinensis.,*

Common - *Plagiogyria communis, Loxogramme involuta., Thelepteris brunnea, Diplazium stoticzkae* var. *hirsutipes, Athyrium thelyplerioides, Dryopteris splendens.*

Rare - *Diacalpe aspicioides* Bl.

Common ferns of Lebong (1500 metres) (Castanopis - Schim association)

Onychium lucidum, Pteris blumena, Mecodium (hymenophyllum) javonicum, M. exsertum - (common) *Vandenboschia* (Trichomenes) *radicans, Microlepea* (Davallia) *Lhasayana, M. strigosa. M. pilosula, M. Trapeziformis, Pteridium aquilinum, Hypolepis* (Polypodium) *punctata* (common) *Araiostegia* (Davallia) *pulchra., Oleandra wallichii, Alsophila glauca, A. glabra, Polystichum lentum,* P. biaristatum., *Athyrium drepanopterum, Dryoathrium boryanum,* (Very common), *D. maximum,* (very common). *D. viridissimum, D. stenochlamys, D. muricatum, Diplazpsis javanica,* (very common), *Thelepteris xylodes* and *Asplenium cheilosorum.*

Climbing Ferns as Component of Ground Vegetation:

Lygodium flexuosum and *L. japonicum,* are very common in plains and foot hills especially in damp areas. *Stenoclaena palustris* is the commonest fern of the moist forests and often climbs to tree top.

Thicket forming species are as follows:-

These are tough ferns and good soil binders. Common species are *Pteridium aqilium Kuhu, Gleichenia glauca Hook* and *Dicranopteris.*

D. GRASSES OF FOREST FLOOR AND GRASSLAND

Extensive study has been made on the eco-botaincal aspects of grasses of plains and hills in the study area. Various aspects of dominance, sociability, indicator value etc. have been described in detail under the heading " Ecological Status of the Grasses and Grassland".

Forest Types and Grassland Types:

Indicators to parallelism between woodlands and grasslands are meagre in this state, although Hele (1911) and Troup (1921) made some observations on parallelism. Factors like moisture, moisture-holding capacity of soil rocks, soils, biotic intervention and light conditions mostly govern the distribution of grasses and grassland types and, therefore, it is not easy to establish a correlation with the woodland types. Grasses being generally light demanders gradually disappear with the spread of a forest canopy, though a few shade tolerant species may persist. They have a wide range of adaptability and take different forms according to environmental conditions. This is the reason why many species manage to survive or flourish over a wide range of conditions. A few grasses, however can adapt themselves only in special conditions, such as *Porteresia coarctata* in the freshly formed estuarine mud banks, *Leersia* and *Hygroryza* in the marshes, *Microstegium ciliatum* and *Narenga porphyrocoma* in best quality Sal forests, *Capilipedium assimile* in poorer types of Sal forests in the foot hills and a preponderance of the spcies of *Aristida, Heteropogon, Tragus, Dichanthium* and *Chloris* in the drier laterite tract.

Hole (1911) and Troup (1921) found some parallelism between woodland and grassland which covers the forests of the State as follows:

Erianthus munia in *Accacia catechu* forests, *Narenga prophyro-*

coma in Sal forests, *Saccharum procerum* and *Phragmites karka* in *Albizzia-Bombax ceiba* forests.

In a clear-felled area grasses do come in but they gradually vanish with invasion of evergreen species and when planted trees form close canopy. It is thus apparent that in a forest climate grasses cannot represent any particular woodland type although some correlation is possible under broad ecological groups. Even these groups are transitory because with change in light and edaphic conition one type of grass vanishes giving place to another.

Some Ecological Features of Grasses and Grassland of Study Area:

(i) Savannah like formation:

The occurrence of vast tracts of coarse tall grasses on the river banks of North Bengal rivers are owing to regular flood that have killed the tree forests of could not reach successional climax owing to various factors. Grazing and fire have kept such areas in a state of temporary equilibrium. The savannah like grasslands in the districts of Jalpaiguri and adjoining areas on the bank of Brahmaputra had been the home of wide varities of wild life species in the past. In fact longe populations of tigers, deer, rhinos, wild buffaloes, bisons etc. used to live in such grassy tracts in the past. Although devastating flood over years have transferred such grassy tracts into vast tracts of sand silt and boulders some low lying Savannahs still exist. Several wild life Sanctuaries and National parks now testify to the biological and ecological importance and potentiality of such grassy tracts where many surviving wild life species have taken refuge.

There are vast tracts of Khair *(Accacia catechu)* and Sissoo *(Dalbergia sissoo)* forests on the bank of principal hilly rivers. In such forests, many species of the genera *Saccharum, Erianthus, Phragmites, Arundo, Sclerostachya, Hemarthria, Rottboellia, Themeda, Neyraudia* thrive. These species occur either in pure, gregarious or in mixed formations over wide range of areas. Such grasses form dense thickets which make it difficult for even elephants to penetrate them. Rhinos form distinct tunnels in them through which they graze. Interspersed in these thickets of coarse grasses lie many softer fodder grasses and herbs that grow either sporadically or in small isolated colonies.

(ii) Grazing grounds and lawns:

Fire, grazing and other biotic interference affect the grazing land. Under extreme pressure many tall grasses also form turf.

Imperata cylindrica, Chrysopogon aciculatus, Cynodon dactylon Cyperus, brevifolia (=Kyllinga triceps) have been found very common in West Bengal and Assam pastures. *Imperata* and *Chrysopogon* associates are the commonest of all. *Chrysopogon* possess a woody root stock and prostrate creeping stem which roots abundantly at the nodes. This is a favourite fodder grass and is capable to withstand intense trampling. *Imperata* is a tougher grass than *Chrysopogon*. The lawn formed by *Cynodon* is delicate, picturesque and soothing to the eyes.

(iii) Wet mixed forests:

Once the area is cleared of forests, the debris are burnt, soil is prepared by pulverizing, and protection is afforded by erecting fencing, several soft species of fodder grasses invade such area. These are *Paspalum conjugatum, Setaria palmifolia, Panicum* species, *Eragrosits* species and *Elytrophorus* species. These grasses persist so long the planted seedlings continue to grow and do not form dense thick crown to create permanent shade on ground floor, *Paspalum conjugatum* grow gregariously in the early years of plantations. *Elytrophorus* with its globose spicate inflorescence occurs sporadically. *Scripus erectus* (a sedge) also occurs sporadically in such areas. *Imperata cylindrica* sometimes invade such areas gregariously, especially where the plantation is damaged by fire. *Dactyloctenium* species occurs sporadically. *Setaria palmaefolia* is a very common and distinct shade loving grass of Forest floor. It has other associates viz., *Centotheca lappacea, Cyrtococcum oxyphyllum, Oplismenus compositus, Dactyloctenium aegyptium, Paspalidium flavidum* and *Panicum* species, *P. humidorum, P. brevifolium, P. sarmentosum, Echinochloa colonum, Sacciolepis indica, Paspalum longifolium* etc.

(iv) Grasses of the Forest Road Flanks:

Besides *Imperata cylindrica* and *Chrysopogon aciculatus*, there are some grasses that are conspicuous during the rains. They are *Setaria glauca, Paspalidium flavidum, Echinochloa crusgallia, Microstegium ciliatum* etc.

(v) Dry mixed forests:

Microstegium ciliatum, Narenga porphyrocoma, Pogonatherum species and *Ichaemum* species are commonest in the dry mixed forests. In Bhabar tracts there are some sprinkling of *Arundinella decempedalis, Capillipedium assimile* and *Saccharum* species.

(vi) Grasses of Sandy River Bank:

On the sandy bank, inundated by flood once or twice a year, *Brachiaria distachya, Digitaria preslii* are common decombent grasses.

Cymbopogon jwarancusa is also a very conspicuous grass for its size, shape and smell of leave. Vast tracts are covered by *Saccharum spontaneum, S. procerum*, and species of *Imperate, Arundo* and *Phragmites*. Other grasses are *Cynodon dactylon, Digitaria longifolia, Mnesithea* leavis, *Sacciolepis indica* etc.

In such tract two sub-types of tall grasses have been identified viz; one high level and another low level. High level genera are:-

Saccharum and *Erianthus* primarily. Low level genera are:-

Neyraudia, Phragmites, Arundo, Vetiveria and *Thysanolaena* etc.

(vii) Thatch Grass:

The leves of *Imperata cylindrica* is one of the best thatch grass; others are *Narenga porphyrocoma, Coelorhachis stricta, Themeda species, Sclerostachya fusca* and *Apluda mutica* etc.

Grasses as Indicator of Soil Conditions:

During the study of articficial and natural regeneration of Sal and its ecology by pioneer workers it was observed that *Narenga porphyrocoma* was the typical grass on soil suitable for Sal. This grass was followed by *S. arundinaceum, Arundinella* clarkei, *Erianthus fastigiatus, Andropogon nardus* and certain other species although they occur only in the openings but rarely grow gregariously. Area too moist or otherwise unsuitable for sal is generally indicated by the presence of *Imperata cylindrica and Themeda gigantea*. The invasion of the evergreen type of shrubs and herbs in so rapid due to present day fire protection that these grasses can only be found growing sporadically, and they are rarely found to form any definite association.

Saccharum spontaneum which has a tremendous capacity to colonise recently formed alluvium characterises the vegetation of moist low lying growing unsuitable for sal. It is often associated with *Phragmites karka* and *Erianthus ravennae*. *Khair-sissoo* forests of low alluvial land are characterised by *Saccharum procerum, Erianthus elephantinus, Themeda gigantea, Phragmites karka* and other coarse grasses. In moist pockers annual soft grasses occur.

The fresh deposits of sand and shingle which are only inundated during the rains are typified by some of the above xerophytic grasses and also *Saccharum munja, S. spontaneum* and *Triraphis madagascariensis*. *Cymbopogon jwarancusa*, a tufted gras may also be found growing on sandy soil along with *Digitaria preslii* and a few others. *Capillepedium assimile* represents poor soil in the foot hills of Himalayas where sal and other species are shunted in growth.

Grasses and Grass-Land Types of Moist-Deciduous Forests:

The grasses of this region may be broadly clasified into the types described below:-

In the marshes and moist areas *Hygrorhiza aristata* and *Leersia hexandra* are perennial floating grasses which root profusely at the nodes and may be seen occurring with *Enhydra fluctuans, Jussiueu udscendens Limnanthemum cristatum, Hydrilla* and other aquatic weeds.

Along with these grasses, at the margin of the water, the following species occur:-

Panicum interruptum, P. myurus, *P. paludosum, Echinochloa crus-galli, Sacciolepis indica, S. myosuroides, Apluda mutica, Arundinella benegalensis, A decempedalis, Brachiaria distachya, B. remosa* and sometimes *Rottboellia exaltata. Ischamemum regosum, Dactyloctenium aegyptium, Isachne miliacea* and I, *australis.*

Arundinella species are reed-like, robust grasses and are perennial. *Brachiaria distachya* is a soft grass and has spikes spreading horizontally, whereas *B. ramosa* has slender culms and spike fascicled, and leaves slightly pubescent. All these species occur at the initial stages of grassland formation in the marshy localities which, when left uninterrupted by fire, grazing and other biotic factors, get invaded by taller and coarser grasses like *Phragmites karka, Arundo donax, Vetiveria zizanioides, Coix lacryma-jobi* and *Erianthus ravennae.*

A common sub-type is that of *Typha* and *Phragmites* which may be seen in the marshy and low lying areas annually inundated by fllood. They may sometimes be found associated with species of *Alpinia* and giant sedges. In the rice fields the common grasses that may be seen are *Eragrostis unioloides, E. japonica, Elytrophorus spicatus, Eleusine indica* besides many sedges.

Microstegium Narega type:

Microstegium ciliatum is the extensively occurring grass of Dry Sal forests and can grow in partial shade and is gregarious. *Narenga porphyrocoma* which occurs in good sal area, however grows in gaps in the midst of *M. ciliatum* where sufficient light falls on the forest floor. This grass can also be seen growing in dry mixed miscellaneous forests. In blanks *Narenga porphyrocoma* can be found growing with *S. arundinaceum, Arundinella clarkei, Erianthus fastigiatus, Andropogon nardus* and sometimes in moister clayey pockets with *Themeda gignatea* and *Imperata arundinacea.*

Chrysopogon-Imperata type:

The *Chrysopogon-Imperata* association is the common sight in all grazing grounds, foot paths, road sides and in lawns with sprinklings of

Cynodon dactylon, Sporobolus diander, Eleusine species and and *Eragrostis* species *Chrysopogon* is very tough and unaffected by heavy grazing. *Desmostachya bipinnata* which is a common and troublesome grass in the south and central parts of this State is only a sporadic species in this region.

Paspalum-Seteria-Panicum type:

The pioneer genera in the wet mixed type of forests and especially in the clearings are *Oplismenus, Paspalum, Panicum, Eragrostis* and *Eleusine*; but they are gradually ousted by dominating and shade-loving genera like *Panicum and Setaria*. In the more clear-felled areas protected from grazing, *Paspalum distichum, Eragrostis unioloides, E. japonica* and *Eleusine indica* along with *Dactyloctenium aegyptum, Digitaria setigera* and *Crytococcum oxyphyllum* come in as pioner species and are followed by *Setaria* and various species of *Panicum* but they cannot keep pace with the invading evergreen herbs and shrubs. *Setaria palmaefolia* and *Panicum* species do, however, survive to some extent in the competition. *Setaria* sometimes occurs gregariously as ground vegetation in the shade. *Panicum* species which occur sporadically with *Setaria P. humidorum., P. brevifolium, P. patens, P. montanum, P. psilopodium, P. proliferum, P. sarmentosum, P. trypheron, P .colonum, P. ramosum, P.i ndicum,* var *gracile, P. setigerum* and *P. punctitum, Oplismenus* is a very widely distributed and shade-loving soft grass whose purple and viscid awn offers its best method of indentification.

Paspalum scrobiculatum sometime becomes locally dominant over large areas in the plantations and road sides. This grass has maximum frequency of occurrence in the first two years of a plantation but gradually becomes custed by invading evergreen species. Other shade loving grasses are *Centotheca lappacea* and *Alloteropsis cimicina. Centotheca* has well advanced adaptation for dispersal of seeds. Tips of pendant spikeless catch very readily in cloth or fur and sometimes a whole panicle comes away from the plant. In the forest margin, however, occurs *Setaria glauca, Paspalidium flavidum, Echinochloa crus-gulli, Imperata cylindrica, Chloris incompleta, Cynodon dactylon, Chrysopogon aciculatus, Eleusine* indica and other common grasses.

Arundo-Phragmites type:

This type can be seen all over the division in the swamps, swamp margine and stream banks and is the typical grassland association of the Teri low ladds. Either *Phragmites karka* or *Arundo donax* may be locally dominant and intermixed with various other grasses.

Narenga-Themeda-Vetiveria Type:

An association of the above type is quite common along the road sides and railway clearing in the forests on clayey soil. These areas are burnt annually and fire appears to stimulate their growth and spread. Typical dominant grasses are *Saccharum fuscum, S. spontaneum, S. munja, Narenga porphyrocoma, Themeda gigantea, T. Villosa and Vetiveria zizanioides*. Associated with these are *Erianthus ravennae* and *Imperata cylindrica*. Repeated fires, heavy grazing and interference by men tend to encourage *Vetiveria zizanioides* against its associates.

Other grasses that occur sporadically are:

Apocopis royleana. Andropogon schoenanthus, A. brevifolius, A. assimilis, Digitaria pruriens, Perotis indica, Polytoca bracteata (Sheath and leaves hirsute), *Pogonatherum saccharoideum* (rare) *Thysanolaea maxima* (rather common), *Eragrostis nigra, Erianthus longisetosus and Egragrostis coarctata.*

One remarkable feature in the tea garden areas in the neighbourhood of the forests of this region is the well maintained polot of thatch grass. The species of obvious choice for such thatch plots is *Imperata cylindrica* (large leaved form) although sporadically *Apluda mutica, Ischaemum rugosum, Rottboellia exaltata* etc. occur. In winter most species of *Saccharum* and *Erianthus* throw out silky white, pinkish white, pinkish silvery, mauve-purple, vulvous, reddish brown, black-violet coloured flowers and the grasslands of this region offer a great atraction to the observers.

Common Grasses of the Terai:

The grasses and grasslands of Terai do not differ much in their types and characters from those of the grasses of Duars. The only difference is that extensive Savannah like formations are absent in the Terai and the extent of the tract is small.

The Sal forests of Kurseong division adjoining Baikunthapur Raj Estate forests re sometimes affected by fires and as a result coarse grasses may be found mostly in open places. The commonest of them are *Imperata cylindrica, Narenga porphyrocoma, Saccharm arundinaceum, Arundinella decempedalis, Eulalia fastigiata, Themeda arundinacea, T. villosa, Andropogon nardus* and *Vetiveria zizanioides.* Very common in the month of November is the colourful display of lowering heads of *Themeda arunidinacea and Narenga porphyrocoma with* sprinklings *of Cymbopogon, Rottboellia and Coelorhachis.* They are also common along the Railway lines running through forests and other lands. As usual *Vetiveria* is very common along the road and railway borrow pits and margins of stagnant pools. *Chrysopogon aciculatus characterise* the open grazed areas and foot paths. *Cymbopogon nardus* and *Arundinella* species are quite common in Bagdogra, Pankhabari and Sukna ranges, especially in areas adjoining foot hills. Common grassland types are as follows:-

1. —*Hygrorhiza-Leersia-Panicum* type.
2. —*Paspalum-Setaria Panicum* type.
3. —*Chrysopogon-Imperata* type.
4. —*Arundo-Phragmites* type.
5. —*Nagenga-Themeda-Vetiveria* type.

The type No. 1 may be found in marshy areas adjoining Jalpaiguri district. The commonest species are *Hygrorhiza aristata, Leersia hexandra, Panicum repens, P. paludosum, P. pseudointerrupta* and *Sacciolepis interrupta*. Other grasses which occur in temporary marshes or occur sporadically along with the aforesaid grasses are *Echinochloa crus-galli, Sacciolepis indica, S. myosuroides, Paspalum distichum* (common in the temporary pools in Bagdogra range), *Paspalidium punctatum, Brachiaria distachya, B. ramosa, Rottobellia exaltata, Dactyloctenium aegyptium.*

In moist habitat *Apluda mutica, Isachne miliacea, I. australis, Ischaemum rugosum, Chionachne koenigi, Paspalum scrobiculatum, Echinochloa colonum, Eragrostis gangetica* etc. are commonly found.

The type No. 2 typifies the areas which have been clearfelled and protected from grazing and also forest areas which are shady, moist and protected from the fire and grazing. The pioneer genera in such areas are *Paspalum, Setaria, Panicum, Eragrostis* and *Eleusine*. Paspalum conjugatum comes in profusion in the early rains and overwhelms *P. scrobiculatum, Brachiaria distachya, Eragrostis unioloides, E. japonica, Eleusine indica* and other grasses whose aggressiveness is less than that of the first named species. Gradually, when the forest speceis grow and evergreen herbs and shrubs invade such areas, more exacting grasses like *Setaria palmaefolia, Crytococeum oxyphyllum, Oplismenus compositus* and a few *panicum* species come in and oust them. These grasses can survive very well in shade and are well represented all the wet mixed forests. *Setaria* and *Oplismenus* in fact thrive in all types of forests upto 2134 m. and are very important for the forest officers to note. Other grasses that are associated with them and occur sporadically are *Dactyloctenium aegyptium, Panicum perakense, P. brevifolium, P. mentanum, P. psilopodium, P. proliferum, P. sarmentosum, P. triferon, Brachiaria ramosa, Cyrtococcum patens, Brachiaria setigera, Sacciolepis indica* and *Elytrophorus spicatus.*

In forest margins can be seen *Setaria glauca, Cynodon dactylon*

Paspalidum flavidum, Imperata cylindrica, Eleusine indica and *Chrysopogon aciculatus.*

The type No.3 flourishes mostly in all grazing grounds and lawns. Most dominant of the grasses are *Chrysopogon aciculatus* and *Imperata cylindrica* which thrive well when grazed or trampled. Other sporadically occurring grasses are *Sporobolus dianeer, Cynodon dactylon Eragrostis* species, *Arundinella bengalensis* and *Demostachya bipinnata.* Among the sedges, *Kyllinga triceps, Fimbristylis* species and *Bulbostylis* species sometimes occur amidst the above mentioned grasses very sporadically.

The type N. 4 is mostly dominant in marshy areas. *Arundonax* and *Phragmites karka* occur in various patterns of admixture sometimes in assocation with *Typha elephantina* or *Calamus tenuis* with *Alpinia allughas* or sometimes in pure formations. In the riverain areas they are quite common and their admixture with *Saccharum spontaneum, S. procerum, Narenga porphyrcoma* and *Saccharum elephantinum* are very interesting.

The type No. 5 occurs along the roadsides and railway clearings in the forest especially on clayey soil. As the areas get burnt annually their incidence increases. Common grasses of such areas are *Saccharum spontaneum, Narenga porphyrocoma, Erianthus munja, Themeda gigantea, T. villosa* and *Vetiveria zizanioides.*

Other sporadically occurring grasses of such areas where soil is less clayey are:-

Cymbopogon clandestinus, Capillipedium assimele, Perotis indica, Politoca digitata, Thysanolina maxima, Eragrostis nigra, Erianthus longisetosus and Eragrostis coarctata.

Common Grasses of the Lower Hill Forests of all States (305 - 915 m. Altitude):

Most dominant species of the lower hills are *Arundinella bengalensis, A. wallichii, A decampedalis, A. nepalensis, Cymbopogon martinii, C. hookeri, C. flexuosus* and the genera *Erianthus, Saccharum, Capillipedium, Themeda, Pogonatherum, Imperata, Microstegium* and *Dendrocalamus hamiltonii* (Bamboo). Less common are *Neyraudia*, a few *Saccharum* species and *Pseudostachyum polymorphum (bamboo), Cymbopogon, Arundinella* may be distinct type in the Himalayan foot hills.

Dendrocalamus hamiltonii is the commonest bamboo of the lower hills and covers extensive areas in sal forests of Kurseong and Kalimpong divisions. It can grow straight upto 16m. but is generally straggling and a menace to sal forests of the foot hills, Patches of *Imperata cylindrica* and *Capillipedium assimile* can be seen in the sal forests of the foot hills. Sal forests of Sukna and Tista valley ranges are characterised by the thick growth of *Pogonatherum panicum* and *P. crinitum, Microstegium ciliatum* is more common in the sal forests of Sukna and Bagdogra ranges (Terai). These are all gregariously growing grasses and their identification is quite easy and worth knowing in the field by foresters. *Erianthus* species are not generally gregarious except in repeatedly burnt areas and their occurrence does not strike the eyes of foresters except in the flowering season (late rains and after rains) when they become very showy. Common associate species are *Eulalia fastigata, Eraianthus hookeri* and *E. ravennae. Pogonatherum* species have a capacity to grow on steep cuttings of hill slopes and can be best seen along the Tista valley and hill cart roads. They can grow on almost sheer rock and extend into the middle hills. Commonly they may be found associated with *Thysanolaena maxima* on such steep slopes. *Cymbopogon nardus* is common in the lower hills in Sukna and Pankhabari ranges although it extends down to Boiranti block of Bagdogra Range. *Neyraudia reynaudiana* is a sporadically occurring grass of the plains and mostly of lower hills of Kurseong division and is also quite common along the cart road leading to Darjeelong. Less sporadically occurring species of this zone are *Saccharum spontaneum, Arundo donax, Phragmitis karka, Erianthus munja, Themeda arundinacea, Imperata cylindrica* and *Vetiveria zizanioides.*

Notes on other grasses occurring in the zone:

Paspalum dilatatum has a wide distribution like *Setaria palmefolia, Isachne albens, Oplismenus compositus* and *Imperata cylindrica.* While *Paspalum dilatatum* extends upto 1829 m., *Setaria* and *oplismenus* occur from plains upto about 2439 m.; *Cynodon Chrysopogon* and *Imperata* may be noticed sometimes upto 1524 m. In marshes *Caelchne simpliciuscula* occurs sporadically. Of the *Eragrostis* species *E.nigra* and *E. coarctata* generally extend upto middle hills, whereas *E. tenella, E. diarrhena, E. unioloides* and *E. tremula* extend upto 610 m. or so. They may sometimes be found with *Arundinella intricata* and *A. bengalensis* upto 610 m. to 915 m. in the hills. One may thus find *Andropogon ischemum* at 305 m., *Pollinia articulata* at 610 m., *Garnotia stricta* at 610 m. and a few *Panicum* species higher up.

Common Grasses the Middle Hill Forests of all States (915 -1829 m. Altitude):

Common genra of this zone are *Arundinella, Erianthus, Miscanthus, Agrostis, Oplismenus, Setaria* and several bamboos *(Dendrocalamus* and *Arundinaria,* species).

Among the forest grasses that grow in shade, *Setaria palmaefolia, Isachne albens* and *Oplismenus compositus* are common all over the hills although several species of shade loving *Panicum* and *Eragrostis* grow in this zone. Sometimes they form thick carpets on the forest floor ousting other evergreen herbs and shrubs. Most showy during the month of August, September and October are the flowering panicles of *Erianthus fulvus* and *Miscanthus nepalensis*. They have perhaps the best looking inflorescences among the middle upper and alpine hill grasses. The small mauve-purple inflorescence of *Erianthus* that fade to fulvous colour are very attractive. The nodding corymbose-fastigiate, fulvous-haired panicles of *Miscanthus nepalensis* soothe the eyes of the vistors of Darjeeling travelling along the cart road when the flower heads peep out growm the thickets of the common spreading *Gleichenia glauca* and *G.l inearis* (fern) jungle, *Erianthus* is common in waste places and may extend upto 3048 m. In the tea estate waste lands adjoining Pugriangbong block it has been found growing with *Bothriochloa intermedia, Setaria viridis* and some other sedges. *Eulalia fastigiata* very rarely extends upto 2134 m. *Agrostis micrantha* and *A. filipes* are soft and slender grasses and are found between 1220 - 1829 m. although they may extend upto the alpine zone.

The species of *Pogonatherum* have their uppermost distribution in Kurseong division upto 2134 m.

Digitaria sanguinalis is common at 1829 m. and can be identified by its bluish black inflorescence and may be found often in association with *Isachne* and *Muehlenbergia* in the forest margins and also in shade.

Sporadically occurring other grasses are mentioned below:-

Arundo donax, Arthraxon lancifolous, Eragrostis nigra, E. diarrhena, Eulalia mollis, Garnotia stricta, Keeloria cristata., Hackelochla granularis (occurs on rock sometime with *Pogonatherum), Muehlenergia huegelli, Neyaudia reynaudiana, Setaria glauca, S. viridis, Sporoblus elalngatus, Thysanolaena maxima, Themeda hookeri, Paspalum dilatatum, Erianthus hookeri, Eragrostis coarctata, Eleusine coracana* (cultivated), *Hordeum vulgare* (cultivated) and *Oryza sativa* (cultivated).

Common Grasses of Temperate or Upper Hill Forests (1829-2744 m.):

Conspicuous temperate genera of this region are *Poa. Agrostis Helictotichon, Muehlenbergia, Danthonia, Dactylis, Miscanthus, Arthraxon* and *Arundinaria.*

In the plantations and at forest margin softer and shade loving grasses are *Oplismenus compositus, Digitaria sanguinalis, Setaria palmaefolia* and *Isachne albens*. All these grasses are found in indentical areas in the plains. At 1900 m., the plains species of *Pogonatherum crinitum, Neyraudi reynaudiana, Setaria glauca, sacciolepis indica* are sometimes found. All these indicate that these grasses primarily of plains have capacity to resist forest and extreme cold.

White the incidence of occurrence of *Oplismenus, Isachne, Pagonatherum, Neyraduia, Setaria, Sacciolepis, Digitaria* etc. gradually decline the following species become prominent.

Agrostis filipes, Poa annua, Arthraxon lancifolius, Dactylis glomerata, Miscanthus nepalensis, Helicototrichon virescens, Muelhenbergia huegelli, M, ciliatum & M. vimineum.

Poa annua is very common from 1829 m. upwards, It is a very slender grass, common in grassland and along road sides; its flowers flush in early July. It ascends up to 3048 m. or a little over, *Arthraxon lancifolius* is an annual slender gregarious grass of moist places and is common in Senchal pasture. *Pogonatherum crinitum* and *Capillipedium parviflorum* may be found on rocks in the Senchal and Ghoom regions throwing their flowers up from September to November. *Miscanthus nepalensis* is best developed between 2134 and 2439 m. although it extends upto 2744 m. *Eulalia mollis,* a slender and tufted grass also extends in Senchal along with *Arthraxon, Isachne albens, I. sikkimensis, Setaria palmaefolia, S. glauca, Digitaria cruciata* and D. *bifasciculta* also occur along with soft herbs and shrubs along road sides. Sporadically mixed with them occur a species of *Muehlenbergia* and *Digitaria sanguinalis.*

Other grasses that have been recorded from this zone are - *Arundinella hookeri, Eragrostis nigra* (quite common in oepn places at 1829-2134 m.), *Garnotia stricta, Holcus lanatus, Trisetum spicatum, Anthoxanthus hookeri, Hierochloa flexuosa, Keoleria cristata, Muehlenbergia huegelii, M. ciliatum, M. vimineum, Phalaris arundinacea* (a big perennial grass) *Pseudoraphis brunoniana, Sacciolepis indica and Trisetum spicatum.*

People trekking in the Tiger hill and Tonglu areas find big grass land where most of the above grasses throw out their flowers during the rains; the combined beauty of their golden flower heads (many sedges also occur here and flower at the same time) along with the white, yellow and purple-mauve flowers of *Anaphalis contorta, Gnephalium leutoalbum* and *Labiatae* against the background of green grasses assumes a strikingly colourful feature. Scattered amidst these flowers peep out scores of big sized puffballs *(Lycoperdon* species) throwing clouds of dust like deep brown spores when slightly disturbed and these can never escape the naturalist's eyes.

Common Grasses of Alpine Hill Forests (Above 2700 M. Altitude):

Principal grassland type in high temperate Himalayan hills is *Duyeuxi - Deschampsia* type.

Strikingly beautiful are the green Apline grasslands where myriads of *Anemones, Fragarias, Petentillas, Primulas, Aconitums* and scores of other flowering plants threw out their exquisitely tinted flowers during the rains. Most of the grasses of this region flower during and after the rains, and the hill sides and meadows are then set ablaze with manifold colours. Some of these grasses are annual, some biennial and others perennial. Some have decorative flower heads; some from carpets. *Arundinaria* species are wide spread in the lower elevations whereas species of *Poa, Festuca, Deschampsia* and *Duyeuxia* dominate the Alpine grasslands.

In the Darjeeling district such grasslands do not extend upto the snowline; nevertheless they constitute much coverted feeding grounds for cattle and sheep during the Summer months. Pioneer botanists have observed that several species of grasses of the this region show strongly European characteristics, *Deschampsia caespitosa, Bromus mollis, Brachypodium sykvaticum, Poa annua, Festuca rubra, Holcus lanatus* and *Dactylis glomerata* have world-wide distribution. Typical European genera are *Deyeuxia Helictotrichon, Brachypodium* and *Bromus.*

The common genera that characterise the grasslands of this region are:-

Agropyron, Anthoxanthum, Bromus, Catabrosa, Deyeuxia, Deschampsia, Danthonia, Elymus, Eragrostis, Festuca, Glyceria, Hierochloa, Helictrotrichon, Kaoleria, Oryzopsis, Phleum, Poa, Stipa, Trisetum, Arundinaria and Thamnocalmus species.

The following are among the important grasses that have been recorded from this alpine zone:-

Agrostis pilosula, A. nervosa, A. sikkimensis, Agropyron longearistatus, A. thoroldianum, Anthistiria hookeri, Anthoxanthum hookeri. Brachipodium sylvaticum, Bromus asper, B. tectorius, B. himalaicus, Calamagrostis tripolifera, C. emodensis.

Danthonia cachemyriana var. *miner, D. Jacquemontii, Deschampsia caespitosa, Deyeuxia scabrescens, D. pulchella, Elymus mutans, E. dasystachys, Eragrostis nigra. Festuca cumminsii. F. valesiaca, F. polycolea, F. rubra. F. undata. Glyceria tonglensis, Oryzopsis a-equighemn's, Hierochloa, flexuosa, Heliprotrichor virescens, Phleum alpinum, Polytoca, bracteata, Poa khasiana, P. flexuosa P. attenuata, P. sikkimensks, P. nemoralisx, P. himalayana, P. ludens, P. tibetica, P. alpina,* var *pratensis., Polypogon species. Stipa roylei, S. purpurea, S. mongholica, S. arthoraphium, Sporobolous elongatus.*

Species of *Danthonia* occur extensively, in Tenglu, Phalut and Sandakphu. *Calamogrostis, Helictratrichon* and *Trisetum* extend from tiger Hills to Phalut.

Deyeuxia, Agrostis, Oryzopsis, Stipa, Hierochlea, Poa, Glyceria, and *Bromus* are of common occurrence amongst the grasslands of Tonglu, Phalut and S. Andakphu. These are intermixed with many species of sedges. Most of these plants flower during August to November.

Figure 6 shows the altitudinal distribution of grasses in the study area.

Ecological Status of the Grasses and Grasslands

There is no naturally stabilised grassland of Savannah type in the study area. Whatever grasslands exist today mainly result from fires either accompanied or not accompanied by shifting cultivation, grazing, and excessive cutting. The vast stretches of Savannah like formations that are found all over the study area are certainly the results of fire, grazing and flood. If those limiting factors were to be withdrawn there would be a succession of vegetation towards an evergreen type. Various indications of this can be found in the forests. In the dry mixed forests, grasses occur in suitable pockets which are mainly caused by edaphic variations. Some of these pockets of grasslands have been in existence for over a century and there has been no large scale invasion of evergreen or deciduous species succeeding them. The density of the stocking, thickly matter roots, annual fires and to some extent grazing have ensured this stability. In spite of these facts, the grassland of these regions is only relatively stable and cannot be called a climax formation. Such Savannah like formations are common in Terai and Bhabar areas, in river valley swamps and burnt Sal forests of Assam. In the riverain forests of *Accacia catechu and Dalbergia sissoo*, tall grasses generally invade, and once burnt, they persist, grow vigorously and suppress seedlings of the forest tree species. These conditions have lead to the formation of vast areas of tall grasses in the riverain forests of Terai and Bhabar zones.

In the alpine areas of the Himalayas extensive grasslands occur which are the result of devastating fire succeded by grazing.

Thus it may be seen that the grasses and grasslands of the study areas are the results of birotic and edaphic factors leading to an edaphic or bioedaphic climax and there is no true climatic climax grassland. (Fig - 6 gives altitudinal distribution of grasses and sedges)

E. SPORADIC OCCURRENCE (About 10 to 40% Frequency) OF HERBS IN GRASSLAND TRACTS:

There are limited areas of high level and low level Savannah like formations and grassy tracts. Some herbaceous vegetations are associated with such vegetation and they occur sporadically. Some of the species that were recorded in various plots are as follows:-

(a) Herbs of tall reeds:

Alysicarpus monilifer, Atylosia scarabaeoides, Cassia tora, Clitoria ternatea, Desmodium triflorum, Medicago denticulata, Melilotus alba, Momosa pudica, Pueraria species, *Tephorosia candida, Uraria picta, Vicia hirsuta.*

Some of the remaining herbaceous species are *abutilon indicum Achyrananthes sessilis, Argemone mexicana, Blumea membranacea, Euphorbia hirta, Hyptis brevipes, Kyllinga brevifolia* (Sedge), *Leucas aspera, Borreria articulata, Cardiospermum halicacabum, Carex condensate* (Sedge), *Centella asiatica, Cleome viscosa, Datura metel, digera muricata Eichornia cressipes* (Monocot), *Ocimum sanctum, Phyla nodiflora, Phyllanthus maderaspatensis, Polygonum faccidum, Pycreus nitens* (Sedge), *Scirpus erectus* (Sedge), *Tridax procumbens, Vernonia cinerea and Xanthium indicum.*

(b) Herbs of drier Foot hill tract:

Atylosia scarabaeoides, Cassia tora, Crotalaria prostrata, Desmodium polycarpum, Indigofera cordifolia, I. linifolia, Ageratum conyzoides (herbs), *Asparagas fulicinus, Borreria hispida, Cannabis sativa, Carex cruciata* (Sedge), *Centella asiatica, Cissampelos pareira, Cyanotes barbata, Cyperus riveus* (Sedges), *Euphorbia hirta, Gnaphalium luteoalbum, Leucas aspera, Morina longifolia, Oxalis corniculata, Phyllanthus maderaspatensis, Polygala crotalarioides, Senecio nudicaulis, Thalictrum foliolosum, Thymus serpyllum, Vicea indica.*

(c) Temperate Alpine Himalaya Grassland:

Herbaceous vegetation form of about 50 species of which *Indigofera heterantha, Lathyrus pratensis, Anaphalis cuneifolia, Anemone obtusefolia, Capsella bursa-pastoris, Ganphalium leutoalbum, Inula cappa, Nasturtium officinale, Plantago major, Taraxacum officinale, Thalictrum foliolosum, Carex species* - 12 species (Sedge) are conspicuous.

(d) Jalpaiguri - Siliguri Road side grassland tract:

Centella asiatica, *Desmodium* species, *D. triflorum, Equisetum* species (Fern allies), *Moghonia* species, *Typha* species, *Vernonia cinera, Abutilon indicum, Amaranthus* species, *Argemone maxicana, Atylosia scarabaeoides, Cassia tora, Cleome viscosa, Clitoria ternatea, Crotalaria* species, *Croton bonplandianum, Datura metel, Digera muricata, Euphorbia hirta, leucus aspera, Mikania cordata* etc.

(e) Temperate hills - Grassland tracts of Darjeeling and Kalimpong:

Polygala mollis, Trifolium repens, T. tomentosum, Polygonum species, *Ageratum conyzoides, Borreria articultata, Cannabis sativa, Cro-*

talaria species, *Desmodium polycarpum*, *Alysicarpus* species, *Pueraria* species, *Mimosa* species, *Urena lobata* etc.

Shrubs-as Fodder and Host Plant:

Most of the species are browned and grazed by animals. But all of them are not palatable. Some of the families and genera of shrubs are palatable and are of fodder value as follows:-

Capparis, Flacourtia, Casearia, Saurauia, Hibiscus, Thespesia, Helecteres, Grewia, Zizyphus, Crotalaria, Derris, Priotropis, Indigofera, Lespideza, Desmodium, Flemingia, Dalberia, Barteria, Pigeum, Rubus, Eriobotryan, Hydrangea, Dichroa, Ribes, Woodfordia, Aralia, Brassiopsis, Viburnum, Morinda, Maesa, Buddleia, Cordia, Erhetia, Callicarpa, Premna, Achyranthus, Polygonum, Piper, Antidesma, Bridelia, Glochidion, Croton, Moraceae (all genera and species), *Urticaceae* (all species).

Shrubs-as A Source of Fruit for Animals and Birds:

Although many other genera come under this category, some common genera are as follows:-

Grewia saurauia, Capparis, Flacourtia, Glycosmis, Evolia, Zizyphus, Mimosa, Acacia, Pigeum, Rubus, Syzygium, Osbeckia, Oxyospors, Melostema, Brassiopsis, Viburnum, Coffea, Cordia, Erhetia, Piper, Litsaea, Machilus, Loranthus, Phyllanthus, Antidesma, Bridelia, Sauropus, Glochidion, Moraceae (all species), *Solanum* species, *Bridelia, Sauropus, Glochidion, Moraceae* (all species), *Solanum* species, *Eurya* species, *Allophyus* sp., *Aphenia* species, *Flemingia* species, *Desmodium* species, *Misoma* species, *Dicroa* species.

F. LIANAS, CLIMBERS AND SHRUBBY CLIMBERS - AS VERTICAL COMPONENTS OF FORESTS:

General Observation

Large woody climbers, a type of 'Life Form' and a biotic association, are remarkable features of the forests owing to their abundance and fantastic forms. Climbers are of various sizes, morphology and life forms add to the complexity of the vertical structure of the forests. The woody climbers or lianas often become large with thick convoluted stems. They show many kind of abnormal cambial activities which result in an irregular outline of the stem giving a rope like appearance due to the partial disintegration of tissues. Their foliage can hardly be identified when they climb tall trees. Often they slip off and fall down from their supports, form great-loops on the ground and make the passage through forests impenetrable. There are delicate climbers and shrubby climbers which number almost the same as woody and shrubby climbers which number almost the same as woody climbers. *Ficus* and *Schefflera* are well known stranglers which anastomose the host giving a beautiful form. Some of them climb by means of spines or thorns as for example the climbing palms (*Calamus* spp.) and the Robus. The root climbers throw out adventitions roots to fasten themselves to crack in the bark (many *Araceae*); they have long branches, very long branches, very long internode, modified leaves or side shoots which are sensitive to touch and serve as a clinging organ. Tropical lianas are said to be longer lived than the temperate ones. The climbers are more common in degraded and secondary forests than in virgin forests but may be numerous in the ecotone Zones. The percentage of woody and shrubby climbers is about 15% of the total number of species in the study once. The proportion of woody to shrubby climbers is about 50% each. Per hectare occurrence may be upto 500 stems.

Large woody climbers are remarkable features of some North Bengal Foot hills, and ecotone zones of evergreen and deciduous forests. In fact monsoon moist-deciduous forests have maximum incidence of climbers per unit area. The present survey indicates in Darjeeling and Jalpaiguri hills, there are about 111 lianas and 105 climbing shrubs. They are also abundant with luxuriant growth and the riverain forests, blanks and fire burnt forest areas.

Climbing plants are usually classified according to their means of attachments to their supports as scramblers, twiners, root climbers and tendril climbers.

Most common climbing plant families are Apocynaceae, Asclepiadaceae, Leguminosae, Combretaceae, Menispermaceae, Convulvulaceae, Titaceae, Ranunculaceae, Schizandraceae. Anonaceae, Lardizabalaceae, Actinidiaceae, Sterculiaceae, Celastraceae, Rhamanceae and Araliaceae, Among the monocotyledonous climbers Plamae occupies the most important place. The *calamus* and *Simlax* species play a large part in the physiognomy of the forest. Crowns are well spread and conspicuous. Araceae has two species *(Scindapsus* and *Raphidophora)* of epiphytes. Gnetaceae (Gnetum scandens) has a single specie of large climber. There are a few climbing ferns such as *Stenoclaena, Lygodium* and *Polypodium* species They compete with trees for space and light and cause their crowns become misshapen. Twiners have much constricting effects and leave sprial grooves on the stems and barks of host plants. The weight of climbers often breaks the branches of supporting trees.

It is not possible to study entire physiography of climbers, as climbers are regarded as weeds and they are removed from the forests. A special type of climber mass has been described by Cowan as 'Convolvulous Mictum" with large scattered trees all over and herbaceous and semi-herbaceous creeper beneath. The principal genera are *Argereya, Porana, Impmoea, Millettia, Smilax, Mucuna, Luffa, Padaerua,* and

Mikania, There are several scandent *Capparis, Dalbergia, Homlskioldia, Zizyphus, Munronia* and other species in such jungle.

The small climbers always pass their lives in shade and they do not impart any significant character to the forest composition. This type is represented by *Pothos, Lygodium, Piper* and several other species.

Stranglers are special groups which begin their lives as epiphytes and later send down roots to the soil becoming independent plants often killing tree that supported them. *Ficus* and *Schefflera* are typical genera of such stranglers. There are more than forty (40) species of Ficus in North Bengal.

3. Climbers contributing to conspicuous physiognomy of the Forests.

(i) Colouration of conspicuous flower - (Genera)

1. White flowers:- *Beaumentia, Chonemorapha, Thunburgia, Porana, Clematics, Schizandra, Millettia, Jasminum, Vallaris, Ichnocarpus, Biswaria.*
2. Red flowers:- *Holmskioldia, Rivea, Thunburgia, Coccinea, Holboellia.*
3. Purple flowers:- *Mucuna imbricata, M. monosperma, Dalbergia stipulacea, Tinospora malabarica, Derris, Paederia, Porana grandiflora.*
4. Yellow flowers:- *Naravelia, Desmos (Uvaria), Stephania, Gissus discolor, Caesalpinia crista, Edgaria.*

(ii) Very common and conspicuous genera:

Millettia, Mucana, Entada, Spatholobus, Uvaria, Stephania, Melodorum, Tinospora, Tetrastigma, Ampilocissus, Cissus, Dalbergia, Bauhinia, Acacia, Combretum, Mimosa, Ichnocarpus, Argyreia, Erycipe, Ipomoea, Smilax, Mimosa, Dioscorea, Raphidophora, Pothos, Scindapsus, Stenoclaena, Lygodium, Calamus, Daemonorops and *Perthenocissus.*

(iii) Common genera

Delima, Aspidopterys, Celastrus, Ventilago, Gouania, Caesalpinia, Hodgsonia, Heptapleurum, Vallaris, Porana, Thunbergia, Jasminum, Cryptolepis, Tiliocora and *Marsdenia.*

(iv) Sporadic genera

Uncaria, Beaumontia, Cissampelos and very many other spp.

The woody climbers are very conspicuous in the plains and foot hill forests of the study area. An idea of abundance of species may be had from actual quantitative assessment which may be at places about 500 individuals per ha.

In a plantation in Terai region a 5 ha. plot was laidout and 33 species were identified (refer to "Ground vegetation under plantation").

G. EPIPHYTES - AS VERTICAL COMPONENT OF FORESTS:

General Observation:

Epiphytism is a type of biotic association and a specialised Life Form where the epiphytes grow on other plants wihout deriving any food from inside the supporting plant. The macro vascular & Nonvascular forms are very common in warm and humid tropics, but the non-vascular forms are prevalent in the temperate and alpine zones. The epiphytes derive food from the bark crevices. Roots of epiphytes sometime store water in a tissue called velamen. *Dischidia* sp. have a remarkable adaptation where the leaves fold and from a containerlike structure to hold water. The monsoon temperate East Himalayan forests and other hill forests of the study area where moisture level is high and leaves constantly drip, the incidence of epiphytes is high. Epiphytes consume water economically just as the succulents do. Many orchids have leaf tubers as water reservoirs and also have thick leaves.

The frequency of occurrence of all epiphytic species have been recorded as abundant, common, frequent, sporadic and rare in the analytical synopsis of plant families etc.

Tropical and Temperate Epiphytes

The epiphytes grow in luxuriance in the forest of North Bengal and comprise an intergral part of the composition. It is, however difficult to classify the epiphytes into communities or types on any specific basis. This is because the epiphytes are not selective in choosing their hosts and the choice is not universal. There are special types of life forms that grow on living plants and struggle for light at different heights.

The epiphytes have three types of habitat viz., (i) Tree trunks (ii) Branches and (iii) Crowns (heliophytes). Some are succulants, xerophylous shade tolerent and hygrophilous types.

Type No. (i) Tree trunks - It has maximum incidence of occurrence and comprise about 50% of the epiphytes (mostly nonvascular).

(ii) On Branches - Epiphytes of this type compose about 40% of all epiphytes (mostly vascular)

(iii) On crowns - Epiphytes form 5% of the forms, and the balance 5% comprise of succulent xerophylous shade tolerent and hygrophilous type.

The above types have the morphological forms:-

Caespitose form:- The orchids have this form and they are mainly distributed in the temperate hills. The buds are dormant at the base protion of mother plant close to the substratum and the reproduction is vegetative.

Fascicularis form:- The ferns have this form where the leaves are rosetee and there is no creeping or upright shoot. The dormant buds are close to the subsrata and there is no vegetative reproduction.

Creeping forms:- The shoots are creeping; the shoots or stolons root at junction of leaves and the leaves are well apart. Some ferns belong to this form.

All the above types and forms relate to the non-vascular forms; there are woody vascular epiphytes which are a few in number but very conspicuous at 100 metres in the forests. One form is woody and upright, another form is monocauline or of single stem, and the third form is caespitose, either woody or herbaceous climbing.

Epiphytic Fern:

(i) Epiphytes on tree trunk:

They get only diffused sun light and do not have any effect of wind velocity and exposed environment. Ferns belonging to this type are wiry and they are *Mecodium exsertum, M. javonicum, M. membranaceum, M. polyanthes, Loxogramme chinensis and Hymenophyllum simosianum.*

The fronds dry up in winter and shrivel. *Oleandra wallichii* is conspicuous (at 1400 metres) and grows on rocks as well; the leaves turn yellowish and assume fine attractive landscape.

Some epiphytes growing on tree trunks are branchy. They are *Vittaria ophiopogonoides, Lepisorus loriformis, L. excavatus, Loxogramme lanceolata, P. argutum, P. amoenum, Lepisorus and L. kashyapii.*

(ii) Nest forming epiphytes:

Asplenium nidus known as 'birds nest fern' grow on Sal and miscellaneous trees iin the duars and foothills. This has longer and tougher leaves than *Microsorium punctatum (Polypodium punctatum).* This species sometimes grows in profusion on a tree. Large simple fronds radiate on all directions, collectively forming a basket into which humus collects. *Pseudodrynaria coronans (Polypodium coronans)* grows in thick formations on road side trees in Tista Vally (Fronds are not dimorphic). This is associated with *Vittaria elongata* and rarely of *Lycopodium phlegmaria* of *Lycopodium squarrosum*, and rarely of *Lycopodium phlegmaria.*

(iii) Bracket - forming epiphytes:

Drynaria quercifolia with dimorphic leaves (Sterile leaf, Stiff, scaly and fertile large and green) has this form. This is very common in Duars forests and road-side trees. At elevation above 1100 metres, this species is replaced by *D. prophinqua.* Smith. Other common ferns are *Lepisorus nudus, Pyrrosia stigmosa, Phymatodes lucida, Davallia hullata, Nephrolepis cordifolia* and *Vittaria sikkimensis.*

(iv) Epiphytes of (1500m. - 2400m. growing at foot of the trees.):

The epiphytes that complete their life cycle during the rainy seasons after which they wither are as follows:-

Mecodium exsertum, M. javanicum, M. polyanthos, Hymenophyllum simosianus, Vandenboschia auriculata, V. radicans, Microsorium hancocki M. membranaceum and Loxogramme chinensis. Common epiphyte on trees and rocks at 1500 - 2000 metres are *Oleandra nerriformis* and *O. wallichii. Microsorium* and *Loxogramme* are filmy ferns.

(v) Epiphytes growing on branches in shade are as follows:

Vittaria ophiopogonoides, Lepisorus loriformis, Loxogramme lanceolata, Asplenium ensiforme, Polypodium microrhizoma, P. argutum, P. amoenum, Lepisorus exeavatus (Lepisorus species also occur at the tree tops.)

(vi) Mossy forest epiphytes are as follows:

Polypodium argutum, P. amoenum, Lepisorus excavatus, Vittaria flexuasa, Phymatodes griffithiana, Lepisorus loriformis, Asplenium laciniatum Don., Asplenium enseforme and *A. planicaule.*

(vii) Epiphytes of Rhododendron-Conifer forests (2500-3600 M.) are as follows:

Oleandra wallichii, Polypodium argutum, P. lachnopus, Lapisorus excavatus, L. kashyapii, L. longifolius., Arthomeria himalayansis, A. lechmani, A. wallichiana, Vittaria himalayansis, Ctenopteris subfalcata and *Phymatodes ebinipes.*

(viii) Epiphytes of Rhododendron-Conifer forests (above 300 metres):

Phymatodes oxyloba, Polypodium lachnopus and *Araiostegia dareiformis.* In the pure *Abies-Rhododendron campanulatum* forests *Phymatodes malacodon, P. strachyi, P. erythrocarpa, Polypodium atkinsoni* and *Lepisorus nudus* grow.

Above this altitude ferns are no feature of landscape.

(ix) Vascular epiphytes in Temperate hills:

Conspicuous vascular epiphytes which are widely distributed in the hills are as follows:-

Pentapterigium serpens, Aeschynanthus sikkimensis, Gonathanthus sarmentosus, Roscoea spicata, Globba species, *Lysionatus serrata, Loxostigma griffithii, Aeschynanthus ramosissima, A.* species, *Pentapanax leschenaultii* (climbing), *Scindapsus officinalis* (creeping), *Rapidophora glauca* (creeping) and *Pothos scandens* (creeping).

Rosette leaved and fasciculariform *Asplenium* and *Microsorium* specis grow on the stem of *Shorea robusta* and are very conspicuous. Widely occurring and less conspicuous ferns are *Vittaria elongata, V. longifolia and Polypodium* species.

Conspicuous orchids of caespitosa form are represented by *Malaxix rufilabris*. Single stem orchids that grow on *Shorea* are *Rhynchostylis retusa* and *Doritis wightii.*

(x) Vascular epiphytes on branches:

Vascular epiphytes on branches:

Vasculr epiphytes that grow on branches are *Hymenodictyon flaccidum, H. parasiticus* (white bracts), *Agapetes saligna, A. auriculata, Pentapterigum hookeri, Rhododendron edgworthi, R. dalhousiae, Fagara obovata and Acer hookeri.*

Many species of *Ficus* initially start as epiphytes and subsequently grow as independent trees.

(xi) Some common epiphytic families:

Common families are Ericaceae, Vaccinaceae, Rubiaceae, Gesneraceae, Asclepiadaceae, Aceraceae, Araceae, Scitaminae and Orchidaceae.

(xii) Special features of epiphytes:

Epiphytes thrive on shade in an atmosphere of variable humidity. Epiphytes, especially those on high branches and on crowns face wind blasts. They seem to differ in their microclimatic and humus requirement, as some are found in shades and some in open while others avoid strong light and heavy shades. Density of crown and stratification also control their distribution. Hoya species prefer swampy and wet forests, *Asplenium nidus* prefers tree trunks to branches; *Pseudo-drynaria* prefers the junction of stem and branches. Mature and old tree trunks and branches have more of epiphytes than the young; Bamboo stem has only lichens as epiphytes. It is possible to correlate the epiphytes with the tree associations as done in the cases of grasses. It is also possible to classify them under shade epiphytes, sun epiphytes, hygrophylous and xerophilous epiphytes. Epiphytes have remarkable structural specialisation. Several epiphytes grow as terrestrial herbs and shrubs.

(xiii) Other Non-vascular epiphytes:

Most of the lichens, hepatics, mosses and some algae are epiphytes. Temperate forests are rich in such flora. Those plants are useful from the view point of soil and water conservation, animal and bird food and aesthetics of the forests.

(xiv) Special observations:

In suitable open habitation (on soil and bare rock) many epiphytes can grow successfully on the ground. The roots of epiphytes in many cases are nesting places of arboreal ants. A gignatic epiphyte is *Wrightia gignatea*. In temperate forests 40% of the trees bear epiphytes on them. A single tree may have more than 20 species of epiphytes. A square kilometre of forest in many areas will have more than 100 species.

In several advanced countries epiphytes have been studied in relation to air pollution. It is possible to count the surface area occupied, total number of such species, frequency of each species per bole or quadrat, height reached by each species above ground level, vitality and fertility of the epiphytes and to get an idea about the influence of epiphytes on environment.

(xv) Epiphytic Orchids:

Sal Forests in plains:

Malaxis rufilabris Reichb., *Vanda roxburghii* R.Br., *Bulbophyllum careyanum* Spreng.,

Foot hills 800 m.:

Eria flava Lindl., Phalaenopsis panishii Reichb., Oberonia prainiana King & Pantl., Dendrobium pierardi Roxb.,

At 1,300 m.:

Dendrobium longicornu Lindl., *D. cathcartii Hook., Eria coronaria* Lindl., *Otochilus alba* Lindl., *O. fusca* Lindl.

At Sureil 2.300 m. :

Dendrobium chrysanthum Wall., *Eria paniculata* Lindl., *Liparis perpusilla* Hook., *Bulbophyllum odoratissimum* Lindl., *Dendrobium nobile* Lindl., *Eria bractescens* Lindl., *Eulophia nuda* Lindl., *Otochilus porrecta* Lindl.,

At Singalila Forests:

Bulbophyllum careyanum Spr., *B. eublepharum* Reichb., *B. leopardinum* King & Prantt., *B. striatum* Reichb.,

Calanthe species. *Coelogyne corymbosa* Lindl., *C. flavida* Wall., *C. ochracea* Lindl., *C. prolifera* Lindl., *Dendrobium amoenum* Wall., *D. densiflorum* Wall., *D. eriiflorum* Wall., *D. hookerianum* Lindl., *D. porphyrochilum* Lindl., *Eria convallarioides* Lindl., *E. dasyphylla* Par & Reichb., *E. graminifolia* Lindl.,

The following genera of Orchids have been identified of which First five genera have several species.

Coelogyne	-	All pseudobulbous epiphytic orchid. A very large genus.
Dendrobium	-	Epiphytic. A very large genus.
Eria	-	Epiphytic. A large genus.
Currhopetalum	-	A large genus.
Bulbophyllum	-	A large genus.

Others:

Ione - Epiphytic. A large genus. *Panisea* - Tufted orchids. *Pholodota* - Epiphytic. *Otochilus* - Epiphytic. *Sarcopodium* - Epiphytic. *Malaxis* - Several species.

Liparis - Several species. *Oberonia* - Several species.

H. TREE VEGETATION IN PALINS:

(1) Belt Transect in Plains:

(i) Wet Mixed Forests (Table A and Histogram 4 & 15)

At Moraghat :

The soil is moist and the sub-soil water is high. The forests have a density of 250 trees per hectare. There is preponderance of species of *Lauraceae* and *Meliaceae* besides *Jambosa formosa, Artocarpus chaplasa, Talauma hodgsoni, Glochidion* species, *Castanopsis tribuloides* and *Dysoxylum* species. In fresh water swamps *Podocarpus neriifolia* occurs over limited areas. *Acronychia laurifolia*, a small evergreen tree, is very extensive. In the top storey *Shorea* is associated with the evergreen *Artocarpus chaplasa*.

At Lataguri:

The density of crop is 212 per hectare. There is patchy concentration of Machelia montana, a rare species, in the forests (edaphic pockets). Other associates of this species are *Lophopetalum fimbriatum, Elaeocarpus aristatus. E. rugosus, E. varunua, Dysoxylum benectariferum, Aphanamixis polystachia, Amoora wallichii,* several *Lauraceous* species, *Olea dioica, Styrax serrulatum* and *Turpinia pomifera*.

The vertical column in the statement of analysis gives the list of tree species in different canopy layers. Some forest sites indicate edaphic type and has been described separately.

At Damanpur:

The density of crop is 225 per hectare. There is preponderance of *Meliosma simplicifolia, Casaeria varoca, Syzygium cuminl and Jambosa praecox, Vatica lancaefolia,* an associate species of evergreen forests of Assam is found in the area only in the whole of North Bengal.

At Apalchand:

The density of crop is 195 per hectare. Predominant spcies are *Syzygium praecox, Macaranga denticulata, Dillenia indica, Drimycarpus racemosus, Garuga pinnata, Ficus* species *Sloanea sterculacea* and *Meliosma simplicifolia.* The presence of *Garuga* and *Drimycarpus* in large numbers shows that these deciduous species have gradually crept in the evergreen belt. *Shorea* is absent from this area.

At Rajabhatkhawa:

The density of crop is 275 per hectare which is very high. There is preponderance of *Meliosma pinnata, Dillenia pentagyna, Macaranga denticulata, Bauhinia purpurea, Castanopsis tribuloides* and *Eriolaena wallichii.*

At Udlabani:

The density of crop is 200 per hectare. Principal species are *Dillenia pentagyna, Meliosma simplicifolia, Premna benagalensis, Syzygium* species, *Syrax serrulatum, Macaranga denticulata, Ilex godajam, Litsaea monopetala, Machilus gamblei.*

(ii) Scattered Shorea robusta Forests (Table B and Diagram 3, Histogram 16).

At Moraghat - The density of crop is 200 per hectare. Number of species per hectare is less than the Wet Mixed type. The species of high density are *Artocarpus chaplasa, Toona cillata, Dillenia pentagyna, Gmelina agborea, Garuga pinnata, Ilex godajam, Macaranga denticulata* and *Premna* species. *Shorea robusta* comprises 25% of the crop.

Artocarpus chaplasa, Ilex gadajam, Actinodaphne obvata, Litsaea species, *Machilus* species, *Talauma hodgsoni* and *Mangifera sylvatica* show wet mixed pockets in the transect; *Careya arborea, Terminalia belerica* and *Shorea* show the mixture of dry mixed type.

At Lataguri:

The density of crop is 168 per hectare. *Artocarpus chaplasha, Elaeocarpus varunua, Eurya Cavinervis. Ilex godajam* indicate the wet mixed character of the crop. But gradually the number of deciduous species and their density increase. Main difference with the wet mixed type is in the percentage composition of *Shorea robusta* which is 28.4% in this type and in the absence of *Michelia montana.*

At Damanpur (Jamguri forests):

The density of crop is 325 per hectare. Shorea robusta has a percentage of 26.7%. There is high density of *Premna, Meliosma simplicifolia* and *Casearia vareca.* Striking feature of the forest is the complete absence of *Vatical lancaefolia.*

At Rajabhatkhawa:

The density of crop is 325 per hectare. The percentage of *Shorea robusta* is 9.6%. *Schima wallichii* has a higher then the wetter type. *Aphanamixis polystachia, Vitax heterophylla, Elaeocarpus arunua* and *Bauhinia purpurea* have a very high percentage of occurrence. In fact such a high concentration of *Vitex* does not occur elsewhere is North Bengal.

At Jainty:

The density of crop is 362 per hecare. The percentage of *Shorea robusta* is 31.5% *Aphanamixis polystachia* has 13.0% occurrence. Other associates are *Terminalia belerica, Bridelia retusa, Dillenia pentagyna, Mallotus albus* and *Schima* wallichii.

At Bengdubi:

The density of crop is 362 per hactare. The percentage of *Shorea* is 13.8% and *Schima wallichii* 18.2%. Other prominent species are *Talauma hodgsonii, Symplocos spicata, S. ferrugenia, Eurya japonica, Ilex godajam, Ternstroemia gymnenthera, Endospedrmum chinense* and *Homalium zeylanicum.*

Regeneration:

Common are *Aphanamixis, Acronychia, Schima, Sterospermum, Dillenia, Mallotus, Litsaea, Machilus, Glochidion* species etc.

(iii) Pure Shore forests (Table C and Diagram 4, Histogram 17).

The density of crop is 296 per hectare consisting of 26 species. Total number of tree species in this type is 77 and is less than all other types. This zone extends over large area and is the representative type of forest that is found in the plains.

The distribution of the diameter class of *Shorea* shows that most of the trees are middle aged to mature and there is dearth of young aged crop. The graph (A) also shows that the associates of *Shorea* are mostly young aged crop. This leads to the conclusion that mature crop has either been exploited or the miscellaneous evergreen crop have invaded the mature forests after the fire protection was enforced and there was complete absence of regeneration. Another indication is evident that the forests are primarily of *Shorea robusta* with a few associates having little bearing on the general type of vegetation and some miscellaneous forests located over limited areas forming different edaphic types.

Scrutinizing the figure of the transects it is revealed that Moraghat and Lataguri (Lower Tondu) forests hade more of evergreen associates of *Shorea* than the other types. *Acronychia laurifolia, Actinodaphne angustifolia, Evodea miliaefolia, Chisocheton paniculatus, Artocarpus lakoocha, Baccaurea sapida* occur mostly in Jamguri forests. Species common to all the types are *Aphanamixis polystachia, Callicarpa arborea, Careya arborea, Dysoxylum binecteriferum, Premna* species, *Syzygium cuminii, Lannea coromendelica* and *Macaraga denticulata.*

Histogram Nos. 1 - 8 and statement No. VI show density of different species in different types of forests. Histogram Nos. 12-18 show the canopy class distribution of predominant species in percentage; they also show the distribution in percentage of Dipterocarpaceae, Meliaceae and Lauraceae - the three most well distributed families in different forest types.

(iv) Dry-Mixed forests (Table D and Diagram 2. Histogram **18)**

Dhumchi has one of the highest number of trees per hectare and the area has 70 species in the transect Crop density next in order are at Chapramari, Kharbari, Bhutri, Jainti, Boiranti and Udlabari. The following species are found all over in this type:

Callicarpa arborea, Dillenia pentagyna, Premna species *Garuga pinnata, Syzygium* species, *Sterculia villosa, Stereospermum chelonoides, Lagerstroemia perviflora, Careya arborea, Lannea cormendelica, Mallotus philippinensis, Oroxylum indicum, Wrightia tomentosa* and *Schima wallichii.*

Average height of the trees in this type other than *Shorea* is almost half of Pure *Shorea* type in the top canopy; the reason is ascribed to shallow and sandy loam soiil. The top storey species are *Terminalia belerica, Lagerstroemia parviflora, Stereospermum chelonoides* and

Sterculia villosa; the middle storey species are *Aphanamixis polystachia, Dillenia pentagyna, Litsaea polyantha, Mollotus philippinensis*. There are four distinct layers of vegetation against five in other types. Different tree associatings are:

Top canopy	(1)	*Terminalia - Sterculia - Lagerstroemia* association.
	(2)	*Terminalia - Syzygium - Stereospermum* association.
Second canopy		*Aphanamixis - Dillenia - Syzygium* association.
Third canopy		*Wrightia - Premna - Careya* association.

In the Histogram Nos. 1 - 8 and statement No. 1 percentage of different species in different types of forests have been shown. Histograms 12 - 18 show the canopy class distribution of predominant species in percentage; it also shows the distribution in percentage of Dipterocarpaceae, Meliaceae and Lauraceae.

Frequency (percentage) of some of the species are:

Dysoxylum binecteriferum 100%, *Orxylum indicum* 100% *Wrightia tomentosa* 85%, *Shorea robusta* 85%, *Premana bengalensis* 85%, *Lannea coromendelica* 85%, *Premna integrifolia* 85%, *Litsaea polyantha* 85%, *Litsaea sebifera* 85% *Dillenia pentagyna* 85% and *Callicarpa arborea* 85%.

Analyis of canopy classes is as follows:

Canopy layers	Number of species	Percentage (%)
Top storey	36	34.6
Middle storey	44	42.3
Lower storey	24	23.1
Total	104	100,0%

Histogram No. 2 represents frequency distribution of different species.

The analysis of ground flora has been done in detail in Bhutri, Jainti, Champramari and other forests. In the ground *Ervatamia divericata, Urena lobata, Imperata cylindrica, Dioscorea* species, *Asparages racemosa, Leea* species and regeneration of *Terminalia belerica, Litsaea* species, *Lagerstroemia parviflora, Shorea robusta, Wrightia tomentosa* and *Dillenia pentagyna* are very common.

(2) Random Plots: (Table E)

(i) Pure Shorea including Scattered Shorea and Wet-Mixed forests:

Many small plots in each of the forests at Nilpara and Chilapata, Moraghat and Tondu, Baxaduar, Rajabhatkhawa and Damanpur (Jumguri) and Dalkajhar were studied. Vast areas under this type of forests were analysed. The results have been tabulated and their Frequency and percentage of species calculated.

All the forests are in the plains and consist primarily of pure *Shorea robusta*. In the enumeration economically important species have been counted; uneconomical species have been registered by plus signs.

At Nilpara and Chilapata Forests:

This study has been made over extensive areas where the percentage of *Shorea* is 47.2%. The canopy classification is same as those in the transects. *Shorea* in this vast area has the following associates:

Aphanamixis polystachia 3.8%, *Amoora wallichii* 7.1%, *Castanopsis tribuloides* 2.6%, *Lophopetalum fimbriatum* 5.1%, *Lagerstroemia perviflora* 1.3%, *Sterculia villosa* 2.0%, *Michelia champaca* 3.2%, *Schima wallichii* 4.6%. *Syzygium cuminii* and *S.* species 5.4%, *Terminalia belerica* 1.7%, *Lauraceous, Meliaceous* and the third storey tree species have not been counted but their presence recorded by sign; test checks, however, have been made in the field and the resuts found to tally with the figures of transects.

Concentration of *Lophopetalum fimbriatum* is very high at places and the association of this species with *Shorea* may be ascribed to edaphic variation. The association of *Michelia champaca* with *Shorea robusta* is also prominent at places (Edaphic variation)

At Moraghat and Tondu forests:

The forests extend over large areas in the plains were *Shorea robusta* is 57.5%. This is the richest *Shorea* forest in the whole of North Bengal. Density and the quality of crop is very high.

Principal associates of *Shorea* with species percentage are as follows:-

Aphanamixis polystachia 3.7%, *Amoora wallichii* 0.9% *Castanopsis tribuloides* 3.1%, *Sterculia villosa* 2.3%, *Schima wallichii* 8.3% *Syzygium* species 3.4%, *Terminalia belerica* 3.0%. Other species have similar distribution and concetration as in the transect lines. In the Moraghat, Tondu, Chilapata and Nilpara forests there are isloated areas where *Artocarpus chaplasha* occurs in association with *Shorea. Lophopetlum* has resricted distribution in this zone.

Results of analysis of many plots in Sursuti, Medla and Borohati blocks of Lower Tondu Range are as follows:-

Shorea robusta 50.2%, *Aphanamixis polystachia* 7.6%, *Castanopsis tribuloides* 4.6%, *Schima wallichii* 4.5%, *Dillenia pentagyna* 3.3%, *Syzygium* species 2.7%, *Premna* species 2.7%, *Terminalia belerica* 2.4%, *Machilus* species 1.8%, *Elaecoarpus* species 1.5%, *Lagerstoemia parviflora* 1.1%, *Sterculia villosa* 0.5%.

In the Punjhora, Hillajhora and Bhokai-Mardi blocks of Upper Tondu forests the crop has the following percentage of occurrence:-

Shorea robusta 57.3%, *Schima wallichii* 10.6%, *Sterculia villosa* 6.1%, *Terminalia belerica* 4.7%, *Syzygium* species 2.8%, *Machilus specis* 2.1%, *Dillenia pentagyna* 1.6%, *Castanopsis tribuloides* 1.3%, *Terminalia tomentosa, Bauhinia* species, *Aphanamixis polystachia, Lannea coromendelica* 0.9%.

South Moraghat forests has 60.6%, of *Shorea robusta*, while the other species have very small percentage of occurrence.

At Buxa Duar forests:

The forests of this zone is in the Bhabar tract where the soil is very well drained and the undegrowth is comparatively sparse. *Shorea robusta* has a percentage of 45.4% and is of good quality. Other associates and their percentage of occurrence are as follows:-

Aphanamixis polystachia 12.5%, *Amoora wallichii* 1.8%, *Lagerstroemia parviflora* 3.7%, *Sterculia villosa* 5.5%, *Schima wallichii* 6.2%, *Syzygium* species 3.7%, *Terminalia belerica* 9.6%, *Garuga pinnata* 3.7%.

A slight change in composition is indicated in the increase in the percentage of *Terminalia belerica* and *Aphanamixis polystachia.*

Analysis of ground flora shows a striking similarity to Jainty areas which has been analysed in the Chapter of ground vegetation.

At Rajabhatkhawa and Damanpur:

Most of the forests are low lying and wet and the percentage of *Shorea* in the wet soil goes down. *Shorea robusta* has a percentage of 32.7% while the percentage of other species are as follows:-

Aphanamixis polystachia 6.5%, *Castanopsis tribuloides* 7.5%, *Lagerstroemia parviflora* 3.7%, *Sterculia villosa* 6.2%, *Schima wallichii* 2.3%, *Syzygium* species 5.8%, *Terminalia belerica* 3.7% and *Amoora wallichii* 0.5%.

The quality of *Shorea* in Damnpur areas is bad and the percentage of *Syzygium* and evergreen moisture loving species predominate.

There are many evergreen pockets where *Syzygium* form almost pure crop and is sometimes mixed with *Elaeocarpus varunaua, E. aris-*

tatus, Sloanea sterculiacea, Turpinia pomifera, Meliosma pinnata and *M. simplicifolia.*

***At Dalkjhar forests*:**

The forests of Dalkajhar is in Terai region of the District of Darjeeling. Innumerable small streams originate and most of the forests have a very high water level. The percentage of occurrence of *Shorea robusta* is 50.6% and those of other associates are *Schima wallichii* 10.6%, *Terminalia belerica* 3.2%, *Lophopetalum fimbriatum* 0.4%, *Lagerstroemia parviflora* 2.4%, *Aphanamixis polystachia* 1.2%, *Castanopsis tribuloides* 1.0%, *Amoora wallichii* 0.8%, Sterculia villosa 0.9%.

There are few very rare (local) species which occur with *Shorea* and such type has been described under Edaphic Forest types. Apart from this, *Acronychia laurifolia, Carallia lucida, Talauma hodgsoni Magnolia pterocarpa, Mastixia pentendra* and *Calamus* species occur.

The ground vegetation has fair representatives of families *Labiateae, Compositeae, Scrophulariaceae, Acanthaceae* and *Leguminosae* are very wide spread. *Cycas pectinata* forests of Dalkjhar, Tukriajhar and in Tista Valley forests. This species is vanishing from the forests of North Bengal. *Podocarpus neriifolia* and associate of *Shorea* in the wet mixed forests is also fast vanishing from the forest of Dalkajhar and Moraghat; it also occurs in Lopchu and Buxa hills.

(ii) Drv-Mixed forest (Table F)

Many plots in each of the forests at Sipchu, Bhutri and Pana. Kalabari and Nipania were studied and the resuls have been tabulated showing Frequency, Density, Percentage of occurrence of species and canopy class.

These Dry mixed forests are situated on sandy loam slopes in Bhabar areas. While Kalabari and Nipania are to the west of the river Tista, Sipchu, Pana and Bhutri are to the east.

At Sipchu:

These forests lie either on old river beds or on the sancy and sandy loam soil overlain by boulders. *Shorea robusa* is practically absent and many deciduous species occur as a dominant crop. The crop is comparable to the Dry mixed types of the transects. There are 34 species of economic importance. Species of higher percentage occurrence are *Tetrameles nudiflora* (13%), *Terminalia belerica* 7.5%, *Stereospermum chelonoides* 10.8%, *Chinnamomum cecedodaphne* 7.5%, *Gmelina arborea* 5.0%, *Aphanamixis polystachia* 4.1%, *Sapium baccatum* 3.8%, *Schima wallichii* 3.4%, *Albizzia lucida* 4.2%, *Castanopsis tribuloides*

2.8%, *Bauhinia" species* 2.8%, *Amoora wallichii* 2.6%, *Alstonia sholaris* 2.8%.

Gmelina arborea a very useful plant has its best development in this type of forests.

There is one distinct association in this type viz., *Tetrameles nudiflora - Stereospermum chelonoides - Terminalia belerica* in the top storey; in the storey below *Aphanamixis polystachi - Gmelina arborea - Albizzia lucida* form a distinct association. *Cinnamomum cecedodaphne* is an unusual occurrence in the locality and may be due to some moist pockets in the entire dry rea.

The ground flora is not at all different from Chapramari forests which has been described under dry mixed forests in the anlysis of ground vegetation.

At Bhutri and Pana block:

These blocks also occur in same latitude as Sipchu in the gently sloping areas adjacent to the hills. There are 42 species of little or some importance. This crop is comparable to the dry mixed forests at Sipchu. Species of high density percentage are:-

Aphanamixis polystachia 16.8%, *Gmelina arborea* 10.1%, *Terminalia belerica* 9.4%, *Tetrameles nudiflora* 5.8%, *Stereospermum chelonoides* 4.0%, *Sterculia villosa* 4.5%, *Bauhinia* species 3.9%, *Albizzia lucida* 4.2%, *Schima wallichii* 1.9%.

Gmelina arborea has a very high percentage of occurrence. Top storey association is formed by *Terminalia belerica - Tetnameles nudiflora - Stereospermum chelonoides*. *Sterculia villosa* does not have uniform distribution. In the second storey *Aphanamixis polystachia - Gmelina arborea-Albizzia lucida* association is conspicuous.

The ground flora is not at all different from Chapramari forests which has been described under dry mixed forest in the analysis of ground vegetation.

At Nipania and Kolabari forests:

The vegetation of these areas are contiguous to Nepal and Bihar forests and the composition of crop have fair number of species of those areas. The following are percentage of occurrence of several species of those forests:-

Lagerstroemia parviflora 19.5%, *Terminalia belerica* 13.8%, *Premna bengalensis* 8.7%, *Adina cordifolia* 6.8%, *Castanopsis tribuloides* 6.0%, *Garuga pinnata* 5.1%, *Dillenia pentagyna* 4.0%, *Stereospermum chelonoides* 5.9%, *Sterculia villosa* 6.2% and *Albizzia lucida* 2.5%.

There are 31 economically important species in these forests. The presence of *Adina cordifolia* is a special feature of this area Principal association is *Lagerstromia parviflora - Terminalia belerica - Stereospermum chelonoides.* In the second storey *Premna species, Garuga pinnata - Dillenia pentagyna* association is prominent. The abundance of occurrence of *Castanopsis tribuloides* in this zone is a special feature of this locality.

The ground vegetation is similar to the forests described elsewhere. The crown class of the trees is similar to the Dry mixed forests of the Linear Transects.

I. TREE VEGETATION IN THE HILLS

(i) Shorea forest in Kurseong foot hills: (Table H)

The sampling covers more han 10% of the Shorea and miscellaneous forests of the plain forests of entire Duars and Terai and the figures of analyses are, therefore, true representation of the crop of the area.

The forests of the areas have been studied under three groups as follows:-

Many plots in each of the forests at Champta, Singimari, Bamanpokri, Khairbani, Punding, Jogijhora and Koklong were studied. The results have been tabulated and their frequency and percentage of species calculated.

The topography in slightly slopping hills and adjoining undulating areas bear *Shorea robusta* forests.

Shorea Robusta Forests in Foot Hills.

At Chamta :

There are 44 economically important species in this block. Shorea forms 30.1% of the composition and is the predominant species. The crown classification is similar to those enumerated in the linear transects. The associates forms smaller percentage as follows:-

Terminalia belerica 12.6%, *Dillenia pentagyna* 7.5%, *Aphanamixis polystachia* 5.5%, *Lagerstroemia parviflora* 5.8%, *Garuga pinnata* 3.5%, *Sterculia villosa* 1.9%, *Syzygium* species 1.6%, *Hymenodictyon excelsum* 4.4%, *Elaeocarpus* species 3.0%, *Talauma hodgsoni* 4.2%, *Schima wallichii* 0.9%.

The number species of less importance is 33. The middle storey has about 45 species, top and lower storey have about one third.

At Singimari:

There are 44 economically important species in the block. *Shorea robusta* forms 56% of the crop and is predominant. Crown classification is same as already described. The associates form the following percentage:-

Aphanamixis polystachina 7.8%, *Terminalia belerica* 3.6% *Talauma hodgsoni* 3.0%, *Castanopsis tribuloides* 3.0%, *Dillenia pentagyna* 1.3%, Schima wallichii 0.8%, *Garuga pinnata* 0.8%, *Lagerstroemia parviflora* 0.4%, *Sterculia villosa* 0.8%, *Syzygium* species 0.3%.

At Bamanpokri and Khairbani:

There are 36 economically important species; *Shorea robusta* forms about 44.2%, *Schima wallichii* 5.8%, *Garuga pinnata* 4.7%, *Terminalia belerica* 3.8%, *Gmelina arborea* 2.1%, *Ailanthus grandis* 2.6%, *Careya arborea* 2.0%, *Chikrasia tabularis* 1.2% and *Adina cordifolia* 1.7% of composition.

Schima wallichii is conspicuous here which is not so in the two previous areas. Adina cordifolia is an associate of *Shorea* is a special feature of the forests west of Bamanpokri. *Adina* is completely absent from the forests east of Tista.

At Punding, Jogijhora and Koklong:

There are 34 economically important species. The percentage of occurrence of *Shorea robusta* is 46.3%. Next in order of abundance are *Terminalia tomentosa* and *T. belerica* with 9.4% and 2.6% respectively. Other species are *Adina cordifolia* 5.3%, *Schima wallichii* 6.0%, *Sterculia villosa* 4.5%, *Garuga pinnata* 2.5%, *Aphanamixis polystachia* 2.2%, *Amoora wallichii* 2.0% and *Dillenia pentagyna* 2.1%.

The ares shows *Adina cordifolia* which reaches its eastern most limit; *Terminalia tomentosa* has a very heavy concentration probably highest anywhere in North Bengal.

(ii) Broad leaved and conifer forests of Kalimpong hills (Table - I):

Geologically the rock and soil of the foot hills of Kalimpong, Sub-Division differ very much from the foot hills of Darjeeling and Kurseong Sub-Divisions. The rock and soil have been described in the appropriate chapters. Rainfall also differs. The Jalacca valley gets highest rainfall of the District, and the vegetation combined with the effect of the soil and rock is somewhat different from other areas.

(i) Belt Transect at Eastnar (350 metres):

Analysis shows the presence of 76 species in an area of about 2.53 hectares. This line is representative of the forests of the foot hills of the south-eastern part of the Sub-Division. Rainfall sometimes reaches 8-00 millimetres a year. Principal species are *Jambosa formosa, Phoebe hainesiana, P. attenuata, Beilschmiedia clarkei* and *B. roxburghiana,*

Species of secondary importance are *Cinnamomum obtusefolium, C. impressinerium, Dysoxylum procerum, D. binecteri ferum D. hamiltonii, Gynocardia odorata, Litsaea khasyana, L. salicifolia, L. panamonja, Machilus villosa, Ostodes paniculatus, Polyalthia seminarum, Styrax serrulatum, Syzygium cumini, Talauma hodgsoni, Turpinia pomifera, Dendrocalamus hamiltonii* and *Bauhinia purpurea.*

The forest may be classified into the following types:-

(a) Top storey association - *Syzygium-Phoebe-Beilschmeidia*

(b) Second storey association - *Ostodes-Cinnamomum-Dysoxylum.*

In this storey there is well dispersed *Dendrocalamus* bushes scattered all over.

Other vegetation of the locality are *Castanopsis indica, Michelia champaca, Bauhinia purpurea, Alstonia scholaris, Ficus* species and *Dillenia pentagyna.*

There are some gigantic climbers in the ground flora of which *Uvaria, Baumontia, Ipomea* and *Bauhinia* are common genera. *Capillepedium assimele, Calamus erectus, C. acanthospathus, C. leptospadix Camellia drupifera, Casearia vareca, Desmodium, Flemingia* and *Crotalaria* species commonly occur in the area.

There are several species of ferns and the species are same as those mentioned in the foot hills elsewhere.

(2) Random plots at Tista Valley Forests:

Analysis of many plots in the Tista valley forests reveal the presence of some species endemic to this valley forests. *Ulmus lancaefolia* is worth mentioning in this respect. The herb and fern flora are very rich and there are several species confined to the area. The following tree associations are common.

(i) *Shorea-Terminalia-Tetrameles* association; this is very often mixed with fair amount of *Garuga pinnata, Duabanga grandiflora,* and *Toona ciliata.* In the second storey there are *Callicarpa arborea* and *Spondiusmangifera.* The under growth is composed of *Capillepedium, Microstegeum, Leea, Desmodium* etc.

(ii) *Ahorea-Schima-Terminalia* association with sprinklings of *Toona ciliata.* In the second storey *Garuga pinnata, Castanopsis tribuloides, Spondius mangifera,* and *Dendricalamus hamiltonii* are common. The undergrowth is same as the former type.

(iii) *Terminalia-Schima-Ailanthus* association with sprinklings of *Garuga pinnata, Toona ciliata* and *Castanopsis tribuloides.* This type occurs prominently in the inner valleys where the soil is moist and deep.

In the second storey *Magnolia pterocarpa, Syzygium kurzii, Ostodes paniculata* are very common. The ground flora is dense with preponderance of *Urticaeous* species, *Piper* species, *Polygonum* species, Ferns and grasses.

The ferns, grasses, shrubs and herbs of the Tista valley have been described elsewhere in this work.

(3) Random Plots at Chel Forests:

The presence of *Shorea* in this zone is indicative of a transitional zone. Principal association is *Terminalia-Schima-Bauhinia. Terminalia belerica* has a very low percentage while concentration of *T.tomentosa* is high. These species have also wide distribution in the plain forests.

Rainfall of this area is about 5000 millimtres. *Stereospermum*, which is rather common in Tista valley has a poor occurrence in this region.

Principal associates of the above species are *Syzygium* species, *Sterculia villosa, Talauma hodgsoni, Terminalia myriocarpa, Actinodaphne obovata, Ailanthus grandis, Albizzia lucida, Beilschmiedia roxburghiana, Castanopsis tribuloides, Toona ciliata, Chikrasia tabularis, Cinnamomum obtusefolium Dillenia pentagyna, Duabanga grandiflora, Engelhardtia spicata, Sloanea* species, *Eurya cavinervis, Gmelina arborea, Lagerstroemia parviflora* and *Lannea coromendelica.*

The ground vegetation has abundance of bamboo regeneration. There are plenty of climbers which have been described separately.

(4) Random plots at Eastnar forests (900 metres):

The type of forests found in this area is same as described in the linear Transects. *Phoebe* and *Syzygium* are still the dominant species. In the second storey the following are the principal species:-

Macaranga pustulata, Dysoxylum procerum, Beilschmiedia species *Brassipsis* species, *Duabanga grandiflora,* and *Ostodes paniculatus.*

Common climbers are *Mikania cordata, Pueraria sikkimensis, Bauhinia vahlii, Ipomea* species, *Ichnocarpus frutscens, Mucuna* species *Millettia* species and *Spatholobus* species.

Among the shrubs *Sida cordifolia, Urena lobata, Eupatorium odoratum, Desmodium* and *Crotalaria* species are common.

(5) Random plots at Churanti (1000 - 1500 metres):

These areas show a perfect association of *Castanopsis tribuloides, C.indica* and *Schima wallichii*. There are few *Castanopsis* indica. The soil in this area is clayey loam over-lying Sikkim Geniss. Principal associates are *Terminalia tomentosa, Syzygium formosum, Litsaea* and *Machilus* species, *Mallotus philippinensis, Mangifera silvatica, Garuga pinnata* and *Sloanea* species.

In the second storey and shrubby layer *Boehmeria rugulosa, Ficus cunia, Callicarpa arborea, Trichillia connarioides, Grewia vestita, Dendrocalamus hamiltonii, Meliosma simplicifolia* are conspicuous.

Of the ground vegetation *Artimesia vulgaris, Maesa chisia, Croton caudatus, Inula cappa, Clerodendrum bracteatum, Dichroa febrifuga* are worth mentioning apart from the list of vegetation under the analysis of ground flora.

(6) Random plots at Eastnar (1400 metres):

At this altitude *Castanopsis tribuloides-Engelhardtia spicata* association is predominant and conspicuous. *Cinnamomum* species have a very high percentage and are mixed with *Michelia, Nyssa* and *Acer*.

In the second storey *Eurya, Symplocos, Litsaea* and *Machilus* species occur in great abundance.

(7) Random plots at Westner (1600 metres):

In the top storey *Castanopsis tribuloides - Nyssa javanica,* and *Acer* species *(A. thompsonii, A. campbellii, A. lavaegatum)* form association.

Mixed with this association there are equal proportions of mixture of *Beilschmedia, Ailanthus, Cinnamomum, Toona, Echinocarpus, Engelhardtia, Garuga, Michelia* and *Quercus* species.

In the second storey *Eurya* and *Symplocos* species are very common.

(8) Random plots at Bokim (1800 metres):

Two different associations are conspicuous viz.,

(i) *Michelia - Castanopsis - Engelhardtia* Association

(ii) *Castanopsis - Schima - Acer* association. It has wider altitudinal coverage).

The association in the top and second storeys are same as above.

(9) Random plots at Iche (200 metres):

The following is the principal association of vegetation:-

(i) *Micehlia - Acer - Engelhardtia* association with mixture of *Nyssa, Eriobotrya, Prunus, Castanopsis* and *Beilschmiedia* species.

(10) Random plots at Rissum (2200 metres):

In the lower altitude the following association is very conspicuous:-

(i) *Machilus - Michelia* association is common. Other associates are *Acer* speceis, *Beilschmeidia* species, *Litsaea* species (most common in the second storey), Nyssa species and *Prunus nepalensis.* The percentage of *Quercus* species increase with the increase in altitude.

(11) Random plots at Lava (2400 - 2800 metres):

Machilus - Quercus association is common at this altitude. Quercus species gradually increase and in many localities form pure crop.

Higher up, extensive forests of *Rhododendron* occur either pure or mixed with *Tsuga dumosa and Abies densa* associations. Common associates are *Sorbus cuspida, Andromeda villosa, Viburnum cordifolium* and bamboos. There are extensive areas of bamboo (described separately) which established themselves after the destruction of *Rhododendron* forests years back by fire.

In the *Rhododendron* forests common shrubs are *Ardisia macrocarpa Actinodaphne sikkimensis, Arundinria* species, *Berbris* species, *Boehmeria polystachia, Eurya cavinervis, Pilea smilacifolia, Leycesteria formosa* and *Girardinia palmata.* Figures 1 and 3 show altitudinal distribution of species.

(iii) Broad Leaved and conifer forests of Darjeeling hills: (Table - J)

The vegetation of the foot hills upto 600 metres has already been analysed while describig the *Shorea robusta* forests of the foot hills. The following are the omposition of the forests from 600 metres and above:-

(1) At 600 - 975 metres:

The species recorded in this zone are same as areas of similar altitude at Jagijhora, Punding & Bamonputri. The principal association is that of *Shorea-Terminalia-Schima.* Common species of this zone are *Ailanthus grandis, Albizzia stipulata, Betula cylindrostachs, Castanopsis indica, C. Tribuloides, Cordia myxa, Chukrasia tabularis, Cinnamomum cecedodaphne, Duabanga grandiflora, Engelhardtia spicata, Evodia fraxinifolia, Jambosa kurzii, Litsaea* and *Machilus* species, *Mugnolia pterocarpa, Ostodes paniculatus, Schima wallichii, Syzygium cumini, Spondius axillaris* and *Toona microcarpa.*

In the lower storey *Dendrocalamus hamiltonii* is distributed all over the zone. *Ardisia, Boehmeria, Casearea, Ficus, Macropanex* and *Gynocardia* species are conspicuous in the second storey.

Musa hookeri is evenly distributed in the moist slopes. There are plenty of grasses of which *Imperata cylindrica, Capillipedium assimele, Arundinella* species and *Cymbopogon* species are common. Detailed description of the herbs, shrubs and climbers have been made in the chapter on ground flora.

(2) At 975-1300 metres:

Principal association is that of *Castanopsis-Schima.* Common tree species of this zone are *Schima wallichii, Castanopsis tribuloides, Cinnamomum cecidodaphne, C. impressionervium, Dysoxylum procerum, Engelhardtia spicata, Eurya acuminata, E. cavunerns, Evodia fraxinifolia Glochidion thomsonii, Litsaea* and *Machilus* species, *Magnolia pterocarpa, Ostodes paniculatus, Quercus spicata, Rhus insignis, Saurauia nepalensis, Syzygium* species, *Symplocos* species, *Spondius axillaris* and *Toona* species

The transition zones (ecotone zone) of various vegetation association are very interesting. The ecotone areas reflect vast assemblage of species some of which very conspicuous. Most conspicuous in the lower storey is *Macropanax undulatum.* There are many species common to Lower Hill Forests. In this and the former zones the incidence of giant climbers is very high. In shady and moist areas the following species occur abundantly

Urtica parviflora, U. dioica, Laportea crenulata, Girardinia plamata, Pilea anysophylla, P. smilacifolia, Elatostema liniolatum, Boehmeria rugulosa, Pouzolzia viminea, P. zeylania, Lecanthus wightii, Begonia species, *Oplismenus* species, *Globba* species and *Hedychium* species.

Number of genera and species of fern increase with increase in altitude. Several species of tree ferns start coming up, and are very conspicuous. Rich ground flora of these forests have been described in appropriate chapters.

(3) Transects at Lower and Upper Babukhola (1600-2350 metres):

This is a belt transect cutting the hills slope from 2300 metres down to 1600 metres. The vegetation in the transect represents the true picture of the entire hill slope. There are three distinct zones of vegetation viz., the lower being dominated by *Spondius axillaris, Guruga gamblei* and *Castanopsis tribuloides;* the higher being dominated by *Beilschmledia sikkimensis* and *Castanopsis tribuloides* and the highest zone being dominated by *Acer campbellii, Michelia doltsopa, Machilus gammieana, M. odoratissima,* and *Quercus lamellosa.*

There are spinklings of *Beilschmiedia sikkimensis Exbucklandia populanea, Cinnamomum* species and *Evodia* species.

Entire areas has *Symplocos* and *Eurya* as almost pure undergrowth. These two genera with several species have a wide distribution all over the temperate hills; in the upper zone these are mixed with several species of Viburnum. This transect which covers about 1.8 hectares has 33 tree species (above 10 centimetre diameter at breast height).

The ground flora of this zone has been described in appropriate chapters.

(4) Analysis of Vegetation in Lower and Upper Babukhola (1300-2350 m.):

In the lower altitude of the Lower Babukhola the composition is as follows:-

At 1300 metres *Engelhardtia spicata* is very predominant and conspicuous. *Castnopsis tribuloides* is an associate of *Engelhardtia* in this zone though the analytical data do not show such prominence. At this altitude *Duabanga grandiflora, Tetrameles nudiflora, Terminelia myriocarpa* represent the top storey species; at higher altitude *Toona febrifuga, Castanopsis trubuloides, Acer* and *Litsaea* species predominate over limited areas. *Nyssa - Castanopsis* association also exists. Higher up *Michelia doltsopa, Machilus edule, M.gammieana, M. odoratissima, Quercus* species and *Litsaea* species predominate.

All over the forests *Eurya, Cavinervis, E. acuminata, Glochidion thomsonii,* and *Symplocos* form almost pure second storey species, *Viburnum* is wide spread in this zone.

Ground flora is more is dense and there is heavy incidence ferns and species of *Urticaceae, Lauraceae* (Regeneration), *Compositae, Leguminosae, Labiataex, Scrophulariaceae,* and *Gramineae.*

In the Upper Babukhola the composition of crop is as follows:-

In the lower level *Engelhardtia spicata, Tonna febrifuga,* and *Castanopsis tribuloides* form extensive forests. This association is mixed up with sprinklings of *simingtonia populnea.* Further up *Acer* species come in and are in association with *Castonopsis* and *Litsaea elongata.* At the highest level *Machilus gammieana, M. edulis, M. odoratissima, Litsaea elongata, Michelia doltsopa,* and *Quercus* species form conspicuous association.

In the second storey *Eurya acuminata, E. japonica, Glochidion* species *Viburnum* species, *Turpinia nepalensis, Brassiopsis* species, *Saurauia* species, *Ficus* species are conspicuous.

Ground vegetation is dense and there is preponderance of Ferns, grasses sedges, mosses and some lichens on the rocks and on trees. These have been described in the appropriate chapter.

(5) At Mahaldaram (1600 - 2100 metres) :

This zone can be put under three distinct types of associations as follows:-

(i) Lower zone - *Castanopsis - Engelhardita - Beilschmeidia* association.

(ii) Middle zone - *Castanopsis - Nyssa - Garuga* association.

(iii) Upper zone - *Machilus - Quercus* association.

Mahaldaram forests are very dense and are situated on the northern slope. At 1600 - 1950 metres *Quercus lamellosa, Q.lineata, Q.fenestrata, machilus gammieana, M. odoratissima, M. edulis, Cinnamomum impressinervum, Michelia doltsopa, M. cathcartii, Sloanea dasycarpa, Elaeocarpus lancaefolius, Acer campbellii, Nyssa javanica, Beilschmeidia sikkimensis*, and *Meliosma wallichii* are found in the top storey. In the second storey *Symplocos theifolia, S. phyllocalyx, Eurya cavinervis Evodia fraxinifolia, Eriobotrya petiolata, Litsaea elongata, Pentapanax leschenaultii, Viburnum stellulatum, Myrsine semiserrata* and *Acer laevigatum* are conspicuous.

Epiphytic ferns in this region are *Mecodium exertum, M. javanicum M. polyanthos, Hymenophyllum simosianum, Vandenboschia auriculata, V. radicans, Microsorium hancocki, M.membranaceum and Loxogramme chinensis.* On the rocks are found *Oleandra neriiformis, O. wallichii* etc.

The epiphytes on the branches are *Vittaria ophiopogonoides, Lepisorus loriformis, Loxogramme lanceolata, Asplenium ensiforme, Polypodium microrhizoma, P. argutum, P. amoenum* and *Lepisorus exacavatus.*

Ground flora has been listed in appropriate chapter.

(6) At Sukiapokri, Rishihat and Balasan (1600 - 1950 meters):

The following types are distinguished:-

(i) Lower zone - *Castanopsis - Engelhardtia - Beiloschmiedia* association.
(ii) Middle zone - *Castanopsis - Garuga - Nyssa* association.
(iii) Higher zone - *Acer - Michelia - Machilus - Quercus* association.

The associates in the upper storey is same as other types of similar altitude. The second storey species is also same as other types. Some of the common ferns of the forests are as follows:-

Mecodium javonicum, Vandenboschia pyxidifera, Dryopteris splendens, Woodwordia radicans, Vandenboschia auriculata, Denstaedtia scabra, Acrophorus stipellatus, Cheilanthes albomarginatus, Anogramme microphylla, Thelepteris repens, Asplenium normale and *Pteris aspericaulis.*

(7) At Rissop, Gaddikhana and Chhatakour (1950 - 2600 metres)

* Lower zone - *Engelhardtia spicata - Castanopsis - Beilschmiedia* association.
* Higher zone - *Castanopsis - Garuga - Nyssa* association.
* Higher zone *Acer - Michelia - Machilus - Quercus.*

There is preponderance of *Quercus* in the higher zone along with *Lauraceous* species; there are also three species of *Michelia* viz., *M. doltsopa Alchimandra, M. lanuginosa* and *Sloanea dasycarpa* etc.

In the middle storey *Symplocos* and *Eurya* species are common every where along with *Evodia, Eriobotrya, Pentapanax, Viburnum, Myrsine* and *Litsaea.*

Ground flora has been described in the appropriate chapter. Fern flora that are conspicuous are:-

Polypodium argutum, P. amoenum, Lepisorus excavatus, Vittaria flexuosa, Phymatodes, griffi thiana, Lepisorus loriformis, Asplenium laciniatum, A. enseformis, A. planicaule, Pteridium aquilinum, Gleichenia glauca, Dicranopteris linearis, Diplazium stolickzae.

Of the tree ferns *Cyathea spinulosa, Hemitalia decipiens, H. brunoniana* are common.

Commonest ferns are *Mecodium exertum, Asplenium tenuifolium, A. cheilosorum, Dryoathyrium boryanum* and *Loxogramme chinensis.*

Less common are *Plagiogyria communis, Loxogramme involuta* and *Thelepteris brunnea.*

(8) At Singalila Forests (2500 - 2800 meters) :

In Rimbic, Siri, Ramam, Sabarkum, Sandakphu and Phalut blocks species of Lauraceae, *Michelia, Acer*, and *Quercus* species dominate the tree forest vegetation. Of the Lauraceous species *Beilschmiedia roxbuyrghiana, Cinnamomum impressinervium, Litsaea elongata, L. sericea, Machilus gammieana, M. odoratissima, Cinnanomum glanduliferum* species are dominating.

Composition of forests do not differ from those described in this chapter in the foregoing paragraphs. *Litsaea elongata* has a very high density.

The under growth is very dense and their is abundance of *Arundinaria* generation all over the forest floor. Several edaphic types have been isolated and described under Edaphic Forests.

In Ramam blocks there are patches of pure *Tsuga dumosa* forests. Such forests have *Quercus, Acer* and *Michelia* species in the top storey. In the middle storey there is abundance of *Rhododendron, Brassiopsis, Symplocos, Viburnum,* and *Arundinaria* species.

In Rimbik there are pure crops of *Quercus* consisting of all the species. In the second storey there is no difference in the occurrence of species. *Daphne cannabina, Viburnum* species, *Polygonum molle* and other species occur on forest floor. Figure 2 shows altitudinal distribution of tree species.

J. ANALYTICAL SYNOPSIS OF VARIOUS PLANTS RECORDED IN THE QUADRATS AND TRANSECTS

(i) Analytical synopsis of tree Flora

Various tables, statements and the present analytical synopsis pertain to qualitative and quantitative analysis of vegetation falling inside quadrats and transects and in adjoining areas and give a cross section of vegetation in different types of forests. Enumeration of trees has been made familywise in this section. This study has not been intended for listing all the trees occurring in the study area and not purely a botanical emuneration. No botanical description therefore has been aimed at, but fidelity of species, their distribution and time of flowering and fruiting has been incorporated. Numericals following flowering and fruiting indicate months (viz., 1, 2, 3 means January, February and March). Most conspicuous and widely occurring trees have been marked with asterisks (*).

DILLENIACEAE

Dillenia Linn.-

**D. indica* Linn.-A gregarious evergreen tree of wet-mixed forests. *Flower*-white. Common.

Occurrence:- Wet-mixed and wet-Sal forests.

**D. pentagyna* Roxb.-A very widely occurring tree of varied types of forests. *Flower*-yellow, very common.

Occurrence:-All over the plains and foot hills.

MAGNOLIACEAE

Magnolia Linn.-

M. pterocarpa Roxb.-A medium sized evergreen tree of lower middle foot hills. Young parts tawney pilose. *Flower*-white. Common.

Occurrence:-Lower and middle hill forests.

M. sphenocarpa Roxb.-An evergreen tree upto 25 cm. in height, 1.5 m. in girth. Sporadic.

Occurrence:- Foot hills.

**M. campbellii* Hook. f. & Th.-A tree of upper temperate Himalayan hills. Flower-large white or rosy.

Occurrence:- At 2300-2700 m. Sporadic to common in Singalila Forests.

Talauma Juss.-

**T. hodgsoni* Hook. f. & Th.-It is a tall tree. *Flower* large white. Very common.

Occurrence:- In foot hills Terai and Duars.

Alchimandra Dandy

**Alchimandra cathcartii* Dandy (*M. cathcartii* Hook. f. & Throats)-A tall tree of temperate forests. *Flower*-white. Common.

Occurrence:- Middle hill forests 1500-2100 m.

Michelia Linn

**M. doltsopa* Buch Ex Dc Ham. (*M. excelsa* (Wall Bl.)-A tall tree of upper temperate forests. *Flower*-white, 4-5, *Fruits*-10-11 Common.

Occurrence:- Upper temperate hills.

M. velutina Dc

(*M. lanuginosa* Wall)-A tall tree of middle hill forests. *Flower*-creamy white, 8-9, *Fruit*-9-11.

Occurrence:- Temperate hill forests 1600 m.

**M. champaca* Linn.-A very tall tree. *Flower*-yellow, 4-6, *Fruit*-9. Common in Sal and Foot hill forests.

Occurrence:- All over Sal and Foot hills mixed forests.

M. oblonga Wall.-A medium sized tree. Sporadic. *Flower*-2-3, *Fruit* 8-10. Rare

Occurrence:- Sal forests.

M. montana Bl.-A medium sized tree. *Flower*-7-9, *Fruit* 8-9 Rare.

Occurrence:- Sal forests of Lower Tondu Forests.

ANONACEAE

Polyalthia Bl.

P. longifolia Benth & Hook. f.-An evergreen tree. *Flower*-3-5, *Fruit*-7-9.

Occurrence:- Cultivated.

P. simiarum Benth.-A large tree upto 30 m. in height and 2-2.5 m. in girth. Common. *Flower*-4-7, *Fruit*-7.

Occurrence:- Foot hills forests; also in Sal forests.

P. obliqua Hk. f. & Throats.-A small tree Rare.

Occurrence:- Lataguri Forests.

Miliusa Lesch.-

M. roxburghiana (Wall) Hook. f. & Th.-Recorded from foot hills, Sal and mixed forests.

Alphonsea Hook. f. & Th.-

A. ventricosa Hook. f. & Throats.-A tall straight evergreen treo of Sal forests. Sporadic.

Occurrence:-Sal forests and in foot hills forests.

MENISPERMACEAE

Cocculus Dc.-

C. laurifolius Dc.-A small tree branches pendulous.

Occurrence:- Riverain forests Common.

CAPPARIDACEAE

Crataeva Linn.-

C. *nurvala* Buch-Ham (*C. unilocularis* Ham.)- A medium sized tree of riverain forests; also of foot hills forests. Very common. *Flower* 3-4, *Fruit* 9-8.

Occurrence:- Mixed forests all over.

FLACOURTIACEAE (SAMYDACEAE)

Homalium Jacq.-

H. zeylanicum Benth.-A tall tree, whitish bark. Rare.

Occurrence:- Bamanpokri forests.

Gynocardia R. Br.-

**G. odorata* R. Br.-A small evergreen tree. Very common. *Flower*-5, *Fruit* 11-1.

Occurrence:- Cool damp place in Sal forests and also in foot hills.

Flacourtia Commex L'Herit

F. jangomas (Lour) Raeusch.-A small tree. Sporadic.

Occurrence:- Riverain forests.

Casearia Jacq.

**C. glomerata* Roxb.-A deciduous small tree. *Flower* 3-4, *Fruit*-yellow, 6-8. Common.

Occurrence:- In marginal forests.

**C. graveolens* Dalz.-A very common shrub in Sal forests and Middle hill forests. *Flower*-8, *Fruit* 3-4. Also found in the lower hills forests.

**C. vareca* Roxb.-A very common shrub occurs in Sal forests and Middle hill forests in Terai and Duars; also found in the lower hill forests, very common.

C. kurzii Clarke.-Climbing shrub or small tree of the Lower hill forests. Sporadic.

Occurrence:- Foot hill forests.

Xylosma Frost.-

X. longifolia Clos.-A small tree with strong axillary spine.

Occurrence:- Riverain forests.

GUTTIFERAE

Mesua Linn

M. ferrea Linn.-An evergreen medium sized tree with dense crown. *Flower*-white, 2-4, *fruit* 8-9. Rare.

Occurrence:- Khumain and Mal forests.

M. floribuda (Wall) Kost

(*K. floribunda* Wall.) A large tall evergreen tree. Sporadic.

Occurrence:- Wet-mixed forests.

Garcinia Linn.

G. xanthochymus Hook. f. ex Anders- A small or middle sized tree. *Flower*-white, *fruit*-edible, 3-4; Sporadic.

Occurrence:- Evergreen forests.

G. cowa Roxb. - A middle sized tree with drooping branches Sporadic. *Flower* 2-5, *Fruit* 5-6.

Occurrence:- Wet-mixed forests.

G. cornea Linn - A tree of the foot hills. Flower-6.

Occurrence:- Wet-mixed forests in foot hills.

G. stipulata T. -A medium sized evergreen tree, sporadic *Flower* 8-9, *Fruit* 1-2.

Occurrence:- Wet-mixed forests.

Calophyllum Linn

C. polyanthum Wall.-A medium sized to large evergreen glabrous tree. Sporadic. *Flower* 2-3, *Fruit*-11.

Occurrence:- Wet-mixed forests.

TERNSTROEMIACEAE (THEACEAE)

Ternstroemia Mutis ex Linn.-

T. gymnanthera (W&A) Sprague (T. japonica Thunb,)-A tree with a very limited distribution. Rare.

Occurrence:- Bagdogra forests at Bengdubi (Endemic).

Schima Reinw. ex Bl

**S wallichii* (DC) Korth-A large tree upto 26 m. in height; *Flower*-white, *Fruit* 11-12, very common.

Occurrence:- Sal forests and foot hills forests.

Eurya Thunb

**E. cavinervis* Vesque (*E. japonica* Thunb)-It is often found as secondary growth in these forests and in old cultivation, mixed with. Very common. *Flower* 7-8, *Fruit*-11.

Other species of the genus:-

E. symplocina Bl.-A small tree. *Flowers*-7, *Fruit* 11-12. Rare.

Occurrence:- Tista Valley.

**E. acuminata* Dc.-Very common in North Bengal Sal Forest and in hills upto 1500 m.

Gordonia Ellis

G. excelsa Bl (*G. dipterosperma* Kurz.)-A large tree of middle hills for 1300 to 1900 m. *Flower*-3, *Fruit*-12.

Occurrence:- Mixed broad leaved forests, Riang.

SAURAUJACEAE

Saurauja Willd.-

**S. nepaulensis* Dc.-A small tree 8 m. tall in temperate hills 1300-2200 m. Comes up as secondary growth. Very common.

Occurrence:- Taklah, Tindharia and elsewhere.

S. griffithii Dyer.-A small tree from 1300 m to 1900 m. Leaves flocculent tomentose beneath.

Occurrence:- All over temperate hills in village sites from 1300 m to 1900 m.

S. fasciculata Wall.-A very small tree 5 m. of 600-1600 m. Common. Leaves ferrugenous tomentose beneath. Common.

Occurrence:- In villages in temperate hills upto 1600 m.

S. roxburghii Wall.-A small tree 9 m. of 1900 m. down to plain. Very common.

Occurrence:- In village sites from plains upto 1900 m.

S. panduana Wall.-A small tree at 1300 m. recorded from Darjeeling hills. Sporadic.

Occurrence:- In village sites.

DIPTEROCARPACEAE

Shorea Roxb. ex Gaertn

**S. robusta* Gaertn.-A large deciduous tree. Gregarious. *Flower* 2-3, *Fruit* 5-6, very common, abundantly occurring over wide areas.

Occurrence:- All over upto 900 m.

Vatica Linn.-

V. lancaefolia Bl.-A medium sized tree. Rare.

Occurrence:- Damanpur forests.

MALVACEAE

Kydia Roxb.

K. calycina Roxb.-A moderate sized fast-growing deciduous tree. *Flower* 9-10, *Fruit*, 12-2.

Occurrence:- Foot hills; sporadic.

BOMBACACEAE

Bombax Linn.

**B. ceiba* Linn (*B. malabaricum* Dc.)-(The silk cotton tree.) A lofty deciduous tree. This is the best match wood species of North-East India and is being raised in plantation. *Flower*-1-2, *Fruit* 4-5. It has been heavily exploited. Common.

Occurrence:- All over; especially in riverain forests.

STERCULIACEAE

Sterculia Linn.

S. alata Roxb (Pterygota alata R. Br.)-A tall tree upto 45 m. height and 3 m. in girth with a high narrow conical crown. Fruit-follicle 1-5. Common.

Occurrence:- Foot hills mixed forests.

**S. villosa* Roxb. ex. Smith-A widely occurring moderate sized deciduous tree. A very common tree of Sal and dry type forests. *Flower*-1-2, *Fruit* 3-4.

Firmiana Mars.

F. colorata Roxb.-A spreading deciduous tree upto about 13 m. in height.

Occurrence:- Raimatang and Damanpur forests.

Pterospermum Schreb. (Linn)

**P. acerifolium* Wild.-An evergreen tree upto 25 m. in height and 1.5 m. in girth. *Flower* 2-4, very common.

Occurrence:- Sal forests but primarily in foot hills.

TILIACEAE

Grewia Linn.-

G. subinaegualis Dc (*G. asiatica* Linn.)-A middle sized tree. Leaf base cordate.

Occurrence:- Sevoke and Pesoke forests.

G. disperma Rottl.-A small tree. *Flower* 6-8, *Fruit* 10-11, cyme 3-flowered.

Occurrence:- Dry mixed forests of sipclus.

G. optiva Drumm ex Burrett (*G. oppositifolia* Roxb.)-A middle sized tree. Drupe 2-4 deeply lobed.

Occurrence:- Dry-mixed forests, riverain forests.

G. calophylla Kurz.-A moderate-sized tree. *Flower* foetid smelling. *Fruit*-orange yellow when ripe.

Occurrence:- Dry-mixed forests.

ELAEOCARPACEAE

Elaeocarpus Linn.-

E. floribundus Bl.-A small tree 15 m. in height. *Flower*-white. *Fruit* - edible tastes acidic. Rare.

Occurrence:- Sal forests.

E. tectorius (Lour) Poir (*E. robustus* Roxb.)-A middle-sized tree upto 20 m. in height and 2 m. in girth. Sporadic.

Occurrence:- Wet type Sal Forests, foot hills.

**E. lanceaefolius* Roxb.-A tall evergreen tree of temperate forests. *Flowers*—white. *Fruit* - edible. Common.

Occurrence:- Temperate hills 1500-2200 m.

E. aristatus Roxb.-A tall tree (35 m.) with girth of 2.5 cm. Leave crowded at the end of the branch. Sporadic.

Occurrence:- Sal forests, foot hill forests.

E. rugosus Roxb.-A lofty deciduous tree upto 33 m. or more in height and 4 m. in girth. Sporadic.

Occurrence:- Sal forests, foot hills forests.

**E. varunna* Buch Ham.ex Master-A small tree very common in Sal and Miscellaneous forests in N. Bengal.

Occurrence:- All types of Sal forests.

Other species.-

E. sikkimensis Mast.-A small tree upto 1600 m. *Flower*-5, *Fruit*-7, edible. Common.

Occurrence:- Kurseong forests.

Sloanea Linn.-

**S. dasycarpa* Benth.-A large tree upto 25 m. in height, apparently evergreen. Capsule covered with dense deciduous bristles. Common.

Occurrence:- Temperate hills 1500-2200 m.

S. sterculiaceus (Benth) Rohder & Wilson-Very large tree. *Flower* 2-3, *fruit*-11-12, armed with sharp spines of chocolate colour.

Occurrence:- All over wet type Sal and Miscellaneous forests.

S. tomentosus (Benth) Rehder and Wilson. (= *Echinocarpus tomentosus* Benth)-Large tree of the lower and middle hill forests from 1400-1600 m. *Flower*-3, *Fruit*-9.

Occurrence:- Temperate forests.

RUTACEAE

Achronychia Forst.-

A. pedunculata (Linn) Miq. (=*A. laurifolia* B1)-Common in Wet type Sal forests. *Fruit* - edible white.

Occurrence:- Moraghat, Bagdogra

Evodia J.R. & G. Forst.-

E. lunankenda (Gaortn) Mess (=*E. roxburghiana* Benth)-A tree up to 16 m. in height and nearly 1 m. in girth. Leaves 3-foliate. Sporadic.

Occurrence:- Kalimpong foot hills, Samsing

E. fraxinifolia (D.Don) Hook.F.- Small to moderate sized tree. Leaves opposite, imparipinnate. *Fruit*-red, edible.

Occurrence:- Common in Sal and miscellaneous forest.

E. meliaefolia Benth.-A medium sized tree. *Flower* 7-8, *fruit* 11-2, red.

Occurrence:- Common in foot hills miscellaneous forests.

Fagara Schrib (=*Zanthoxylum* Linn.-

F. budrunga Wall.- An evergreen tree upto 16 m. in height and 1m. in girth. Trunk covered with broad based prickled. Common.

Occurrence:- Wet type Sal forests.

F. alata Roxb.-A small, strongly aromatic tree attaining about 8 m. Stem with conical prickles. Sporadic.

Occurrence:- Foot hills.

F. ovalifolia Wight.- A small aromatic tree, sometimes unarmed Leaves 3-foliolate. Fruit red.

Occurrence:- Temperate Lower hills.

Muraya Linn.-

M. paniculata (L) Tack *M. exotica* Linn.-A pretty shrub or small tree upto 6 m. in height. Leaves 3-9 foliolate. *Flower*-white. Sporadic.

Occurrence:- Foot hills.

M. Koenigii spreng- A small Tree of terai forests. Also occurs in foot hills. All parts have oily small. .Leaves used in curries all over India. *Flower* 2-5. *Fruit* 5-7.

Limonia Linn.-

L. alata W&A (*L. acidissima* Linn.) -A small tree. Leaves imparipinnate. Flower white, fruit black. Rare.

Occurrence:- Foot hills.

Micromelum Bl.-

**M. internerrimum* Buch Ham W&A (*M. pubescens* Bl.)-A very common tree. Leaflet 7-15. *Flower*-white fragrant. *Fruit*-yellow.

Occurrence:- Very common in Wet Sal and Miscellaneous forests.

SIMARUBACEAE

Ailanthus Desf.

* *A. integrifolia* Lam (*A. grandis* Prain) -A lofty tree upto 40 m. tall with a cylindrical bole. Fruit winged at both ends. Leaves 60-100 cm. Long. Very Common.

Occurrence:- Foot hills miscellaneous forests.

Picrasama Bl.-

P. javanica Bl. (P. nepalensis Benn).-A middle sized tree. Leaflets 3-7. Bark bitter. *Fruit* - black, pulpy seated on a disco.

Occurrence:- Sporadic in Sal forests; also in foot hills. Sevoke.

BURSERACEAE

Garuga Roxb.

**G. pinnata* Roxb.-A middle sized deciduous tree. Leaflets 11-19, serrate, turns red before falling. Very common.

Occurrence:- Sal forests and miscellaneous forests.

G. gamblei King.-A tall tree (22 m.) of temperate hills, spreading crown, often buttressed at the base. Very common in Babukhda block at 1300 m.

Occurrence:- Temperate hills (Lower and middle).

Canarium Linn.

C. bengalense Roxb.-A medium sized tree with a girth upto 3 m. Blaze give resinous small. Leaves 30-90 cm. Long. Sporadic.

Occurrence:- Sal forests (Wet type), Buxa.

C. sikkimense King.-A very tall tree. Leaflets 5-pair serrate. Blaze orange. *Flower*-7, *fruit* 3-4.

Occurrence:- Temperate hills, Darjeeling sporadic.

MELIACEAE

Aphanamixis (= *Amoora* Roxb) Bl.-

**A. polystachia* (Wall) Parker (*Amoora rohituka* W&A)-A small tree, evergreen. Leaves upto 1 m., 4-7 leaflets. Capsule 3-valved, yellow. Very abundant.

**Amoora spectabilis* Miq. (= *A. wallichii* Roxb).-A tall handsome tree with considerable trunk and a high crown. Blaze yields milky juice. Very common.

Occurrence:- Sal forests all over.

Azadirachta A. Juss.

A. indica A. Juss.-A middle sized graceful tree, planted in homestead land. Common.

Melia Linn.-

M. azedarach Linn.-A small or middle sized deciduous tree. Leaves 2-pinnate. *Flower* - lilac-blue, *Fruit* - yellow.

Occurrence:- Scattered in drytype Sporadic.

M. composita Willd.-A large spreading deciduous tree upto 20 m. in height and 2 m. in girth. Leaves 2-3 pinnate. Drupe brown edible. Sporadic.

Occurrence:- Foot hills.

Trichillia P. Brownie (= Heynea Roxb.)

**T. connaroides* (= *H. trijuga* Roxb.)- A handsome tree upto 13 m. in height and 1 m. in girth, evergreen. Stem vertically furrowed. Common.

Occurrence:- In Sal forests.

Dysoxylum Bl.-

**D. hamiltonii* Hiern.-A fairly large evergreen tree. Leaflet 10-18, drooping. Capsule 4 seeded, yellow. Sporadic.

Occurrence:- In Sal forests (wet type).

**D. binectariferum* (Roxb) Hook. f. ex Bedd.-A tree upto 16 m. in height and 1 m. in girth. Leaflet 5-9. *Fruit* - 5-valved, seeds dark purple, very abundant.

Occurrence:- In all types of forests in plains and foot hills.

D. lobara (Buch-Ham) Merr.

**D. procerum* Hiern.-Sub Himalayan tract. Common in Kalimpong foot hills, Chel, Dalimkat.

Walsura Roxb.-

W. tabulate Hiern.-Assam, foot hills of Himalayas, Sikkim, Darjeeling. Leaflet 3-5, sporadic to rare.

Occurrence:- Raimatang hills.

Chukrasia A. Juss.

**C. tabularis* Andr.-Jressieu.-A tall handsome tree. Leaflets 10-24, terminal leaflet pink. Capsule is of the size of hens egg. *Flower*-5, *Fruit*-12.

Occurrence:- Sal forests and foot hill forests. Very common.

Toona M. Rocm. (=*Cedrela* Linn.)

T. surani (Bl) Merr (*C. febriguga* C. Dc.)-A middle sized tree upto 16 m. in height. Leaflet entire; bark exfoliating. Common.

Occurrence:- Temeperate hills.

**T. ciliata* Roem (=*Cedrela toona* Roxb,.)- A large handsome tree. Leaflet 9 pairs; seeds winged at both ends. Very common. *Flower*-12, *Fruit*-2.

Occurrence:- In Sal, foot hill forests.

T. microcarpa (C. Dc.) Harm *(cedrelamicrocarpa)*- A tree of foot hills. Leaflets 9 pairs. Common.

Occurrence:- Foot hills forests.

Chisocheton Bl.

**C. paniculatus* (Roxb) Hiern.- A small tree scarcely exceeding 13 m. and 2 m. in girth. Leaflet 5-12 pairs. Terminal leaflet aborted. *Fruits* globose, dark orange red 5 cm. diameter.

Occurrence:- In sal forests all over.

ILICACEAE
(AQUIFOLIACEAE)

Ilex Linn.-

I. godajam Coleb. ex Hook-A large deciduous tree with spreading branches over 30 m. in height. Bark has thick horizontal wrinkles. Blaże red. Common.

Occurrence:- Sal forests wet type.

Other species sporadicallly occurring are:-

I. odorata Buch-Ham ex D. Don. Hook. f.-Singalila range, Darjeeling.

I. fragilis Hook. f.-Singalila forests, Darjeeling.
I. sikkimensis Kurz.-Singalila forests, Darjeeling.
I. intricata Hook. f.-Singalila forests, Darjeeling.

CELASTRACEAE

Lophopetalum Wight.-

L. fimbriatum Wight-A large tree attaining nearly 30 m. in height. Leaves opposite. Blaze exudes red gum. Flower red, *Fruit* woody 3-celled Common.

Occurrence:- Sal forests of Buxa, Chilapata, Moranghat.

Cassine Linn (*Elaeodendron* Jack)

C. glaucum Pers.-A small tree. Leaves serrulate. Fruit drupaceous. Sporadic.

Occurrence:- Foot hills.

Kurrimia Wall.-

K. pulcherrima Wall.-A tree upto 25 m. Young leaves red. *Flower* white, solitary or paired. *Fruit* - yellow arillous. Sporadic to rare.

Occurrence:- Foot hills.

RHAMNACEAE

Zizyphus Mill

Z. mauritiana Ham. (=*Z. jujuba* Lamk.)- A tree or shrub. *Flower*-fascicled, cyme sessible. Drupe red.

Occurrence:- Common all over drier tracts.

Hovenia Thunb.-

H. acerba Lindle (*H. duleis* Thunb)-A dedicuous tree 30 m. tall. Leaves upto 18 cm., obliquely ovate, serrate, 3-nerved. Sporadic to rare.

Occurrence:- Sal and miscellaneous forests.

STAPHYLEACEAE

Turpinia Vent.-

T. cochinchinensis auct non Merr. (*T. nepalensis* Wall ex Wt)-in the middle and upper hill forests from 1219-1234 n, Very common beyond Rissum at 1829 m. Leaflet 5, *Fruit*-13 m.

Occurrence:- Temperate hills, Middle and Upper.

**T. pomifera* (Roxb) DC- In the mixed hill forests, in the Terai. Duars and Lower hill forests. Leaflet 7, *Fruit*-3 cm. Very common.

Occurrence:- In Sal and foot hill forests.

ACERACEAE

Acer Linn.-

A. oblongum Wall ex DC.-A middle-sized deciduous tree. Leaves have 3 basal nerves. Sporadic.

Occurrence:- Foot hills.

A. laevigatum Wall.-A middle-sized deciduous tree. Sporadic.

Occurrence:- Temperate hills.

A. sikkimenses Mig.-A medium sized tree. Sporadic.

Occurrence:- Temperate:- Temperate forests.

**A. campbellii* Hook. f. & Th.-A tall widely occurring deciduous tree. Leaves 5-7 cut, petiole reddish, Very common.
Occurrence:- Temperate hills at 1500-1900 m.

A. pectinatum Wll. ex Pax-A tall deciduous tree. Leaves 3-5 lobed, with a deeply cordate base.
Occurrence:- Temperate forests.

HIPPOCASTANACEAE

Aesculus Linn.-

**A. assamica* Griff; (*A. punduana* Wall.)- A small tree of shaded stream bank in duars. Leaves opposite, digitate, leaflet sessile. *Flower* leathery capsule.

Occurrence:- Sal forests and miscellaneous forests of plains.

SAPINDACEAE

Aphania Bl. (=*Sapindus* Plum.)-

A. detergens (Roxb) Radlk.-A medium sized tree of Lower hills. Leaves 1-3 foliolate. Foot hills. Sporadic.

Erioglossum (Roxb) Bl.-

E. rubiginosum Bl.-A small tree rather rare. Leaves 10-4, *flower*-white, drupe yellow to reddish. Edible.

Occurrence:- Scattered in dry forests.

SABIACEAE

Meliosma Bl.-

**M. simplicifolia* (Roxb) Walp.-A small tree 6 m. tall. Leaves simple upto 35 cm. Panicle 40 cm. long. *Flower* white. Very Common.

Occurrence:- Wet type sal forests.

**M. wallichii* Planch ex Hook. f.- A tall tree of 20 m. tall, in the temperate forests. Leaflet 5-13. Drupe red when ripe.

Occurrence:- Temperate forests 1800 m.

**M. pinnata* Roxb.-Common throughout the forests of N. Bengal and Assam plains. A small tree of 8 m. tall. Leaflets cuspidate serrate.

Occurrence:- All over wet type Sal forests.

ANACARDIACEAE

Rhus Linn.-

R. hookeri Shani & Bahadur *R. insignis* Hook. f.)-A tree of 10 m. tall; branch tips rusty villous. Common.

Occurrence:- Foot hills.

R. javanica Linn.- (=*R. semialata* Murr.)- A small tree, petiole winged. Leaflet 4-6 pairs, serrate. Fruit-edible.

Occurrence:- Foot hills. Sporadic.

R. succedanea Linn.-A tree of 16 m. tall. Leaflet 3-6 pairs. Common.

Occurrence:- Foot hills.

Semecarpus Linn. f.-

**S. anacardium* Linn. f.- In sal and other forests, sporadic. Leaves upto 60 cm, by 30 cm. Drupe seated on fleshy receptacle. Sporadic.

Occurrence:-In Sal forests.

Mangifera Linn.-

**M. sylvatica* Roxb ex Wall-A lofty evergreen tree with a dense globose crown. *Flower*-white, *Fruit*-long, small.

Occurrence:- Sal and foot hill forest.

Drimycarpus Hook. f. (Roxb)

D. racemosus Hook. f.-All over the forests of plains but very sporadic. Moderate sized tree, upto 20 m. Blaze gives acrid sap turning black. Leaves slightly wavy. Sporadic.

Occurrence:- Foot hills miscellaneous forests.

Lannea A. Rich.

**L. coromandelica* (Houtt) Merr (*Odina Wodier* Roxb)- A small to middle-sized deciduous tree. Leaflet entire, drupe 2 cm. long, red when ripe. Very common.

Occurrence:- Sal forests; foot hill forests.

Spondias Linn.-

**S. pinnata* (Linn) Kurz (=*S. mangifera* Willd)-A middle-sized tree with a pleasant aromatic smell. Leaflet broad. Drupe 4-5 cm. long. Sporadic.

Occurrence:- Foot hills forests.

Choemospondias Burtt & Hill

C. axillaris (Roxb) (*S. axillaris*) Roxb.)-A small evergreen tree. Leaflet 7-17, drupe ablong, yellow when ripe, sporadic.

Occurrence:- Temperate forests.

FABACEAE

Millettia Wt. & Arn.-

M. glancescens Kurz (*M. prainii* Dunn)-A small tree of foot hills. Leaflet 9-11, *Flower*-purple. *Fruit*-Pod with wars. Sporadic.

Occurrence:- Foot hills near Sukna; sporadic.

Erythrina Linn.-

**E. variegata* Linn (*E. indica* Lam.)-A moderate sized tree. Leaflet entire broader than longer. Pod 1-8 seeded.

E. arborescens Roxb.- A small tree with a low spreading crown. Leaflet upto 15 cm. Pod curved 20 by 2 cm.; narrow at both ends 4-6 seeded. Common.

Occurrence:- Foot hills.

Butea Roxb ex Willd

**B. monosperma* (Lamk) Kuntz. Taub. (B. frondosa Koen ex Roxb.)- A small to medium sized tree. Leaflets 3. *Flower*-orange red. Common.

Occurrence:- In waste land; in burnt forests.

Dalbergia Linn. f.-

**D. sissoo* ex Dc. Roxb.-A middle sized gregarious and deciduous tree. Leaflets 3-5, rhomboid, cuspidate, pod 1-3 seeded. *Flower*-3-4, *fruit* 10-11.

Occurrence: Gregarious tree in riverain tract.

**D. lanceolaria* Linn.- A large deciduous tree. Leaflet 7-17. Pods 1-3 seeded. Common.

Occurrence:- Foot hills.

**D. assamica* Benth.-A middle-sized deciduous tree attaining 25 m. in height and 2 m. in girth. Leaflets 15-21. Pod 1-2 seeded. Common.

Occurrence:- In foot hills mixed forests.

D. sericea G. Don.-A middle-sized tree upto 25 m. in height and 2m. in girth, often fluted in old age. Leaflets 17-27. Pods 1-3 seeded.

Occurrence:- Foot hills mixed forests sporadic.

D. latifolia Roxb.- A large deciduous tree. Leaflet 3-7, broad elliptic. Pod 1-4 seeded. Rare .

Occurrence:- Foot hills in Pankhabari forest.

Derris Lour.-

D. robusta Benth.-A middle-sized deciduous tree. Leaflets upto 25. Pod 1-5 seeded. Sporadic.

Occurrence:- Foot hills

Ougeinia Benth.-

O. oujeinensis (Roxb.) Hochr. (*O. dalbergioides* Benth).- A small and rather crooked tree in the area. Blaze exclude red juice. *Flower*-mauve. Rare.

Occurrence:- In Khairbani Sal forests.

Sesbania scopoli

S. sesban (L) Merr (*S. aegyptica* Pers.)- A fast growing small tree.

Occurrence:- Cultivated.

CAESALPINIACEAE

Acrocarpus W&A.

**A. fraxinifolius* Arn ex Wight.-A tall deciduous unarmed tree. Leaves bipinnate. *Flower*-Red or orange. Pod 14 cm long, dark brown. Common.

Occurrence:- Foot hills east of Sevoke.

Cassia Linn.

**C. fistula* Linn.-A medium-sized tree with a presented pubescent young shoots. Blaze pink. Leaflet opposite. *Flower*-yellow. Pod 30-60 cm. long, cylindric. Very common.

Occurrence:- All over foot hills and dry mixed forests.

Bauhinia Linn.

**B. malabarica* Roxb.-A moderate sized tree. Blaze pink. Fertile stamen-10. Leaf tastes acid. Common.

**B. purpurea* Linn.-A moderate sized tree. Blaze. biscuit coloured. Leaves cleft half way down. Fertile Stamen-3.

Occurrence:- Dry type forests all over plains and foot hills.

**B. variegata* Linn.-A moderate-sized tree. Blaze pink and white. *Flower*-white. Fertile stamen-5. Leaves cleft one third the way down.

Occurrence:- Dry mixed forests, foot hills. Common.

MIMOSACEAE

Adenanthera Linn.-

A. pavonina Linn.-A handsome middle sized to large unarmed deciduous tree. Leaves bipinnate; pinnae 2-4 pairs. *Flowers*-5 yellow. Seeds pink with black spot. Sporadic.

Occurrence:- Foot hills mixed forests.

Acacia Mill

**A catechu* (Linn f) Willd.- A middle-sized deciduous tree; often gregarious and deciduous. Thorns stipular. *Flower*-elongate spike. Abundant.

Occurrence:- In riverain forests.

Albizzia Durazz.-

**A odoratissima* (Linn f) Benth.- A large deciduous tree with spreading crown. Leaves bipinnate. Leaflets 10-30 pairs. *Flower*-white.

Occurrence:- Riverain and miscellaneous forests of foot hills.

**A. procera* (Roxb) Benth.-A tall handsome sub-deciduous tree, attaining over 25 m in height and 2 m in girth. Bipinnate leaves pinnae 2-12 pairs. Leaflets 6-16 pairs. Very common.

Occurrence:- Riverain forests.

**A. lebbek* Benth.-A large deciduous tree with a spreading crown. Leaflets 3-10 pairs. Pods straw coloured. Very common.

Occurrence:- Riverain and dry mixed forests.

**A. chinensis* (Roxb) Merr (*A. stipulata* Bolvin) - A large handsome flat crowned tree. Pinnae 6-12 pairs, leaflets 20-40 pairs; blaze deep crimson. Common.

Occurrence:- In reverain and msicellaneous forests.

**A. lucidior* (Stend) Neilson & Hara (*A. lucida* Benth.)- A very common tree of riverain and moist deciduous forests. Pinnae 1-2 pairs, leaflets 2, *Flower* 4-5, *Fruit* 8-2.

Occurrence:- Riverain and wet mixed forests.

Pithecolobium Mart.

P. Htero phyllum (Roxb) Macbride (*P. angulatin* Benth) -A small to middle-size tree. Pinnae 2-3 pairs. Leaflets 2-8 twisted spirally.

Occurrence:- Riverain forests, foot hill forests.

P. montanum Benth.-Foot hill forests, rare.
P. bigerminum (Linn) Benth.-Foot hill forests, rare.

ROSACEAE

Prunus Linn.-

P. communis Linn.-A shrub or moderate sized tree, sometimes spinescent. Leaves acuminate, serrate, petiole glandular. *Flower* white in fascicles, *Flower*-2, *Fruit*-6-7.

Occurrence:- Upper temperate forest.

P. napaulensis (Ser) Steud.-A moderate sized to a fairly large deciduous tree attaining about 20 m in height and 1.25 m in girth. *Flower*—white, 10-11, *Fruit* 7-8.

Occurrence:- Upper temperate forests.

Pygeum Gaertn.-

P. acuminatum Coleb.-A large evergreen tree attaining 30 m. in height and 2.5 m in girth. Crused leaves. Stems, fruits smell bitter almond. *Flower*-5, *Fruit* 12-3. Sporadic.

Occurrence:- Sal forests, wet mixed forests.

P. montanum Hk.f.-A medium to tall sized tree. Leaves also give bitter almond smell. *Flower* 10-12, *Fruit*-2.

Occurrence:- Temperate hills 1700 m.

Eriobotrya Lindl.

E. petiolata Hk. f.-A middle sized tree. All parts covered with brown wool. *Flower*-4-5; white, crowded in panicle, *Fruit* 8-9, yellow succulent.-

Occurrence:- Temperate forests 1600-2200 m.

E. hookeriana Decane.-A medium sized tree. In temperate forest.

Occurrence:- Sporadic.

E. bengalensis Hk. f.-A middle-sized trees in temperate forests. Common in temperate forests.

HAMMAMELIDACEAE

Ex bucklandia RW. Pr. (*Bucklandia* Br.)-

E. populnea (R. Br. ex Gribb) RW. Brown-A medium sized to tali handsome tree. Leaves thick polar like, fleshy stipule. *Flower* and *Fruit* 11-12. Sporadic.

Occurrence:- Upper temperate forests, Singalila forests.

RHIZOPHORACEAE

Carallia Roxb.-

**C. - brachiate* (Lour) Merr (C. integerrima D.C.) (Lour) Mer-A large evergreen tree. Medium sized, thick leaved opposite leaves. *Flower*-small, sessile; 2, common.

Occurrence:- In stream bank in Bagdogra and Moraghat forests.

COMBRETACEAE

Terminalia Linn.-

**T. belerica* (Gaertn) Roxb.-A large deciduous tree. Leaves clustered at the end of the branches, petiole long. Spike simple. *Flower*

4-5, *Fruit*-drupe, not winged. Very common.

Occurrence:-All types off forests-Sal, mixed foot hills and riverain forests.

**T. chebula* Retz.-A medium-sized deciduous tree. Fruit-drupe 3-5 cm. long, 5-ribbed. Sporadic.

Occurrence:- In wet mixed forests.

**T. citrina* Roxb.-A medium sized to large deciduous tree, often with buttressed stem attaining 26 m. in height. Leaves not clusted. *Fruit*-5 angled. Sporadic.

Occurrence:- Wet mixed forests.

**T. alata* Heyne ex Roth (*T. tomentosa* (Roxb) W&A)-A meduim sized to large deciduous tree. Bark very conspicuous owing to black deep vertical fissures and smaller horizontal ones. *Fruit*-5 winged, 4-5, common.

Occurrence:- In foot hills and in dry type Sal forests.

**T. myriocarpa* Heurek et Muell.-A very large evergreen tree with branches pendulous. Petiole has 1-2 glands. Drupe-small, winged. Sporadic.

Occurrence:- In foot hills shady places.

MYRTACEAE

Jambosa DC.-

**J. kurzii* Duthie.-A moderate sized to large tree. *Flower*-white pedicillate, 3-6, *Fruit*-globose edible.

Occurrence:- Foot hills and plains.

J. ramosissima (Bl) Wall ex Duthic.-Very common in Kalimpong foot hills. Leaves small, acuminate. *Flower* 2-4, *Fruit* 6-7.

Occurrence:- Sevoke, Kalimpong.

Syzygium Gaertn.-

S. cymosum Leark.-A small tree. *Flower*-white sussile, *Fruit*-bluish black.

Occurrence:- Foot hills.

**S. tetragonum* Wall ex Kurz (*E. tetragona* Wight.)-A medium sized or large evergreen glabrous tree. Leaves sessile. Petals white. *flower* 7-12, *Fruit* red and black, 5,-edible.

**S. operculatum* (Roxb) Neidz. (*E. operculata* Roxb.)- A small or middle sized sometimes a large tree. Leaves turn red before falling. *Flower* 4-5, *Fruit* 6-7.

Occurrence:- Burnt Savannah.

**S. cumini* (Linn) Skeels (*E. jambolana* Lamk)-A large tree, *Flower*-greenish white, sessile 4-5. *Fruit*-black, purple, juicy 6-7.

Occurrence:- Very common in wet mixed Sal and less common in other forests.

S. wallichi (Wt) Walp (*Jambosa praecox* Roxb).-A medium sized tree; Branches quadrangulas, winged, *Flower*—white, 3-6; *Fruit* 8-9.

Occurrence:- Temperate hills.

LECYTHIDACEAE

Careya Roxb.

**C. arborea* Roxb.-A moderate to medium sized deciduous tree. Very common as a small gnarled tree in Savannahs. Oblanceolate. Leaves globose. Blaze red. *Flower* 4-5, *Fruit*-7, many seeded. Very common.

Occurrence:- All types of forests, more concentrated in drier types.

Barringtonia Forst

B. acutangula Gaertn.-A small to medium sized tree. Flower in pendulous raceme. Fruit angular. Not common.

MELASTOMACEAE

Memecylon Linn.

M. plebejum Kurz.-A small tree in Kalimpong foot hills (Samsing) *Flower*-light blue, *Fruit*-dark coloured. Sporadic.

LYTHRACEAE

Largerstroemia Linn.-

**L. parviflora* Roxb.-A large to medium sized deciduous tree. Leaves with sounded base. *Flower*-6 merois, *Fruit*-dry capsule. Very common.

Occurrence:- An associate of Sal and also of miscellaneous forest in plains and foot hills.

L. reginae Roxb. (*L. speciosa* Pers)—A large and rather branchy tree. Flower in terminal panicle. Calyx strongly ribbed. Rare.

Occurrence:- Only a few trees occur in Boroblisha forests.

L. hypoleuca Kurz.-A large deciduous tree from Andamans. Planted in gardens and forests.

SONNERATIACEAE

Duabanga Ham

**D. grandiflora* (Roxb) Walpers (=D. sonneratioides Ham.)-Tall handsome tree with drooping branchlets. Leaves large, opposite, clayx segments 7, capsule 4-8 valved. Common.

Occurrence:- Abundant in foot hills; also in plains Sal forest.

TETRAMELACEAE

Tetrameles R. Br.-

**T. nudiflora* R.Br.-A very tall deciduous tree, buttressed, bark whitish. Capsule coriaceous ribbed.

Occurrence:- In foot hills forest. Abundant.

CORNACEAE

Mastixia Bl.-

M. arborea C.B. Clarke.-Small tree of 10 m. Leaves crowded at the top of a branch. *Flower*-5, green, *Fruit* 8-9. Rare.

Occurrence:- In wet Sal forests.

M. pentandra Bl. - Sal and Wet mixed forests of Bagdogra, Moraghat. Rare.

Nyssa Linn.-

**N. Sessili* flora Hook. f. & Th.-A large tree young parts silky pubescent. Leaves elliptic, closely punctate on both surface. Very common. Berry eaten by birds and Monkeys.

Occurrence:- 1400 m. in the hills and lower temperate forests.

Alangium Lamk.-

**A. chinense* (Lour) Harms Rehder (*A. begonifolium* Roxb)-A tree of foot hills and Sal forests. Branch zigzag. Petals of flower 10-15. *Fruit* beccate. Very common.

Occurrence:- On marginal forests in moist deciduous forests and foot hills.

Cornus Linn.

C. capitata Roxb.-A small tree. Leaves opposite. *Flower*-white or yellow. Sepals-4. Petal-4.

Occurrence:- Sporadic; in temperate hills.

RUBIACEAE

Cephalanthus Linn.-

C. occidentalis Linn.-A tree found in lowlying areas of Latergguri. Not common.

Anthocephalus A. Rich.-

**A. chinensis* (Lam) A. Rich exwall. (A. indicus A. Rich)-A tall tree, branches whorled and horizontal, common all over plains and foot hills. *Flower* head 3-5 cm. diameter, *Fruit* 7-9.

Occurrence:- All types of forests.

Adina Salisb.

A. cordifolia Hk. f.-A large tree, spreading crown, butteressed. Wood yellow. Leaves cordate, petiole red. *Flower*-8, *fruit* 11-12. Rare.

Occurrence:- Turiajhar and Nepania forests.

Hymenodictyon Wall

**H. excelsum* Wall.-A large deciduous tree, bitter bark. Leaves glandular. *Flower*-white 5-6, *fruit*-many seeded, capsule-11.

Occurrence:- Riverain forests.

Wendlandia Bartl. ex DC.

W. excerta (Roxb) Dc.-*W. puberula DC* and *W. Tineforia* (Roxb) DC have been recorded from Sal forests and foot hills forests.

Ixora Linn.

**I. arborea* Roxb ex Sm. (I. parviflora Vahl)-A small tree of Sal forests. *Flower*-white, sub-sessible, incompact 3-5 branched panicle. *Fruit*-black red.

Xeromphis Rajin

**X. uliginosa* (Retz.) Mahesh (R. uliginosa Dc.)-A small tree, armed. Leaves large clustered. *Fruit* 3.5 cm. many seeded. Sporadic.

Occurrence:- Sal forests.

**X. spinosa* (Thunb) (*R. dumetorum* Lamk.)-A small deciduous tree armed with stout horizontal spine. *Flower*-large 3-5, *Fruit* 1-2. Sporadic to common.

Occurrence:- Foot hills.

Randia Linn.

R. fasciculata Dc.-A small tree with short axillary spine. *Flower*-white fragrant 3-4. Berry purplish 4-5.

Occurrence:- Sporadic in foot hills.

R. wallichi Hk.f.-A tree (upto 16 m.) branches unarmed, very stout but flexous. *Flower*- yellowish, fragrant, 5.

Occurrence:- Sporadic in foot hills.

Gardenia Linn.-

**G. campanulata* Roxb.-A tree armed with spinescent abortive branches. *Flower*-greenish white 3-4, *Fruit* 10-11.

Occurrence:- Dry mixed forest. Sporadic.

G. turgida Roxb.-A small deciduous tree. *Flower*—3-5. Fruit—5-7 cm. long.

Occurrence:- Dry mixed forests Sporadic.

Pavetta Linn.

P. indica Linn.-A small tree. Stipules connate in a loose deciduous sheath. *Flower*-white, 2-3, dichotomously branched cyme.

Occurrence:- Sal forests. Sporadic.

ERICACEAE

Enkianthus Laur

**E. deflexus* (Griff) Sehn (*E. himalains* Hk. f. & Th.)-A small tree about 7 m. tall found in temperate forests. from 2700-3300 m. *Flower* 5-7, *Fruit* 8-10. very common.

Occurrence:- Temperate forests.

Lyonia Natt (=*Pieris* D.Don)

**L. ovalifolia* (Wall) Drude-Very common in Darjeeling and Sikkim hills. It ranges from 700 m. to 3300 m. *Flower*-small 1.5 cm, coroa white. Leaves poisonous to cattle. Very common.

Occurrence:- Temperate hills.

Rhododendrons Linn.-

**R. arboreum* Srn.-Commonest of all *Rhododendrons* at 1800 m. to 2500 m. gregarious. *Flower*-red. Commonest of all *Rhododendrons.*

Occurrence:- Long large in temperate hills.

R. grande Wright.-Small tree at 3300 m. *Flower*-white, sometimes gregarious over small area.

Occurrence:- Temperate hills 3200-3600 m.

R. hodogsoni Hook.f.-Small tree. *Flower*-pink or lilac.

Occurrence:- 3300 m. temperate hills. Common.

R. falconeri Hook. f.-A small tree. *Flower* cream, purple spotted on throat. Leaves braod elliptic, rusty tomentose beneath.

Occurrence:- 3300 m. temperate hills.

R. barbatum Wall.-A small tree. *Flower*-brick red, barbata hairs in petiole. Gregarious.

Occurrence: 3300 m.

R. griffithianum Wright-A small tree. *Flower*-white.

Occurrence:- 3000 m. temperate hills.

R. edgeworthii Hook.f.-A small tree. *Flower*-white.

Occurrence:- 3000 m. temperate hills.

R. cinnabarinum Hook. f.-A gregarious tree. Head 5-8 flowered, brick red.

Occurrence:- 3500 m. temperate hills.

SAPOTACEAE

Sideroxylon Linn.

S. hookeri Clarke.- Small tree of 8 m height. Spines on branches. Exdues milky juice. *Flower*- 3-4, *Fruit* 4-5.

Occurrence:- Foot hills miscellaneous forests.

S. gamblei C.B.C.-Big tree of the L and M.H.F. upto 1300 m. Rare.

Occurrence:- Foot hills.

Sarcosperma Hk.f.

S. arboreum Benth.-A large spreading tree of 27 m. tall. Leaves sub-opposite, gland pitted on nerve axiles,. *Flower*-yellow, 12-2 *Fruit* 5-6. Common.

Occurrence:- Tista Vallyey and in foot hills.

Aesandra (=Bassia Linn)

A. butyracea (Roxb) Bahen (=B. butyracea Roxb.)-Big tree of the foot hills and upto 1300m. The tree generally grows very crooked. Sporacdic.

Occurrence:- Tista Valley and foot hills.

EBENACEAE

Diospyros

D. montana Roxb.-A small tree. *Flower* 3-4, fruit 9-11. Blaze light yellow turns green.

Occurrence:- Foot hills in Pankhabari. Rare.

D. embryopteris Pers.-A medium sized tree with short trunk and spreading crown. Evergreen species. *Flower*-white 5-7. Ripe *fruits* globose.

Occurrence:- Sporacdic in plains forests.

SYMPLOCACEAE

Symplocos Jacq. -

S. cochinchinesis (Laur) Moore (*S. spicata* Roxb.)—A small tree of 8 m. high. Leaves upto 15 cm. *Flower*-scented, cream coloured. *Fruit* a drupe 5-6.

Occurrence:- Foot hills upto 1600 m. Common.

**S. lucida* Th. (*S. phyllocalyx* Clarke)-A small tree. Leaves coriaceous serrulate. *Flower*-in compact spike. *Fruit* -long 1.2 cm. Very common.

Occurrence:- Temperate 2500 m.-3000 m.

**S. glomerata* King ex Calrke-Small tree. Leaves glandular serrate. *Flower*-in dense axillary fascicles. Common.

Occurrence:- Temperate forests 2500-3500 m.

S. racemosa Roxb.-Small evergreen tree. *Flower*-yellow. Common.

Occurrence:- Foot hills.

S. racemosa Roxb.-An evergreen tree. Common.

Occurrence:- In plains.

S. theaefolia Ham.-A middle sized tree of 21 m. high. *Flower*-10-3. Common.

Occurrence:- Temperate forests 1800 m.

STYRACEAE

Styrax Linn.

**S. serrulatum* Roxb.-A small tree, bitter bark, blaze yellow. *Fruit*—yellow. Sporadic.

Occurrence:- Plains Sal forests.

S. hookeri Clarke.-A small tree. Flower-yellow. Sporadic.
Occurrence:- Sal forests.

OLEACEAE

Olea Linn.

**O. dioica* Roxb.-A middle size tree of 10 m. high. *Flower* 2-4, *Fruit* 10, drupe, blue when ripe. Sporadic.

Occurrence:-In Sal forests.

Fraxinus Wall.

F. paxiana Lingelsh Var. *Sikkimensis*.-Tree of 16 m. height. Found in cave in Tista Valley near Labdah, near Melli or elsewhere.

Occurrence:- Tista Valley.

APOCYNACEAE

Alstonia Br.

**A. scholaris* Br.-A tall evergreen handsome tree. Branches in whorl, so are the leaves. *Flower*-white 11-1, *Fruit* 1-4. Sporadic.

Occurrence:- All over plains and foot hills.

A. nerlfollia Don.-A small tree. Rare.

Wrighita Br.

**W. tomentosa* R & S.-Common in riverain tract and dry mixed forests. *Flower*-5-6, cream coloured, *Fruit* 12-4, long cylindrical capsule. Very common.

Occurrence:- Riverain forests and dry mixed forests.

W. sikkimensis Gamble.-Small tree of the lower and middle hill forests; upto 1219 m. Rare.

Occurrence:- *W. coccinea* Sims.-Small tree of the lower hills forests and upto 1219 m. *Flower*-5 red, *Fruit* 6-7. Rare.

Occurrence:- Foot hills.

Holarrhena Br.-

H. antidysenterica Wall.-A small deciduous tree. Leaves upto 25 cm. *Flower*-white. Folicle 2-divaricate upto 400 cm. by 1 cm. Very common.

Occurrence:- All over riverain and dry mixed forests.

BIGNONIACEAE

Stereospermum Cham.

**S. personatum* (Hassk) Chatterjee (S. chelonoides Dc.)-A large deciduous tree. *Flower* yellow 4-5, *Fruit* long curved twisting into spirals 5-8.

Occurrence:- Very common in plains and foot hills.

S. suaveolens Dc.-A middle sized tree. *Flower*-red, 4, *Fruit*-11-2. Sporadic.

Occurrence:- Plains and foot hills.

BORAGINACEAE

Ehretia Linn.-

**E. laevis* Roxb.-A small deciduous tree. Leaves upto 15 cm. Drupe 4-ribbed. *Flower*-white. Common.

Occurrence:- Dry mixed forests.

E. acuminata Br.-A small sized tree. *Flower* white, *Fruit*-2 seeded. Common.

Occurrence:- Dry mixed and moist deciduous forests.

E. wallichiana Hk.f.-A medium sized tree. *Flower*-fragrant. *Fruit*-4, 1-seeded, Pyrenes. Sporadic.

Occurrence:- Riverain forests.

Cordia Roxb.-

**C. grandis* Roxb.-A medium sized tree. Leaves upto 20 cm. *Flower*-white, *Fruit*-white, 1-seeded.

Occurrence:- Riverain and dry mixed forests. Common.

**C. obliqua* (=C. dichotoma) (=*C. myxa* Linn.)-A deciduous small tree. *Fruit*-yellow viscid, sweet, when ripe. Common.

Occurrence:- In riverain and dry mixed forests.

VERBENACEAE

Tectona Linn.f.-

**T. grandia* Linn.-A planted species. Leaves upto 60 cm. long. *Flower*-white. *Fruit*-sub-globose drupe. *Flower* 7-8, *Fruit* 2-4.

Occurrence:- In plantation in plains and foot hills.

Gmelina Linn.-

**G. arborea* Linn.-A deciduous tree. A pair of gland at the junction of leaves. *Flower* 2-4, *Fruit* 2-4 glossy, yellow when ripe. Very common.

Occurrence:- All over plains and foot hills.

Premna Linn.

**P. bengalensis* Clarke.-A middle sized evergreen tree often grows a large size.

Occurrence:- Common.

**P. latifolia* Roxb.-A middle sized evergreen tree. Stem irregularly fluted. Corolla white. *Flower* 5-7, *Fruit* 8-11.

Occurrence:- Sal forests. Common.

P. latifolia Var *mucronata* Clarke.-Larger tree 2 m to 3 m in girth.

Occurrence:- Sal forests.

P. interrupta Wall ex Sch.-In Sal forests. Not common.

**P. integrifolia* Linn.-In sal forests. Not common.

Vitex Linn.-

**V. quinata* (Lour) F.N. Will (=*V. heterophylla* Roxb.)-A large tree. Leaves 5-foliolate. Leaflets petioluted.

Occurrence:- Common in Wet mixed forests.

V. peduncularis Wall ex Sch.-A large tree. Leaves 3-foliolate with shining resinous gland on the underside.

Occurrence:- Wet mixed forests.

MYRISTICACEAE

Horsfieldia Willd.-

**H. kingii* (Hook f.) Warb. (=*Myristica* Kingii Hk.f.)-Tree upto 20 m. high. Blaze red. Leaves upto 22 cm.

Occurrence:- All over Sal and miscellaneous forests.

**H. amygdalina* Wall.-A moderate sized glabrous tree.

Occurrence:- Sal forests. Common.

Knema Lour.-

**K. linifolia* Worb.-a middle sized tree.

Occurrence:- Sal forests. Common.

K. angustifolia Worb.-A middle sized tree.

Occurrence:- Sal forests. Sporadic.

LAURACEAE

Cryptocarya R.R. Br.

**C. amygdalina* Nees.-A tree attaining 13 m. in height. *Flower*-3, *fruit*-11, narrow ellipsoid 2.5 cm. long.

Occurrence:- Sal and wet mixed forest. Common.

C. floribunda Nees.- A tree attaining 13 m. height.

Occurrence:- Sal forests. Sporadic.

Beilschmiedia Nees.-

**B. roxburghiana* Nees.-A tree greatly varying in size. *Fruit*-upto 3.5 cm. long, dark purple.

Occurrence:- Foot hills. Common. More in Kalimpong hills upto 1600 m.

B. assamica Meissn. - A large tree. Kalimpong hills. *Flower*-12, *Fruit* 2-3.

Occurrence:- Sporadic; Also occurs in Moraghat forest.

B. sikkimensis King.-Large tree. Leaves opposite. *Flower* 5-6, *Fruit*-ellipsoid, 11.

Occurrence:- Foot hills.

Cinnamomum Blume.-

C. tamala Nees.-A moderate sized tree. Aromatic of high order. Leaves 3-nerved. *Fruit* -succulent, black. common.

Occurrence:- Foot hills, Ghoramara forests (common).

**C. impressinervium* Meissn.-Medium sized tree. *Flower*-7, *Fruit*-12. Nerves deeply impressed on the upper surface.

Occurrence:- Temperate forests 1600-2500 m. Common.

C. caudatum Nees.-A medium sized tree. Leaves alternate, 3-5 nervd. Common.

Occurrence:- Temperate hills 1800-2200 m.

**C. gluaeolens* (Nees) Hard-Marr (*C. cecidodaphne* Meissn.)-A large evergreen tree. Bark furrowed. Wood sweet scented. *Flower* 1-2, *Fruit* 9-10. Sporadic.

Occurrence:- All over Sal forests; foot hill forests.

**C. bejalghata* (Buch Hon) SW. *C. obtusifolium* Nees.)-A tall tree in North Bengal and Assam plains and foot hills. Leaves opposite, 3-nerved. Very common.

Occurrence:- All over plains and upto 1500 m in the temperate hills.

Neocinnanomum Lion Ho

N. Candatum (Cinnamomum Candatum Nees)- A medium sized tree. Leaves allernate 3-5 nerved.

Occurrence:- Common.

Persea Hill

P. odoratissina (Nees) Kosteron (*M. Odoratissima* Ness.)-A medium sized tree. Leaves coridaceous *Flower* 3-5, *Fruit* -ellipsoid or oblong. Very common.

Occurrence:- Gregarious in temperate forests 2000-2300 m.

**P. villosa* (Roxb) Kostern (*M. villosa* Hk. f.)-A large or middle sized tree. Leaves densely tomentose. Perianth segments tomentose or silky. Common.

Occurrence:- In plains Sal forest.

Machilus Nees

**M. edulis* King.-A large tree. *Fruit* - edible 1-3, *Flower*-5. Common.

Occurrence:- Temperate forests 1900-2200 m.

**M. bombycina* King.- A medium sized tree. Leaves glabrous. Leaves silky beneath. Common.

Occurrence:- Plains Sal forests.

**M. gamblei* King.-A medium sized tree. Bark with gall (quite often). *Flower* - 2-3, *Fruit* 3-7.

Occurrence:- Plains and foot hills forest. Common.

**M. gammieana* King.-A large tree, leaf buds covered with large obtuse silky scales.

Occurrence:- Temperate hills. Common.

Phoebe Nees.-

**P. lanceolata* Nees.-A small or middle-sized tree.

Occurrence:- In foot hills miscellaneous forests. Common.

P. attenuata Nees.- Sometimes a very large tree of 35 m. high. *Flower* 3-5, *fruit* 5-7. Undersurface of pedicel rusty tomentose.

Occurrence:- Foot hills in Samsing, Mal, Mo forests.

P. hainesiana Brandis.-A large timber tree, attaining 45 m. height and 6 m. girth. *Flower*-5, *Fruit*-12. Black fleshy.

Occurrence:- Foot hills in Kalimpong (Samsaig). Sporadic.

Actinodaphne Nees.-

**A. reticulata* Meissn.- Evergreen tree. Leaves pubescent lanceolate.

Occurrence:- Darjeeling hills upto 2000 m.

A. angustifolia Nees.- A large or moderate-sized tree. Leaves coriaceous. Cluster of flowers seated on the small cup shaped perianth. Common.

Occurrence:- Plains forest Sal and Miscellaneous.

**A. obovata* Blume.- A large tree. Leaves pubescent.

Occurrence:- Sal forests. Common.

Neolitsea (Benth) Merr.-

N. foliolosa (Nees) Gamble (Litsaea fliolosa Nees)-A tree of 18 m. high. Leaves acuminate, triple nerved. Fruiting pedicel stout, *Flower*-11, *Fruit* 11-12.

Occurrence:- Temperate hills 2000 m.

Litsaea Lamk.-

L. lancifolia Roxb. ex Wall.-A shrub or small tree, aromatic. *Flower*-white, 10, *Fruit* 4-5. Sporadic.

Occurrence:- Foot hills upto 900 m.

L. khasyana Meissin.-A small or middle sized tree. Rare.

Occurrence:-Foot hills.

L. elongata Wall.- Asmall or moderate sized tree of 16 m. Leaves almost in whorls. *Flower* - 9-11, *Fruit* 4-5.

Occurrence:- Foot hills upto 1600 m.

**L. panamonja* Buch. Ham.-A larger tree. *Flower* 3-4, *Fruit* 4-5.

Occurrence:- Common in plains Sal forests.

Other species of the genus:-

L. cubeba (Laur) Pers (*L. citrata* Bl.)-A small deciduous tree. Bark green. Leaves dries black. *Flower*-smell of organge. Sporadic.

Occurrence:- Sal forests.

L. kingii Hook.f.-A small tree 2200-2800 m.

Occurrence:- Temperate hills.

L. sericea Hook.f.-A small tree 2900-3000 m. *Flower* 4-5, *Fruit* 9-10. Leaf bud with rigid scales.

Occurrence:- Temperate hills.

**L. glutinosa* (Lour) Robins (*L. chinensis* Lam)-A deciduous tree of 13 m. high. Blaze turn black. *Flower* 5-6, *Fruit* 7-8. Common.

Occurrence:-Foot hills.

Lindera Thunb.-

**L. assamica* Kurz.-A tall tree. Common on Senchal and Ghumpahar. *Flower* - 9-11, *Fruit* 8-10.

Occurrence:- Temperate hills. Common.

L. pulcherrima (Nees) Benth. - A tree at 2000-2600 m. *Flower*-5, *Fruit* - 9-10, black. Leaves 3-nerved.

Occurrence:- Temperate hills (Uppper).

PROTEACEAE

Helicia Lour.-

H. nilagirica Bedd (*H. erratica* Hk. f).- Fairly common near the Durbin Dhara in Rinkingpon near Kalimpong. A small tree. Leaves serrate. *Flower*-yellowish.

Occurrence:- Foot hills; confined to small area.

SANTALACEAE

Pyrularia Mich.-

P. edulis (Wall) Dc.-A small thorny tree 600-100m. in Himalayas. *Flower* - 3-5, *Fruit* - 7-10, edible.

Occurrence:- Temperate hills 1600 m.

EUPHORBIACEAE

Bischofia Blume.-

**B. javanica* Bl.-A large deciduous tree with tall cylindric trunk and often of great girth 30 m. usually with a spreading crown. Leaves 3- foliolate, serrate. Balze pink. *Flower*- 3-4, *Fruit* - 12-3.

Occurrence:- Very common in Riverain forest.

Bridelia Willd.-

**B. retusa* (Linn) Spreng.-A large or medium sized deciduous tree 20 m. armed with conical spines. *Flower*- 8-9, *Fruit* -10-12. very common.

Occurrence:- All over Sal and miscellaneous forests.

**B. montana* Willd.-A middle sized tree, glabrous. *Flower* in dense axillary clusters.

Occurrence:- Plains Sal forests (drier type).

Emblica Gaertn.-

E. officinalis Gaertn. (*Phyllanthus* emblica Linn.)-A small to medium sized tree, deciduous.

Occurrence:- Dry mixed forests. Sporadic.

Antidesma Linn.-

**A. ghesaembilla* Gaertn.- A small tree. *Flower* and *Fruit* 4-5. *Fruit*-black to red edible. Common.

Occurrence:- Dry type Sal forests.

A. bunius Spreng.-A small tree. Leaves and fruit edible. *Flower* - 5, *Fruit* 9-11. Common.

Occurrence:- Sal forests.

**A. acidium Retz* (*A. diandrum* Roth.)-A small tree. Leaves red when old. Leaves acidic 7-8, *Fruit*-red purple.

Occurrence:- Sal forests.

Gelonium Roxb.-

G. multiflorum A. Juss.-A small tree. Leaves upto 18 cm. *Flower*- greenish yellow. *Fruit* - globose 3-lobed.

Occurrence:- Riverain forests.

Drypetes Vahl.

D. lancifolia (Hk.f.) Paxet & Hoffm.-A large or middle sized tree (26 m.) branchlets lenticellate. *Flower*-5, *Fruit*-11, orange coloured.

Occurrence:- Sal forests. Sporadic.

D. assamicus (Hk.F.) Paxet & Hoffm.-A small or middle-sized evergreen tree of 15 m. height, branches slender, usually pendent. *Flower*- 10-1, *Fruit* -3-4, red when ripe.

Occurrence:- Sal forests.

Glochidion Forst.

**G. multiloculare* Muell.-A shrub or a small tree, chiefly found along roads and streams in the Terai and Duars, near the hills and ascending to 300 m. Leaves reddish.

Occurrence : Common

**G. hirsutum* Muell. Arg.-Small or medium-sized tree of the middle plain forest e.g. Dalka Jhar. Leaves densely white or rusty tomentose.

Occurrence:- Sal forests. Dry type. Sporadic.

G. oblatum Hk.f.-Small tree of the lower hill forests.

Occurrence:-Sal forests dry type.

G. thomsoni Hk.f.-Common in lower hill forest. *Flower* - 4-7, *Fruit* - 9-10.

Occurrence:- Lower hills upto 1600 m.

G. acuminatum Muell. Arg.-It is of the middle and upper hill forest from 1300-2300 m. *Flower* - 4-6, *Fruit*- 10-2.

Occurrence:- Temperate hills (Lower) Common.

G. assamicum Hk. f.-A small tree of the lower hills forest valleys. *Flower* and *Fruit*-3.

Occurrence:- Sporadic.

G. lanceolarium voigt.-It is of the plains forests and foot hill forests upto 1600 m. chiefly in the valleys. *Flower*-12, *Fruit*-3.

Occurrence:- Common.

Croton Linn.-

**C. tiglium* Linn.-Small of medium sized tree of the upper temperate forests from 1500 m. Common about Damsong and in Lulygaon. Common.

Ostodes Bl.-

**O. paniculata* Bl.-A small tree. *Flower* 4-6, *Fruit* 10-11.

Occurrence:- Kalimpong temperate hills.

Endospermum Benth.-

E. chinense Benth.-A tall tree, white bark. Leaves broadly ovate. *Flower* 4-5, *Fruit* 7-8. Sporadic.

Occurrence:- Foot hills.

Sapium P. Br.-

**S. eugeniaefolium* Buch Ham ex Hook.f.-A middle sized tree of 18 m. high. Leaves ovate-elliptic biglandular. *Flower* 5-6, *Fruit* 10-1.

Occurrence:- Dry mixed forests.

S. baccatum Roxb.-A middle sized or large evergreen tree of 25 m. height, glabrous. Common. *Flower* 4-6, *Fruit* 8-11.

Occurrence:- All types of forests.

Aporosa Blume.-

**A. octrandra* (Buch Ham ex Don) Vickery. (*A.roxburghii* Baill.)-A small tree. Sal forests. Leaves 10-15 cm. Spikes of male flower clustered. Very common.

Occurrence:- Sal forests.

Baccaurea Linn.-

B. rami flora Lour (B. Muell.)-An evergreen tree. Leaves upto 20 cm. Flower creamy yellow, berries globular yellow, purple, edible.

Occurrence:- Wet mixed forest.

Trewa Linn.-

**T. nudiflora* Linn.-A middle sized deciduous tree. *Flower* 1-3, *Fruit* 7-8, Woddy smooth.

Occurrence:- Riverain forests.

Mallotus Lour.-

**M. albus* (Roxb) Muell. Arg.- In the middle hills forest upto 1600 m. Leaves peltate, 5-9 nerved *Flower* - 4-7, *Fruit*-8.

Occurrence:- Foot hills. Common.

M. roxburghianus Muell. Arg.-A small tree. Leaves peltate, 9-nerved. *Fruit* 5-7.

M. nepalensis Muell. Arg.-Very common. Leaves peltate, 3-nerved *Flower*-7, *Fruit*-8.

M. philippinensis(Lamb) Muell. Arg.-The tree occurs upto 1300 m. Leaves 3-nerved, underside with red glands. *Flower* 7-11, *Fruit* 1-5 covered with red powder.

Occurrence:- Very common in Sal and foot hills forests.

Macaranga Thouars.-

M. pustulata Kin.-Small tree. Leaves 3-7 nerved. *Flower* 10-11, *Fruit*-3.

Occurrence:- Temperate hills.

**M. gmelinaefolia* King.-A small tree of the lower hill forest valleys. Common along the Tista River. Differs from the preceding by its less broad membranous leaves with broadly cuneate bases, and less distinct nerves.

**M. gamblei* Hk.f.-A small tree. Leaves elliptic. lanceolate. 3 nerved.

Occurrence:- In plains in open land.

M. peltata Muell. Arg. (*M. roxburghii* Wight.)-A small tree. Leaves 9-nerved, *Fruit*-6.

Occurrence:- Plains forests.

**M. denticulata* (Blume) Muell. Arg.-It comes up chiefly on land which has been cleared, forming almost pure forest. Very common. *Flower* 3-4, *Fruit* 4-6.

Occurrence:- All over plains.

ULMACEAE

Ulmus Linn.-

U. lancifolia Roxb. ex Wall.-A tree 20 m. Climbs upto 1300 m. Rare. *Flower* and *Fruit* 10-3.

Occurrence:- Tista Valley.

Celtis Linn.

**C. tetrandra* Roxb.-A large handsome deciduous tree of 30 m. Drupe-orange red. *Flower* 2-3, *Fruit* 10-11.

Occurrence:- Plains, in clearing.

C. timorensis (C. cinnamomea Lindl ex Planch)-A middle sized tree. *Flower*-1, *Fruit* 1-7 Heart wood has disagreeable smell (almost like dried faceal matter).

Occurrence:- Plains forests.

Trema Lour.

**T. orientalis* (Linn) Blume.-Common in the Terai, Duars, and foot hills especially in old cultivation, where it comes up in great quantity. very common. Leaves have 3-basal nerves. *Flower* 3-6.

Occurrence:- In plains.

**T. politoria* Planch.-common in the foot hills, especially on formerly cultivated land. Leaves have 5-basal nerve. *Flower*-6-7, *Fruit* 6-10. Common.

Occurrence:- In plains.

T. lomentosa (Roxb) Hasa (*T. amboinesis* Blume)-Tree of the plains and foot hills, similar to the preceding. *Flower* 3-8.

MORACEAE

Ficus Linn.-

F. tinetoria Forst [F. gibbosa Bl. var. Cuspidifera (Miq) King]-A small or sometimes a large tree, often epiphytic; surrounding the trunks of other trees with a net work of branches. Leaves inequilateral. *Flower* 1-3.

Occurrence:- Plains forest. Sporadic.

F. neterophylla Bl. (*F. urophylla* Wall)-An erect shrub or small tree (15 m.); often epiphytic. Leaves ovate, ovate elliptic. *Flower* 2-3.

Occurrence:- Plains forests. Rare.

**F. bengnalensis* Linn.-A large tree, generally epiphytic. Common as shade tree. Leaves 4-7 nerved thickly chartaceous. Ripe *Fruit* 4-7.

Occurrence:- Plains and foot hills. Common.

**E. drupacea* Thunb vas. pubescens (Roxb) Corno (F. mysorensis Acyne)-A very large epiphytic tree 3-5 m. in girth. Leaves have 12-20 nerves. Receptacle bright orange. *Fruit* 4-8.

Occurrence:- Plains and foot hills. Sporadic.

**F. altissima* Bl.-A enormous spreading tree, epiphytic, with aerial roots (5-6 m. in girth.) Receptacle yellow. Leaves 3-nerved.

Occurrence:- Plains and foot hills. Common.

**F. curtipes* Cornees (*P. obtusifolia* Roxb)-A large tree, glabrous, often epiphytic. Leaves oblanceolate, 3-nerved. Short petiole. Receptacle yellow with red specks.

Occurrence:- All over plains and foot hills. Common.

F. elastica Roxb.-A gigantic tree, quite glabrous.

Occurrence:- Foot hills. Common.

R. maclelandii King (*F. rhododendrifolia* Miq.)-A moderate sized tree, often epiphytic. Leaves elliptic, *Flower* 10-11, *Fruit*-orange spotted-4-5.

Occurrence:- Plains. Sporadic.

F. benjamina Linn.-A large tree, spreading with drooping branches. *Flower* 10-11, *Fruit*-large yellow or red, 4.

Occurrence:- Plains and foot hills. Sporadic.

F. retusa Linn.-A large tree, generally epiphytic, forming a tangled mass with aerial roots and horizontal branches. Receptacle red or yellow.

Occurrence:- Plains and foot hills. Common.

**F. nervosa* Roth.-A very large tree with broad crown, often buttressed. Leaves coriaceous, elliptic. *Flower* 10-12.

Occurrence:- Plain and foot hills. Common.

**F. rumphii* Bl.-A large deciduous tree, often epiphytic, petiole long. Leaves cordate, apex truncate.

Occurrence:- All over plains. Common.

**F. religiosa* Linn.- A large deciduous tree, usually epiphytic, trunk irregularly shaped.

Occurrence:- Very common.

Other species recorded are:-

F. lacor Buch-Ham *F. infectoria* Roxb.-A middle-sized or large deciduous tree, with spreading crown, epiphytic in early stage, sending down a few aerial roots. Common.

F. geniculata Kurz.-A large tree, twigs lenticellate. Sporadic.

F. rubineisa Sm (*F. clavata* Wall)-A small tree or shrub. Sporadic.

F. cyrtophylla Miq. (*F. obscura* Bl.)- A medium sized tree with spreading crown. Sporadic.

**F. hispida* Linn.-A small tree or shrub with hollow branchlets very common.

F. semicordata Buch. Ham (*F. cunia* Ham).-A middle-sized or small tree. Common.

F. hirta Vahl - A small tree or shrub, young branches hollow. Sporadic

F. neriifolia var. *nemoralis* (Wall ex king) (*F.nemoralis* Wall.)- A small hand some tree Sporadic

F. curiculata Laur (*F. roxburghii* Wall)-A middle sized tree with a spreading crown. Common.

F. oligodon Miq. (*F. pomifera* Wall)-A tree. Sporadic.

F. ramosa (F. glomerata Roxb)-A large deciduous tree. Common.

F. glaberrima Bl.-A large tree (25 m) glabrous generally epiphytic, stem usually buttressed Sporadic.

Artocarpus Forst.-

**A. chama* (A. *chaplasha* Roxb.)-A very large deciduous tree (45 m.). *Flower* 5, *Fruit* 7-9. Syncarp globose.

Occurrence:- Wet mixed forests. Very comon.

A. hetrophyllus Lamk (*A. integrifolia* Linn)-A large evergreen tree with large dense crown; young shoots with stiff hairs. Cultivated.

**A. lacuna* Hamitt (*A. lakoocha* Roxb)-A large deciduous tree with a large spreading crown; young parts tomentose. *Flower* 2-5, *Fruit* globose, valvety yellow. Sporadic.

Occurrence:- Wet mixed forests.

Morus Linn.-

M. australis Pair (M. indica Linn.)-A moderate sized deciduous tree. Riverain tract *Flower* 2-5, *fruit*-dark purple, 4-7, edible. Sporadic.

Occurrence:- Riverain forests.

M. macroura Miq (*M. laevigata* Wall)-A large tree (30-40 m.) Leaves shortly accuminate. *Fruit*-yellowish. sporadic.

Occurrence:- Foot hills.

Streblus Lour.-

S. asper Lour.-Small tree, yields milky juice. Leaves dentate rough. *Fruit*-yellow. very common.

Occurrence:- Riverain and dry mixed forests.

JUGLANDACEAE

Juglans Linn.-

J. regia Linn.-A large aromatic deciduous tree, shoots valvety. sporadic to rare. *Flower* 2-4, *Fruit* 7-8.

Occurrence:- Tongly range.

Engelhardtia Lescher.-

E. spicata Bl.-A large deciduous tree (30 m.) often gregarious; base often buttressed. Raceme pendulous, *flower*-11-12, *Fruit*-3-4. Fruit-3-lobed wing. very common.

Occurrence:- Plains to 1800 m.

E. spicata var *Cloebr aokiana* (Lindlex Wall) Koord & Valet (*E. colebrookiana* Lindl)-A small tree, often with spreading crown, sometimes gregarious. *Flower*-2, *Fruit* 4-5.

Occurrence:- Very common.

E. acerifolia Bl.-A large deciduous tree. *flower* 8, *Fruit*-1.

Occurrence:- Temperate hills.

BETULACEAE

Carpinus Linn

C. viminea Wall.-A middle sized tree. Leaves ovate acuminate, doubly serrate. *flower* 1-4, *Fruit* 6-7.

Occurrence:- Singalila forests; rare.

Corylus Linn

C. ferox wall.-A small tree, buds silky, elongated. *Flower*-4, *Fruit*-11.

Occurrence:- Singlila forest, rare.

Alnus Gaertn.-

**A. nepalensis* D. Don.- A large deciduous tree; bark compact, silvery grey, *Flower* 10-12, *Fruit* 2-3. Very common.

Occurrence:- Temperate hills; on landslide and river valleys.

Betula Linn.-

**B. alnoides* Buch Ham ex D. Don.-A moderate sized tree, often gregarious. *Flower* 4-5, *Fruit* 7-8.

Occurrence:- Temperate hills upto 2200 m.

**B. cylindrostachys* Wall.-A very large and handsome tree 30 m. *Flower* and *Fruit*-white.

Occurrence:- Temperate hills upto 1500 m.

**B. utilis* Don.-A moderate sized tree attaining at times 20 m. Bark exfloiating in broad horizontal roll. *Flower* 4-6, *Fruit*-9-10.

Occurrence:- Upper temperate hills 3000 m.

FAGACEAE

Quercus Linn.

**Q. lineata* Bl. var. *thomsoniana* Wenzig.-A large tree. Leaves ovate lanceolate.

**Q. lamellosa* Smith.-A very large tree, evergreen 2000-3000 m Lower surface of leaves glaucous. Large nuts lamellate cups.

Occurrence:- All over temperate forests.

Lithocarpus Blume

L. pachyphylla (Kurz) Rehder (Q. pachyphylla Kurz)-A large tree at 2500-3000 m in Darjeeling Sikkim hills. Frutis covered woody coalesced group of 3-6 acorns.

Occurrence:- All over temperate forests.

L. lancaefolia (Roxb) Rohder (*Q. lancaefolia* Roxb) -A small tree. In acorns cupule is thin and ridged and entirely covers the nut.

Occurrence:- Lower temperate hills.

**L. panestrate* (Roxb) Rohder (*Q. fenestrata*Roxb)-A tall evergreen tree. Ripe acron. Solitary or in trees, much crowded. Rufous pubescent corps.

Occurrence:- Lower temperate hills.

**L. elegans* (Bl.) Hatus ex soepadma *Q. spicata* Smith)-A small tree extends from plains to 2500 m.

Castanopsis Spach.-

**C. indica* A. Dc.-A middle sized or large tree (30 m) Leaves sharply serrate. Fruit-globose. Common.

Occurrence:- All over plains and foot hills.

**C. hystrix* A. Dc.-A large evergreen tree (30 M) stem often fluted.

Occurrence:- Common in plains and foot hills.

C. tribuloides A. Dc.-A middle size, often lage evergreen (26 m), . *flower* 9-11, spines in tufts in *Fruits*. very common.

Occurrence:- All over plains and temperate hills.

**C. tribuloides* A. Dc.-var *ferox* King.-Leaves entire spines fewer and stouter. Nuts 1-3.

Occurrence:- All over plains and foot hills upto 1600 m.

**C. tribuloides* A. DC. var *longispina* King.-Leaves larger acuminate, spines stout and short. .Nuts 1-3.

Occurrence:- Plains and foot hills.

Var echinocarpa King.-Leaves serrate. Involucre. Smaller. Nuts solitary.

Occurrence:- Plains and foot hills.

Var Wattii King.-Involuere small, wooly. Nuts 1-3.

Occurrence:- Plains and foot hills.

SALICACEAE

Salix Linn.-

S. salcoinensis Handel.-Recorded from Singalila range forests Rare.

S. pyrina Wall (=*S. teratpserma* Roxb.):- N. Bengal. Rare.

S. daltoniana Anders:-Tonglu forests (2800 m.) Rare.

Populus Linn.-

P. gamblei Dobe.-Medium-sized tree of the L. and M.H.F. at 600 m. Common in the Dumseng forests. Temperate hills, rare.

P. glauca Haines.-A medium sized tree found on Tonglu at about 900 m. Endemic to Tonglu.

Occurrence:- 3600 m. Tonglu.

(ii) Analytical Synopsis of Shrub Flora

In this section (Analytical Synopsis) listing of Conspicuous and other shrubs falling within quadrats and transects and adjoining areas have been made. A very brief description of each species has been given, but flowering and frusting time and fidelity of each species have been recorded. Numericals following flowering and frusting indicate months (viz.., 1,2,3, means January, February, March and so on) Asterisks (*) indicate conspicuous and widely occurring species.

ANONACEAE

Desmos Linn.-

Desmos *longiflora* Roxb.- An erect glabrous shrub. *Flowers* red 4-5, *fruit* 10-12. Sporadic.

Occurrence:- Foot hills.

Goniothalamus Bl.-

G. *sesquipedalis* Hk.f. & Th. - A branched shrub; grows in shade. *Flowers* greenish yellow, solitary, 5-6, *fruit* 11 - 12.

Occurrence:- In sal forests in shade.

BERBERIDACEAE

Berberis Linn.-

* *B. chitria* Lindl. (=*B. asiatica* Roxb.) - A very pretty shrub with small 1 - 5 toothed leaves 3 - 8 cm. *Flowers* yellow. *Fruit* - berry, purple - blue.

Occurrence:- Upper Temperate hills.

B. wallichiana Dc.- An erect spinous shrub; 2-3m. high; branches sharply angled. *Flowers* yellow bisexual 8-9, *Fruit* 9-10 (fruit turning from red to blue).

Occurrence:- Upper Temperate hills.

B. Concinna Hk.f.
B. insignis Hk.f.. & Th. -
B. angulosa Wall. ex Hk.f. & Th.
— all occur in Singalila Forests. Common.

Mahonia Nutt.-

* *M. Pycnophylla* Fedde.- An erect shrub. *Flowers* yellow, 10-11, *Fruit* 11-1.

Occurrence:- Upper Temperate hills.

LARDIZABALACEAE

Decaisnea H.k.f. -

D. *insignis* Hk.f. - An erect shrub, leaves unequally pinnate, petiole jointed at base. Flowers in raceme, yellow-green. Fruit follicle, edible.

Occurrence:- Singalila Range

CAPPARIDACEAE

Capparis Linn.

C. *olacifolia* Hk.f. & Th. - An erect shrub upto 4 m. high, *flowers* solitary or two from the same leaf-axil, white, 3-4, *Fruit* 12-2.

Occurrence: Plain riverain forests.

C. *pumila* Champs.,
C. *multiflora* Hk.f.. } Occurs in Tista valley forests. -

FLACOURTIACEAE

Casearia Jacq.

C. *graveolens* Dalz.- A very common *shrub, Fruit* yellow; seeds red. *Flower* 8, fruit 3-4. Common.

Occurrence:- Sal forests.

*. C. *vareca* Roxb. - A large shrub. Leaves oblonglanceolate closely serrate, pubescent, *Flowers* greenishgrey, 6-9, *fruit* 10-4. Very common.

Occurrence:- Sal forests.

POLYGALACEAE

Polygala Linn.

P. *arillata* Buch-Ham ex D.Don.- A small shrub; leaves obovate or oblanceolate. *Flowers* lilac or white with purple tips. *Fruit* capsule, margin ciliat. Sporadic.

Occurrence:- Temperate forests at 2000 m.

TAMARICACEAE

Tamarix Linn.

T. *dioica* Roxb ex Rott.- A gregarious shrub upto 5m. in height;

leaves minute, *Flower* pink, 5-merous, *Fruit* capsule, conical, 1 - celled, 3 - valved. Sporadic.

Occurrence:- On sandy river bed.

HYPERICACEAE

Hypericum Linn

* *H. hookerianum* W. & A.- An evergreen shrub, about 2 to 2.5m. height with reddish brown bark. *Flower* yellow in few flowered terminal crymbose cymes. *Fruit* capsule seated in the persistent calyx. Flowers 4-6; Fruit 10 - 11. Common.

Occurrence :- Singalila forests for 2,400 m

* *H. patulum* Thumb.- A small shrub upto 1.5m. in height with reddish-brown, sometimes nearly black bark, Flowers yellow in few flowered terminal cymes, *flowers* 5-7, *fruit* capsule, ovoid; valved persisting for a long time, 9-10. Common.

Occurrence: Singalila forests 2,400 m.

H. chinense Linn.- A cultivated shrub, with ovate leaves, somewhat amplexicaul; capsule-very small. Common.

Occurrence:- Singalila range.

THEACEAE

Camellia Linn.

C. sinensis Lin. Kuntze.- A large evergreen shrub; *Flowers* white upto 3.5 cm. in diameter, solitary or a few together. *Fruit*-capsule, 3-cornered leathery. *Flowers* 12-3; *Fruit* 9-12.

Occurrence:- Cultivated.

C. Kissi Wall (=*C. drupifera* Lour).- A shrub, bark grey outside, light red and places towards inside. *Flowers* white, erect, pendulous terminal, 7-11. *Fruit*-capsule, globose 1.

Occurrence:- Mangrove forests.

Eurya Thumb.

* *E. cavinervis* Thumb.- A shrub; upto 2.5 m. high; ripe fruits blue-black. Abundant.

Occurrence:- All over plains and foot hills.

* *E. acuminata* Dc. - Shrub upto 5 m. high; flowers in buds, white. Abundant.

Occurrence: All over plain and foot hills, it is a dwarf form.

SAURAIUIACEAE

Saurauia Willd.

* *S. roxburghii* Wall.- A very common shrub; flowers urceolate in cymes, white, fading pink. *Fruit*-berry (whitish), ediable. *Flowers* 3-5; *Fruit* 5-8. Common.

Occurrence:- Temperate forests.

MALVACEAE

Sida Linn.

* *S. rhombifolia* Linn. A shrub; branches rough with stellate hairs. *Flowers* yellow or white. Common.

Occurrence:- Plains forests.

* *S. cordifolia* Linn.- A small erect downy shrub, with long spreading pairs on the branches and petioles. *Flowers* small yellow.

Occurrence:- Plain forests.

Urena Linn.

* *U. lobata* Linn.- A tall erect undershrub, 1m. high. *Flowers* bright pink with a darker centre; carpels rounded on the beak with many hooked bristles. *Flowers* and *Fruits* 6-12.

Occurrence:- All over.

STERCULIACEAE

Helicteres Linn.

H. plebeja Kurz.- Small shrub; flowers 6.

Occurrence:- Senoke forests.

H. *spicata* Colsb.- Shrub; leaves ovate, oblong, serrate.

Occurrence:- Sevoke forests.

Melochia Linn.

M. corchorifolia Linn.- A stellately hairing shrub. Leaves 8 cm often plaited, with cordate base. *Flower* in cluster, hispid, bracteole. Petals white or pink.

Occurrence:- Sevoke forests.

TILIACEAE

Grewia Linn.

G. disperma Roxb. (*G. multiflora* non Juss.) - A bushy shrub; flowers white small. Drupe - 1 - 4 lobed, globose, black or purplish, fleshy, glabrescent.

Occurrence:- Dry mixed forests. Sporadic

G. hirsuta Vahl. - A spreading shrub attaining a height of about 3 m. Flowers yellow, polygamous, Drupe-reddish brown and shining when ripe. *Flowers* rainy season. *Fruit* 11 - 1.

Occurrence:- Dry mixed forests. Sporadic

G. sclerophylla Roxb.- A shrub; Drupe-globose, depressed at the top. *Flowers* 5 - 9, *Fruit* 11.

Occurrence:- Dry mixed forests. Sporadic

G. disperma Roth (*G. glabra* B1.) - A shrub.

Occurrence:- Dry mixed forests.

Triumfetta Linn.

T. pilosa Roth.- A shrub 1-2 m. high with stellate hairy stems. *Flowers* congested in lateral cymes. *Fruit* capsule, long, slender, hooked, hispid in the lower half, 4-celled, 8 - seeded Sporadic.

Occurrence:- Dry mixed forests.

* *T. bartramina* Linn. (=*T. rhomboidea* Jacq.) - An under shrub 1 m. high *Fruit* ellipsoid with the spines. Common.

Occurrence:- Dry mixed forests.

LINACEAE

Reinwardtia Dumort

R. trigyna Planch Syn. (=*R. tetragyna* Planch.) - A glabrous undershrub about 1 m. high. *Flowers* bright, golden yellow, solitary. *Fruit* globose capsule.

Occurrence:- Road sides in plains Sporadic

RUTACEAE

Skimmia Thunp.

S. arborescens T. And ex Gamble.- Large shrub, common specially in damp ground & undershade. Flowers white or greenish yellow. polygamous. *Fruit*-subglobose. *Flowers* 4-5; *Fruit* 9-11.

Occurrence:- Upper Temperate forests. sporadic

Fagara (=*Zanthoxylum* Linn.)

E. ovalifolium Wight.- Straggling shrub; unarmed or with a few short straight prickles. *Flowers* white small, *Fruit* almost globular. Sporadic.

Occurrence:- Foot hills.

Glycosmis Correa.-

G. arborea (*G. pentaphylla* Correa.) - A shrub 1.5 m. high, flowers small white; berry upto 1.25 cm. depressed globose, often pinkish, glassy, 1-seeded. Sporadic.

Occurrence:- Plains.

Micromelum Bl.

M. integerrima (Buch-Ham) (*M. pubescens* B1) -A shrub with oily green leaves. *Flower*-white, fragrant. *Fruit* when ripe orange yellow; green when raw. Very common.

Occurrence:- Plain Sal forests.

Murraya Koenig ex Linn.-

M. paniculata (L) Jack (*M. exotica* Linn.) - A hand some shrub. Flowers white-fragrant. Fruit-berry, narrowed at both ends, red or deep orange when ripe. Flowers 3-5, *Fruit* 12 - 1. Sporadic.

Occurrence:- In foot hills.

OCHNACEAE

Ochna Linn.-

Q. pumila.- Buch-Ham ex Dc.- A small undershrub with perennial root stock dying down yearly. Flowers bright yellow. Fruit - drupes upto 1.3 cm. long, greenish when ripe. *Flowers* 3-5, *Fruit* - rainy season. Common.

Occurrence:- On grassay waste land.

MELIACEAE

Munronia Wight

M. pinnata (Wall) Harms) (*M. wallichii* Wt.) - An undershrub about 1 m. in height. Leaves crowded at the ends of the shoot, alternate, imparipinnate. *Flowers* bisexual; *Flowers*-2; *Fruit* capsule depressed. Rare.

Occurrance:- Foot hills.

OLACAGEAE

Olax Linn.

O. nana Wall ex Benth - Undershrub 60 cm. tall. Leaves subsessile 2-lobed. *Fruit* dark red. Sporadic.

Occurrance:- Foot hills.

CELASTRACEAE

Euonymus Linn.-

E. frigidus Wall. ex-Benth- A shrub, 1.5m. high, branchlets 4-angled. *Flower* greenish yellow, 5-6, *Fruit* capsule 4-lobed, 9-11.

Occurrence:- Miscellaneous forests, foot hills.

RHAMNACEAE

Zizyphus Mill

* *Z. mauritiana* Lam. (Z. jujuba (Linn) Gartn - Not very common in North Bengal plains. *Flower* in axillary sessile, cyme. Drupe - 2 cm. Gregarious in some drier areas.

Rhamnus Linn.-

R. nepalensis Wall. - A bushy shrub; *Flowers* very small green. *Fruit* dark-red to back in ripening.

Occurrence:- Temperate forests, rare and Singalila Range.

LEEACEAE

Leea Van royen ex Linn.

L. alata Edgew.- A stiff herbaceous shrub with pinnate leaves, sharply serrate. *Flowers* 7-8; *Fruit* 9-11. Red

Occurrence:- Sporadic in Sal forests.

* *L. guinunsis* D. Don (*L. acuminata* Wall ex clarke.) - A robust herbaceous shrub upto 4 m. high, long acuminate *Flower* red. *Fruit* orange red. Sporadic.

Occurrence:- Sal Forests.

* L. crispa *Van Royen* ex Linn.- A gregarious undershrub. 1.5 m. high, *Flower* 7-8; pale yellow. *Fruit*-1-12 deep grey when ripe.

Occurrence:- Sal forests, Mixed forests. Very common.

L. indica (Burn) Merr. (=*L. sambucina* Willd)- A semiherbaceous shrub upto 4 m. *Flowers* greenish white. -5, *Fruits* - 12. Very common.

Occurrence:- Sal and Miscellaneous forests.

SAPINDACEAE

Aphania Bl.-

A. rubra Radlk.- A shrub attaining 3m. *Flower* brick red. 11-1; *Fruit* Red. 3-4. Sporadic.

Occurrence:- Found in Tukriajhar and Foot hills forests.

FABACEAE

Piptanthus D. Don.-

P. nepalensis D. Don.- A shrub, 3-4.8 m. high. *Flower* yellow in compact short hairy racemes, 3-5, *Fruit* 9 - 11. Sporadic'

Occurrence:- Temperate hills 2,800 m.

Crotalaria Linn

C. cytisoides W & A.- Small about 2.5 m. high much branched shrub. Seeds tar-brown when mature. *Flowers* 4-8, *Fruit* 8 - 10. Sporadic.

Occurrence:- Foot hills. Sporadic.

* *C. tetragona* Andrews.- A shrub with angular stem 2-3m. high. *Flowers* handsome lemon yellow. 9-10; *Fruit* 11-12. Seeds glossy dark brown.

Occurrence:- Sal Forests. Common.

* *C. spectabilis* Roth, *C. sericea* Retz.- A shrub 1.5 - 2.5 m. high. *Flower* and *Fruit* 10-3. Common in Sal forests.

C. alata Buch Ham ex D. Don.- Leaves simple, base unequal, Raceme 2-3 flowered; Corolla pale yellow, Pod stalked. Common in Sal forests.

Indigofera Linn.-

* *I. dosua* Buch-Ham ex D. Don.- A densely grey weedy shrub. Leaflets 25-31. *Flowers* pink purple; 6-7: *Fruit* brown, hairy. 8-9. Sporadic.

Occurrence:- Wet Sal forests.

I. pulchella Roxb.- A shrub 3-4m. high *Flower* 3-4, red or rose-coloured. *Fruit* 8-10. Sporadic.

Occurrence:- Dry mixed forests.

I. hebepetala Benth.- A shrub of upper temperate hills from 1,800-2,700 m.

Tephrosia Perz.-

T. candida Dc. - A weak shrub with angled or grooved stem branches. *Flowers* creamy white, 8-10, *Fruit* -11-12.

Occurrence: - Sometimes found in wild forests near Tea Estate.

Desmodium Desv.-

D. sequax Wall.- A caespetose shrub of 3 m. in high. Calyx broadly funnel shaped. *Flowers* light purple. 9-10, *Fruit* 12-1.

Occurrence:- Sal forests. Sporadic

D. sambuense Dc. - A woody shrub, 1 to 1.5 m. high. Branches angular. Calyx turbinate, hairy, or nearly white. *Flower* 10-11, *Fruit* 11-2.

Occurrence:- Sal forests. Sporadic

* *D. gyroides* Dc.- A shrub 2-6 m. high with softly hairy branches. *Flower* deep purple in short axillary and terminal racemes 10-11, *Fruit* 11-2.

Occurrence :- In Sal and dry mixed forests. Common

Flemingia Roxb- ex Ait

* *F. strobilifera* (Linn) Ait.- An erect shrub 1.5 - 3 m. high. Leaves simple, secondary nerves 4-6 pairs. *Flowers* purplish calyx velvety and gland dotted outside.

Occurrence:- In dry type forests. Common.

F. macrophyela (Willd) Merr.)
(*F. congests* Roxb ex Ait.)- A shrub, 1.5 - 3 m. high. Flowers purplish. Clayx gland dotted. Corolla standard greenish- white, striped with pink. *Flowers* 4-8, *Fruit* 11-2.

Occurrence:- Sal Forests. Sporadic

Aeschynemene Linn.-

A. *indica* Linn. - An annual undershrub with many slender branches. *Flower* yellow.

Occurrence:- In swamps. Common.

A. aspera Linn.- A tall erect shrub with few branches.

Occurrence:- Common in swamps.

CAESALPINIACEAE

Cassia Linn.

*. *C. occidentails* Linn. - A diffuse undershrub of 2.5 m. high. *Flowers* yellow, 7-9; *Fruit* cold season.

Occurrence:- In waste places. Very Common.

* *C. sophera* Linn. - A small undershrub, 1-1.5m. high. *Flowers* cold season.

Occurrence:- In waste places. Very common.

MIMOSACEAE

Mimosa Linn.

M. tubicaulis Sub. sp. *hymalayana* (Gamble) Otashi (*M. himalayana* Gamble) - A large shrub. *Flowers* pink or white 7-9: *Fruit* 11-1.

Occurrence: In open forests. Very common.

Prinsepia Royale.-

P. utilis Royle.- A large shrub, about 3 m. in height. *Flowers* small white, 3-4. *Drupe* purple. Fruits 12. Sporadic.

Occurrence:- Temperate forests.

ROSACEAE

Rubus Linn.

R. sikkimensis Hk.f.. - Small shrub of upper temperate forests. Common.

Occurrence:- Temperate hills.

R. gracilis Roxb.- Low shrub with silky leaflet. Flower pink. *Fruit* yellow or orange. Common.

Occurrence:- Temperate hills.

R. rugosus Smilh (*R. moluccanus* Linn) - A shrub stem of about 1-6 m. high. branches prickly sermentose. Leaves deeply lobed, as broad as long, *Flower* 8-10; *Fruit* 9-10. Common.

Occurrence:- Temperate hills.

R. acuminatus Sm. - A climbing shrub. Leaves, ovate oblong, simple. *Flower* white. *Fruit* scarlet. Sporadic.

Occurrence:- Temperate hills.

Neillia D. Don.-

* *N. thrysiflora* Don. - A branchy shrub with drooping shoot of height 2m. *Flowers* usually in branched terminal or axillary racemes. Petals white, 5-6. *Fruit* 8-11. Common.

Occurrence:- In temperate hills.

Rosa Linn.-

R. damascena Mill.- Flowers scented, pink, sepals reflexed after flowering. Commonest garden rose.

R. sericea Lindl.- A shrub with white flower. Not common in Temperate hills.

Contoneaster Medik

C. nitidus Taeq (*C. distichia* Lge.) - A rigid shrub with almost horizontal branches. Berries bright purple. Sporadic.

Occurrence:- Temperate forests.

C. sandakphurensis Klotz, (*C. microphyllus* Wall Var. gracilis Cowan & Cowan).- Sporadic. Temperate hills.

C. congestus Baker (= *C. microphyllus* Wall). - Sporadic. Temperate hills.

Argimonia

A. pilosa, Leeb var. nepalensis (D. Don) Vakasi (*A.eupatorium* Linn.) - An undershrub with pinnately compound leaves; flowers yellow in racemes.

Occurrence:- Temperate forests. Sporadic

HYDRANGIACEAE

Hydrangea Linn.-

H. heteromala D. Don (=*H. vestita* Wall.) - A tree or large shrub. *Flower* greenish white. Stamens 10.

Occurrence:- Temperate forests. Sporadic

H. robusta Hk.f. & Th.- A large shrub. Leaves upto 22 cm. broad and long, serrate-dentate. *Flowers* blue, panicle corymbose.

Occurrence:-Temperate forests. Sporadic.

Dichroa Lour.-

* *D. febrifuga* Lour.- An erect shrub 1.5 to 2 m. high. Flowers blue in terminal panicles of cymes. *Flowers* 7-8; *Fruits* blue on maturity. 11-12.

Occurrence:- Temperate forests. Very Common.

Ribes Linn

R. glaciale Wall. - An erect shrub 1.5 to 2.5m. high. Often epiphytic. *Flowers* greenish 5-6; Raceme 3-7 cm. long. *Fruit* 7-8.

Occurrence:- Temperate forests. Sporadic.

R. takade D. Dom
(R. acuminatum Wall) - A shrub. *Flowers* 4-6, *Fruit* Berries nearly glabrous, glandular. 7-10. Sporadic.

Occurrence:- Temperate forests.

MELASTOMACEAE

Osbeckia Linn.-

* *O. stellata* Buch-Ham ex. D Don.- Shrub. Flowers red handsome, red 9, *Fruit* 11-1. Very common

Occurrence:- In marshy places and wastes.

O. nepalensis Hook. - A small branched shrub. about 1.6 m. in height. *Flowers* white or red mainly 4-11. *Fruit* campanulate. Not common.

Occurrence:- Temperate hills.

* *O. sikkimensis* criab (*O. orinita* Benth) - A small branched shrub, 1.5 high. *Flowers* purple. Calyx with stiff stellate hairs. *Fruit* suddenly narrowed to a neck. Very common.

Occurrence:- Foot hills and plains.

Melastoma Linn.-

* *M. melabathricum* Linn.- A beautiful shrub upto 4.5m. tall. *Flowers* purple 4 cm. across in cluster. *Fruit* truncate fleshy purple plancentae. Sporadic.

Occurrence : Temperate hills.

M. normale Don. - A shurb about 45 m. hairy, leaves 3-5 basal nerves. *Flower* rose-purple. *Flower* 2-7; *Fruits* 11-12. Sporadic.

Occurrence:- Temperate hills.

Oxyspora Dc.-

O. paniculata (D.Don) Dc.- A shrub attaining 2 m. in height. *Flowers* white calyx teeth triangular, acute, pedical purple. *Flower* and *Fruits* most of the year, mainly 7-3. Sporadic.

Occurrence:- Temperate hills.

LYTHRACEAE

Woodfordia Salisb.-

* *W. fruitcosa* (L) Kutz.- A large spreading shrub. *Flowers* scarlet, 1-3, *Fruits* 45 nearly leafless in 2-3. Very common.

Occurrence:- In foot hills and plains Sal forest.

CORNACEAE

Helwingia Willd.-

H. himalaica Hk.f. - A small evergreen shrubs unisexual flowers are found in umbels on the midrib of the leaves. (epiphyllous inflorescence) *Flowers* 4-5; *Fruits* 6-8, sessile, red when ripe.

Occurrence:- Temperate forests 2,200 m. Not common.

GAMOPETALAE
(CAPRIFOLLACEAE)

Viburnum Linn.

* *V. colebrookianum* Wall ex DC. - A large shrub. Leaves drooping serrate - *Flowers* white, 2-4 *Fruit* orbicular, red, 6. Common.

Occurrence:- Temperate hills 2,000 m.

* *V. erubescens* Wall ex Dc.- A small deciduous tree. *Flowers* white or pale pinkish. 3-5. *Fruit* 8-10. red.

Occurrence:- Temperate hills 1,800- 3,000 m. Common.

* *V. mullaha* Buch-Han ex D. Don (*V. stellulatum* Wall.).- *Fruit* red., pulky.

Occurrence:- Temperate forests. Very Common.

* *V. cylindrocum* Bach-Ham ex D. Don. (=*V. oriaceum* Blume) - Shrub, fruits green. Leaves simple opposite, with tufts of hairs of the nerves benath. Coryl terminal. Sporadic.

Occurrence:- Temperate forests.

RUBIACEAE

Mussaenda Linn.

M. macrophylla Wall. - A large erect shrub with greyish stem. *Flowers* orange coloured; 8-9. *Fruit* - cold season.

Occurrence:- Sal forests. Common.

M. incana Wall. - A small undershrub, clothed with dense hairs. *Flowers* golden or chrome yellow; 5. Berry with adpressed hair. Sporadic.

Occurence: Sal forests.

Mycetia Reinw.-

M. longifolia (Wall) O. Ktze. - An undershrub or small shrub. Flowers yellowish in compound tricho-tomous zigzagly branched subterminal cymes. *Flower* 10-11. *Fruit* Berries snow white cold season.

Occurrence:- Foot hills. Sporadic

Gardenia Linn.-

G. campanulata Roxb.- A shrub. Male flowers fascicled bisexual flowers sessile. Calyx of male flower. *Flowers* 3-4. *Fruit* 10-11. Sporadic.

Occcurrence:- Foot hills.

Randia Linn.-

* *R. tetrasperma* Benth O Hook f. ex, Brandis.- A small erect shrub. *Flowers* pentamerous, fragrant, greenish white. 4-6, *Fruit* 1-2.

Occurrence:- Foot hills. Common.

R. fascicultata (Roxb) D.- A handsome spreading shrub with axillary spines. *Flowers* white. Fruit a berry, purplish black when ripe. Common.

Occurrence:- Foot hills.

R. sikkimensis Hk.f. - A shrub. Leaves opposite, the pairs unequal. Corolla large, silky. *Flowers* 3. *Fruit* 8. Sporadic.

Occurrence:- Foot hills.

Canthium Lamk.-

C. discoccum (Gaertn) Teys & Binn, (*C. didymum* Roxb.) - A large shrub or low branched tree with spreading or drooping branchlets.

Flowers greenish or white. 2-4. *Fruit* 5, black globose. Sporadic.

Occurrence : - Foot hills. sporadic

Ixora Linn.-

I undulata Roxb.- A large evergreen shrub with oblong or lanceolate glossy leaves. *Flowers* white, 4-5. *Fruit* dull purple 8-9.

Occurrence:- Foot hills drier areas. Common.

I. coceinea Linn. - Ornamental shrub with scarlet or yellow flowers.

Occurrence:- Foot hills Sporadic.

Vangueria Juss (= Meyna)

V. spinosa Roxb.- A deciduous shrub. Flowers greenish white. *Flowers* 4-5. *Fruit* yellowish when ripe 7-8. Sporadic.

Occurrence:- Foot hill forests.

Pavetta Linn.-

P. indica Linn.- A glabrous shrub upto 1.5 m. high. Leaves lanceolate to obovate with tapering base. *Flower* white, fragrant. *Fruit* glossy, blackish-green. Sporadic.

Occurrence:- Foot hills, Sal forests.

Coffea Linn

* *C. bengnalenisis* Heyne ex R & S.- A deciduous shrub. *Flowers* white salver shaped, fragrant, 2-5. Drupes subglobose or ovoid, blackish, *Fruit* 10-12.

Occurrence:- In sal forests and Foot hills forests. Very common.

Morinda Linn.-

* *M. citrifolia* Linn.- A large shrub. Leaves elliptic lanceolate. *Flowers* white. Fruit green, succulent, Very common.

Occurrence:- In dry type Sal forests.

* *M. angustifolia* Roxb.- A Large shrub or small tree. *Flowers* fragrant. *Fruits* free, fruiting head elongating oblong, white.

Occurrence:- In dry type Sal forests. Common.

Psychotria Linn.

* *P. calocarpa* Kurz.- An evergreen shrub, reaching upto 2 m. *Flowers* white-yellowish. *Flowers* 5, *Fruit* yellow when ripe ultimately turning red. 10-12.

Occurrence:- Foot hills. Common.

P. erratica Hk.f. - An evergreen erect shrub. *Fruit* succulant reddish-yellow. *Flower* 4-6 small, yellowish white. Flower 4-6 *Fruit* 11-1.

Occurrence:- In foot hills. Common.

ASTERACEAE

Vernonia Schreb.-

V. subsessilis Dc.- A stout undershrub growing in shade. *Flowers* purple, fading white. 10-11, *Fruit* 12. Sporadic.

Occurrence:- All over.

Blumea Dc.-

**B. myiocephala* Dc.- A shrub. *Flowers* 10-11, *Fruit* hairy, ribbed, pappus red,. 2.

Occurrence:- All over. Common.

* *B. balsamifera* Dc.- A common shrub. Receptacle glabrous Pappus reddish. Very aromatic.

Occurrence:- All over. Common.

* *B. aromatica* Dc. An aromatic undershrub of 3 m. high with glandular hairs. Inflorescence a longer spreading panicle of upto 40 cm. in length. *Flowers* white. *Fruit* 10 ribbed. Pappus red.

Occurrence :- All over. Common

Inula Linn.

I. cappa (Buch - Harn ex D. Don) DC. - Aromatic shrub upto 2.5 m. high, branches brownish silky or woolly. *Flowers* 8-11. Fruit 12-2.

Occurrence:- Temperate hills (Open places). Sporadic.

I. eupatorioides Dc.- A wooly shrub of 2 m. high, *Flowers* 11. *Fruit* 2-3.

Occurrence:- Temperate forests in open. Sporadic

Artemisia Linn.-

* A. indica Willd (*A. vulgaris* Linn).- a tall aromatic shrub 2.5 m. high. *Flowers* 10. *Fruit* 3.

Occurrence:- Foot hills and temperate hills. Very common.

ERICACEAE

Gaultheria Linn.-

G. nummularioides D. Don.- A prostrate undershrub with hairy stem. *Flowers* pink or white. Corolla reddish or nearly white. Rare.

Occurrence;- Singalila Forests.

G. hookeri C.B. Clarke.- An erect shrub, about 2 m. high. Stem hairy, *flower* pink or white. *Flowers* 8. *Fruit* - 10. Rare.

Occurrence:- Singalila Forests.

Rhododendron Linn

R. dalhousiae Hk.f. - A shrub. Often growing on rock or as epiphyte. Flowers large white, tinged with rose on the outside. 5-6. *Fruit* 10-11.

Occurrence:- Darjeeling, Senchal 2,200 m. Common.

R. lepidotum Wall ex D. Don. Small aromatic shrub 30-60 cm. high. Common rocks at about 3,500 m. *Flower* yellow, 7-10. *Fruit* 10-11.

Occurrence:- Singalila Range. Sporadic

R. anthopogon D. Don.- Small gregarious aromatic shrub 60 m. high at about 3,500 m. *Flower* white, 6, *Fruit* 10-11.

Occurrence:- Singalila range. Sporadic

R. triflorum Hk.f..- Shrub, 1.5 - 2.5 m. high. Leaves small. *Flower* 5-7, *Fruit* 11.

Occurrence:- Singalila range. Sporadic

Pyrola Linn.

P. rotundifolia Linn.- A shrub. Flower white or rose coloured, in long scapigerous raceme, about 15 cm. long, corolla lobes 5, stamen 10. *Flower* and *Fruit* 4-6. Sporadic

Occurrence:- Upper temperate forests.

DIPENSIACEAE

Diapensia Linn

D. himalaica Hk.f.. - Densely tufted shrub. Leaves very small, flowers white or rose red. Sporadic.

Occurrence:- Upper temperate forests.

PLUMBAGINACEAE

Plumbago Linn

P. zeylanica. Linn.- An undershrub with glabrous small herbaceous branches; very glandular above, flowers white with pretty long tube and clustered in panicled spokes. *Flowers* - 4. *Fruit* - 12. Sporadic.

Occurrence:- In waste places.

MYRSINEAE

Maesa Forest.-

**M. montana* A. Dc. (M. *indica* Wall.).- A shrub 1.5 - 2.5 m. high. *Flowers*.- white, scented in axillary racemes; 12-3. Berries small, pinkish white, sacculent when ripe. *Fruit* 6-7.

Occurrence:- Foot hills upto 1,500 m. Very Common.

**M. chisia* D. Don.- A gregarious shrub of 4 m. high. Fruit white spongy, juicy. Reticulation of the secondary leaf nerves distinct. Very Common.

Occurrence:- In temperate hills.

Ardisia Swartz.-

A. Solanacea Roxb (*A. humilis* Wall).- A large glabrous shrub. *Flowers* waxy, pinkish, stamen yellow; 6-7 *Fruit* black when ripe. *Fruit* cold season.

Occurrence:- In Sal forests. Very Common.

A. involucrata Kurz.- A shrub. Leaves oblong or ovate, bluntly acuminate, gland dote conspicuous calyx-sements waxy, pinkish.

Occurrence:- Plain forests. Common.

* *A. macrocarpa* Wall.- A small low single steamed shrub of undergrowth, grows in shady place. *Flowers* pinkish very pretty shrub either with its wax like flowers in spring or brilliant red berries in winter; 7. *Fruit* bright red dotted, cold weather.

Occurrence:- Temperate forests 2,000 m. Very Common.

APOCYNACEAE

Rauwolfia Linn.-

R. serpentina (Linn) Benth ex Kurz. - Pretty galbrous shrub 30 cm. high. Flower white with pinkish tube. Drupes polished green and finally black.

Occurrence:- Sal forests (dry areas). Sporadic

Tebernaemontana Linn

**T. divericata* (L) R. Br ex R & S.- A very handsome evergreen bushy shrub. Flowers pure white. *Flowers* 2-5, *Fruit* 12.

Occurrence: In Sal forests (drlm type) and Foot hills. Very Common.

ASCLEPIADACEAE

Streptocaulon Wight & Arn.-

S. sylvestre Wight.- A prostrate, handsome, a pubescent shrub. Leaves cordate, tipped with a short cusp. Flower small. Calyx glandular within 5 - partite. Corolla 5 lobed. *Flowers* 5.

Occurrence:- Sal forests Sporadic.

Calotropis R. Br.-

* *C. gigantea* Bl., A shrub. Flowers lavender lilac or bluish purple. Flowers and fruit throughout the year.

Occurrence:- In waste places (drier localities). Common.

* *C. procera* Br.- A shrub. Flowers purplish red, faintly scented.
Occurrence:- In waste places (drier localities). Common.
C. *acia Hamilt*.- A bluish grey shrub with tomentose fleshy leaves. *Flowers* purplish, *Fruit* 12-5.

Occurrence:- In waste places. Not Common.

LOGANIACEAE

Buddleia Linn.-

B. asiatica Lour.- A pretty shrub 4 m. high, Flower in cymes, calyx teethed, triangular, white, sweet scented.

Occurrence: In riverain forests. Common.

B. paniculata:- Wall. - Large much branched shrub with crooked stem. *Flower* tomentose, 2-6.

Occurrence:- Common in foot hills dry forests.

B. macrostachya Benth.- A large shrub 1 m. to 3 m. high. *Flower* white or pink; 9-12. *Fruit* 4.

Occurrence:- Foot hills. Not Common.

SOLANACEAE

Solanum Linn.-

*S. *ereanthernum* D. Don. (S. *verbascifolium* Linn) - A tall erect unarmed shrub 2-6 m. tall *Flowers* white, *Fruit* almost throughout the year; berry yellow, stellate when young.

Occurrence:- Foot hills. Sporadic

S. ferox Linn.- A coarse undershurb, young parts densly brown. *Flowers & Fruit* December.

Occurrence:- Plains dry forests.

S. torvum Swartz.- A tomentose armed shrub 3 m. high. Flowers white with yellow centres. *Flowers* and *Fruit* 9-10.

Occurrence:- In plain in waste places. Common.

S. anguiri Lam. (*S. indicum* Linn.) - A hairy shrub of 3 m. height. *Flower* bluish purple with yellow centre, 6-10. *Fruit* in November

Occurrence :- In wastes.

S. viaram Dunal (*S. khasianum* var Charterjeanum Sengupta Clarke). A stout undershurb 1 m. high. Inflorescence lateral. *Flowers* red 7; *Fruit* 11.

Occurrence:- In Sal Forests. Common in Wet places.

Datura Linn.-

* *D. stramonium* Linn.- A coarse annual, stem green or purplish. *Flowers* and Fruits Rainy season. Common.

Occurrence:- In foot hills.

* *D. metel* Linn (D. fastuosa Linn.) - Stem erect , attains about 2 m. *Flowers* 6. Corolla usually purplish outside,

Occurrence:- Temperate hills 1,500 m. Common.

Cestrum Linn.-

C. aurantiacum Lindl.- }
C. fastuosum } These are cultivated but somewhat naturalised in Darjeeling hills. The flower are attractive yellow and red respectively. Common.

Occurrence:- Temperate hills.

ACANTHACEAE

Eranthemum Linn.-

* E. pulchellum And *E. nervosum* (vahl) R. Br. - Stout handsome shrub 2 m high, with 4 angled scabrous stem. *Flowers* bright blue in dense spikes arranged in long terminal panicles. Calyx whitish. Corolla deep blue.

Occurrence:- Sal forests. Very Common.

Phlogacanthus Ness.-

* *P. thyrsiflorus* (Roxb) Nees.- An evergreen shrub 3 m. high. Flower dark orange. *Flowers* 12-2, *Fruit* 2-4.

Occurrence:- Sal forests. Very Common.

Justicia Linn.-

J. gendarussa Linn.- An undershrub 1.5 m. high. Leaves lanceolate, bluntly acuminate. *Flowers* white purple spot inside 2-5.

Occurrence:- In moist stream bank. Common.

J. adhatoda Linn.-

(*A. vasica* Nees).- An evergreen shrub 1-2 m. high. Flowers white in dense, peduncled spike. *Flowers* 12-4, *Fruit* 2-5. Backs of flower conspicuous.

Occurrence:- In Wastes. Sporadic

VERBENACEAE

Lantana Linn.-

* *L. indica* Roxb.- An erect shrub 1-3m. high, leaves with bulbous based hairs above. Flowers very faintly scented, light purple with yellowish tube occur in close head; 4-5 fruit purple when ripe. *Fruit* 9-10.

Occurrence:- All over. Very Common.

* *L. aculeata* Linn. (= L. camara Linn).- A straggling shrub, very aromatic (black cuneate small). *Flowers* usually oranges, varying from white to dark purple, Drupe greenish blue on maturity. *Flowers* and *Fruit* nearly all the year round.

Occurrence:- All over. Very Common.

Lippia Linn.-

L. geminata. AB & Kunth.- A gregarious strongly aromatic shrub.

Flowers pink, scented. *Fruit* globose with sub-succulent green percarp, when ripe, Common in open marshy areas.

Occurrence:- Plains.

Callicarpa Linn.-

* *C. macrophylla* Vahl. - A stout spreading large shrub. Flowers rose pink coloured. Drupe white. *Flower & fruit* throughout the year.

Occurrence:- Foot hills. Common.

C. longifolia Lamk. var lanceolarium - A shrub upto 5m. high. *Flowers* rose or purple. Drupe white.

Occurrence:- Foot hills. Common.

Premna Linn.-

P. herbacea Roxb.- Small harbaceous undershrub. *Flowers* very small, white or greenish yellow in terminal corymbose panicles; 1-4. Drupe black glossy. *Fruit* 5-7.

Occurrence:- Burnt forest floor. Sporadic.

Vitex Linn.

V. negundo Linn.- A large, strongly scented shrub. 2-4 m. high. *Flowers* lavender to blue. *Drupe* glotose, slightly ribbed, gland-dotted. Common in valleys.

Clerodendrum Linn.-

* *C. viscosum* Gaertn.- A shrub or undershrub with foetid smell. *Flowers* white, 2-5 tinged bluish black, *Fruit* 5-7.

Occurrence:- Everywhere. Very Common.

C colebrookianum Nalp.- A shrub of 4m. high. *Flower* white. *Fruit* turquoise blue when fully ripe.

Occurrence:- In Sal forests. Common.

C. bracteatum Wall.- A large shrub. Leaves ovate. *Flowers* white, scented, Drupe globose, bluish purple, situated on enlarged red-dish sub-coriaceons calyx.

Occurrence:- In Sal forests. Common.

* *C. serratum* Spreng.- A irregularly branched perrenial shrubs. Flowers blue or sometimes pale pink or white. Drupe black when ripe. *Flowers* and *Fruit* 10-5.

Occurrence:- In Sal forests. Common.

C. Japonicum (Thunb) SW. *C. squamatum* Vahl.)- A very handsome shrub upto 2 m. high. *Flowers* scarlet; 6-10. Drupe bluish black. *Flowers* 6-10, *Fruit* cold season.

Occurrence:- In Sal forests. Common.

C. nutans Wall.- A shrub about 4 m. high. *Drupe* deep purple.

Occurrence:- In Sal forests. Common.

C. indrcum (Suin) o Kunfor.- A tall straight stemmed shrub. *Flowers* white or red; 10-5 Drupe blue. *Fruit* 12-8.

Occurrence:- Sal forests. Very Common.

Caryopteris Bunge.-

C. paniculata Clarke.- A spreading shrub with slender and hairy branches. *Flowers* and *Fruit* 12-4.

Occurrence:- In Sal forests. Common.

* *C. odorata* (D.Don) Robin (*C. wallichiana* Schau).- A large erect or spreading shrub *Flowers* fragrant, bluish, lobes, 2-4 Capsule dark blue when ripe. *Fruit* 5-7.

Occurrence:- In Sal forests. Common.

LABIATAE

Plectranthus L. Herit.-

P. striatus Benth.- A pubescent or hairy perennial shrub. *Flowers* 10. Fruit in cold season. Sporadic.

Occurrence:- Temperate hills.

* *P striatus* Buch-Ham ex. D. Don.- A tall erect much branched, aromatic perennial shrubs. Flowers lavender blue in paniculate cymes. Nutlets broadly ellipsoid, yellow. *Fruit* in July.

Occurrence:- Temperate hills. Common.

P. ternifolius Don.- An undershrubs of 1 to 1.5 m. high densly tomentose. *Flowers* white, Crolla, spotted pink in the upper lip.

Occurrence:- Temperate hills, Sporadic.

Pgostemon Desf.-

*P. bengalenisis (Burn & Kuntz *P. plectranthoides* Desf.) - A branchy. shrub.

Occurrence:- Common in Temperate hills.

Colebrookia Smith.-

* *C. oppositifolia* Sm.- An erect tomentose much branched shrub. Calyx with white hairs. Corolla white. Common.

Occurrence:- Common in Temperate hills.

Elsholtiza Willd.-

* *E. fruticosa* (Don) Rehder *(E. polystachya* Benth.)- A pubcesent shrub of 20 cm. high. Flowers minute. Corolla white. *Flowers* 9-10, *Fruit* cold season. Common.

Occurrence:- In temperate hills.

Perilla Linn.-

* *P. frutescens* (L) Britton (*P. ocimodies* Linn.)- A strongly aromatic bush. Flowers small, white. *Flowers* and *Fruit* cold season. Very common.

Occurrence: Foot hills and waste places.

Colquhounia Wall.-

C. coccinea Wall.- A rambling shrub 3 m. high. Flowers orange or reddish. *Fruit* 8-2. Sporadic.

Occurrence:- Temperate hills.

Gomphostemma (Benth) Wall.-

G. parviflorum Wall.- A gregarious stout undershrub 2 m. high *Flowers* 8-9, Cyme fascicled.; *Fruit* 10-12.

Occurrence:- Temperate hills. Common.

AMARANTACEAE

Achyranthes Linn.-

* *A. aspera* Linn.- An erect undershrub, attaining 1 m. Flowers greenish, shiny, stiff, bracts aristate membranous. *Flowers* 6-10, *Fruit* 12,

Occurrence:- In waste places. Very Common.

POLYGONACEAE

Polygonum Linn.

* *P. chinense* Linn.- A shrub of 1.5 m. height. Flowers white, 9-12.

Occurrence:- Temperate hills. Very Common.

PIPERACEAE

Piper Linn.-

P. nepalense Miq. - Small shrub, flowers 8, fruit 10.

Occurrence:- Common in Sal forests.

THYMELEACEAE

Daphne Linn.-

* D. cannabina Wall.- A bushy shrub, 2-3 m. high. *Flowers* white, pale yellow, or purple outside, scented, 8-12. *Fruit* 8-12.

Occurrence:- Temperate hills. Common.

D. involulucrata Wall.- A shrub or small tree. *Flowers* scented, white. Drupe black when ripe.

Occurrence:- Temperate hills. Not Common.

D. sureil Sm. & Cave.- Erect shrub. Branches straight erect. *Flower* 10-1, *Fruit* 1-2.

Occurrence:- Temperate hills. Common.

Edgeworthia Meisn.-

* *E. gardneri* (Wall) Meissn.- Large shrub. branches long, bling; Flower in dense heirs, spherical heads, yellow. 12-6, *Fruit* 6-9. Common.

Occurrence:- In temperate hills. Common.

LORANTHACEAE

Viscum Linn.-

V. monoicum Roxb.- Large parasitic shrub. *Flowers* greenish in 3-flowered fascicled. Berry yellowish. Flowers and fruit cold season.

Occurrence:- Temperate hills. Sporadic

V. articulatim Burn.- A pale green sub-verticellate parasitic shrub. Berry yellow, smooth. Flowers & fruit cold season.

Occurrence:- Temperate hills. Sporadic

V. album Linn.- Parasitic shrub. Berries white and transparent. *Flowers* and *Fruit* 12.

Occurrence:- Temperate hills. Sporadic

SANTALACEAE

Dufrenonya Ch.

D. granulata Hk.f. (Th) Stauffer. (*H. granulata* Hk.f.)- A small parasitic shrub. *Flowers* 7, *Fruit* cold season 11. Small, obovoid.

Occurrence:- Foot hills. Rare.

D. platyphylla (Spr) Stauffer. *(H. heterantha* Hk.f. & Th) A small parasitic shrub. Fruit small, red ellipsoid. *Flowers* 10-1, *Fruit* hot season - 6.

Occurrence:- Foot hills. Rare.

EUPHORBIACEAE

Phyllanthus Linn.-

P. leschenaultii. Muell.- A small shrub. Flowers minute. Male and female often the same axils. Male perianths whitish, elliptic oblong, disc of 6-orbicular glands. Flowers & *Fruit* 6-8.

Occurrence:- Foot hills. Sporadic.

P. clarkei Hk.f. - A rigid shrub with groved branches. Flowers subsolitary, dise of male glandular, disc of female obsucrely lobed. Capsule small globose, reddish. 6-lobed. *Flower & Fruit* 10-1. Sporadic.

Occurrence : - Foot hills forests.

Bridelia Willd.-

* *B. montana* Willd. - A shrub with glabrous, brown twigs. Flower not spicate.

Occurrence:-Sal forests. Common.

* *B. stipularis* (Linn) Bl.- A scandant shrub. *Flowers* greenish yellow. *Fruit* bluish-black and juicy when ripe.

Occurrence:- In all forests. Common.

B. montana var communis - Leaves variables, broadly ovate. *Flowers* monecious, greenish yellow in dense axillary cluster. *Fruit* oblong or fusiform.

Occurrence:- Sal forests. Sporadic.

Sauropus Bl.-

S. androgynus Merrill.- Small shrub with an erect stem. Male flowers reddish green, 9, *Fruit* 11-1.

Occurrence:- In riverain forests. Common.

S. stipitatus Hk.f. - A glabrous shrub. Sporadic.

Occurrence:- Riverain forests sporadic.

S. pubescens Hk.f. - A Small spreading undershrub. 2-ridged branches.

Occurrence:- Riverain forests. Sporadic.

Glochidion Forst.-

G. hirsutum Muell Arg. - A shrub or small tree. Male and female flower in the same cluster. *Flowers* 11-12, *Fruits* 2.

Occurrence:- Sal forests. Sporadic.

* *G. multiloculare* Muell.- A shrub or a small tree. Seeds with fleshy orange red coat.

Occurrence:- Road side wastes. Very Common.

G. lanceolarium (Royb) Voight Muell. - A shrub or a small tree. Leaves coriaceous. *Flowers* pale green. Ovary 6-8 celled, 6-8 lobed.

Occurrence:- In waste places. Common.

Croton Linn.-

* *C. bonplandianum* (*C. sparsiflorus* Morung.).- A small shrub upto 1 m. high. Male flower pedicelled in clusters of 3-4. Female flowers solitary with large gland; petals 0, disc of red glands. *Flowers* 5-7, *Fruits* 7-9.

Occurrence:- In wastes in plains. Very Common.

Bcaliopserumum Bl.-

B. montanum (Willo) Muell.- A branched, stout undershrub. Leaves lanceolate to broadly ovate. *Flowers and Fruits* throughout the year.

Occurrence:- Kalimpong temperate hills. Sporadic.

B. axillare Bl. - A leafy undershrub; Sepals 4-5 forming a globose calyx, petals o. Disc of 4-6 glands. Stamens 10-30. Fruit of tree 2-valved.

Occurrence:- Kalimpong temperate forests. Sporadic.

Homonioa Lour.-

* H. *riparia* Lour.- A large shrub of stony stream bed with numerous erect stem with prominent leaf scars. *Flowers* dio-

ecious, sessile in axillary brancteate spikes, 3-5. *Fruit* rainy season.

Occurrence:- Dry stream bed. Common.

BUXACEAE

Sarcococca Lindl.-

S. hookeriana Baill. (*S. pruniformis* Lindl.).- A small evergreen shrub (3m.) Flower pale yellow, sepals 4 biseriate. *Flowers* 4-5 *Fruit* 6-7, purple when ripe.

Occurrence:- Temperate hills. Sporadic.

MORACEAE

Ficus Linn.-

F. Squamosa Roxb (*F. saemocarpa* Miq)- A low spreading gregarious shrub on rocks. Receptacle 8-10 ribbed, brownish when ripe. *Fruit* 3.

URTICACEAE

Dendrocnide Mnq. (*Laporatea* Gaud.)

* D. schuata (Bl) Chew (*L. crenulata* Gaud).- A herbaceous stemmed shrub. Leaves and young shoots have poisonous, stinging hairs, utricle inflated white. *Flowers* 8-10, fruit fold season.

Occurrence:- Sal forests, moist situation. Common.

Pilea Lindl.-

**P. glaberrima* (Bl) Bl (*P. smilacifolia* Clarke)-A stout undershrub. Leaves 3-nerved stipules interpetiolar. Male flowers minute, sepal 2-4 female flowers tepals 3 rarely 4. *Flowers* 4, *Fruit* rainy season.

Occurence :- Foot hills. Common.

P. hookeriana Weed.-Spreading shrub upto 3-6 m. high. Flowers-9. Common.

Occurrence:- Upper temperate 2,000 m.

P. cordifolia Hk.f.-Large woody undershrub.

Occurrence:- Upper temperate 1,7000-2,600 m. Common.

Elatostema Forst.-

**E. sessile* Forst.-Shrub, usually pubescent. 2-3, m. high Leaves oblanceolate. Receptacle solitary or 2-3, together. *Flowers* -4.

Occurrence :-Temperate hills, 2,000 m. Common.

**E. rupestre* Wedd.-Stout, rigid shrub 2-3 m. high. Leaves elliptic lanceolate, serrate crystoliths crowded. *Flower* -3.

Occurrence:-Temperate hills. Common.

E. Integrifolium (Don) Wedd, *E. sesqulfolium* (Bl Hassk)- A branched undershrub. Leaves oblong-elliptic coudate acuminate. Flowers 3-4. Fruit 6-7.

Occurrence :- Tista valley. Sporadic.

E. lineolatum Wight.- An undershrub or herbaceous perennial. Leaves oblong lanceolate. Receptacle sessile. *Flowers* 5, *Fruit* 7.

Occurrence:- Temperate hills. Sporadic.

Pouzolzia Gauz.-

* *P. sanguinea* (Bl) Merr (*P. viminea* Wedd.) - A large shrub. Leaves alternate, toothed or serrate, base 3 - nerved, which are branchy. Stamen 4. Achene ovoid, angled.

Occurrence :- Foot hills. Common.

Maoutia Wedd.-

M. puya (Hook) Wedd.- A shrub. Flowers minute, in small cymose globose heads; 6-7, *Fruits* 10-11.

Occurrence :- Foot hills. Sporadic

SALICACEAE

Salix Linn -

S. daltoniana Ander.- A small shrub, not common *Flowers* 6, *Fruit* 1-2.

Occurrence :- Upper temperate forests.

(iii) Analytical Synopsis of Herb Flora

In this section (Analytical synopsis) listing of conspicuous and other herbs occurring within quadrats and transects and adjoining areas have been made. A very brief description of each species has been given, but flowering and fruiting time have been indicated by numericals viz., 1, 2, 3 meaning the months January, February, March and so on. Fidelity of each species has been enumerated. Asterisks (*) indicate conspicuous and widely occurring species.

RANUNCULACEAE

Anemone Linn.-

A. revularis Buch-Ham ex Dc.- Leaves lobed. Flower in cymose umbel. Common in shaddy or dampish place between 1,300 m. and 1,500 m. in temperate hills. *Flower* bluish.

Several species have been recorded from Darjeeling of which *A. vitifolia* Buch-Ham ex DC., and *A. obtusiloba* Don Prodr., are common with white, blue and purple flowers in Alpine grass land (Singalila Range).

Thalictrum Linn.-

**T. foliolosum* DC.-At 2500 m. in Darjeeling Himalayas. *Flower* whitish blue.

T. javanicum Bl.-At 1,600 m-1650 m in Darjeeling Himalayas. (Singalia Range).

Ranunculus Linn.-

**R. scleratus* Linn.-Occurs in swampy areas in plainri has wide distribution. *Flower* white or yellow.

R. diffusus DC.-At 1,650-1950 m. in Darjeeling Himalayas (Singalila Range)

Aconitum Linn.

Several species are distinct in Alpine Sikkim/Darjeeling Himalayas of which *A. ferox* Wall (Flower blue), *A. turidum* Hf.K. (Flower dull red), *A. palmatum* Don Proder (Flower Blue) are commonly seen.

Associated species that sporadically occur are—*Caltha gouaniana* Wall.-*Paeonia emodi* Wall, *Cimicifurga foetida* Linn., and *Actea spicata* Linn.

NYMPHEACEAE

Nymphaea Linn.-

**N. nouchali* Burm.f. Everywhere in ponds and ditches. Two forms, one with glabrous, another with pubescent leaves.

**N. rubra* Roxb.-Less common than the white flowered plant though equally widespread.

**N. stellata* Wild.-Everywhere in ponds.

Nelumbo Wild.-

N. nucifera Gaertn.-Everywhere in ponds. A large water plant, leaves peltate. There are two forms, one with pink flowers common, the other with white flowers, less usual.

PAPAVERACEAE

Argemone Linn.-

**A. mexicana* Linn.-Everywhere in waste places. A annual with yellow juice, yellow flowers and crinckly leaves with white veins.

Meconopsis Viguier

M. napaulensis Dc.-Prodr. -Rare, Darjeeling, Phalut 3500 m.

M. villosa (Hook. f.) Taylor- Rare, Darjeeling, Phalut 3500 m.

FUMARIACEAE

Dicentra Bermh.

D. roylei Hk. f. & Th.-Darjeeling Himalayas, in foot hills, temperate hill. *Flower* yellow.

D. macrocarpa Parin (*D. thalictrifolia* Hk. f. & Th.)- Kalapani, 1,700 m. Darjeeling Himalayas.

CRUCIFERAE

Roripa Seop (*Nasturtium* R. Br)

**R. islandica* schinz Q Thell (*N. palustre* Dc.)-Recorded from Kurseong. A small herbs. *Flower* yellow. Sporadic.

R. indica (L) Heirn (*N. montanum* Wall)-Darjeeling and Sikkim Himalayas. Flowers yellow. Sporadic.

R. dubia (Pers.) Hara. [*N. dubia* (Pers) Hara] -Everywhere common. A small herb.

Cardamine Linn.-

C. sulcata Thunb (*C. hirsuta* L. var. sylvatica)-*L. pinnate* 8-20 cm. *Flower* white on wall. Foot hills of Himalayas. Common.

CAPPARIDACEAE

Cleome Linn.-

C. viscosa Linn.-In fields and waste places, everywhere. A viscid herb. *Flower* yellow.

VIOLACEAE

Viola Linn.-

**V. begonifolia* J.E. Smith.- *Flower* lilac. Sal forests in Bagdogra and also in hills. Very common.

V. serpens Wall.-*Flower* bluish white. Common in 1500 to 2000 m.

POLYGALACEAE

Salomonia Lour.-

S. cantoniensis Lour.-A diffuse weed of Tista valley. Sparadic.

Polygala Linn.-

P. linerifolia Willd (*P. glomerata* Lour.)-At 2000 m. Darjeeling hills. Flower greenish. Rare.

P. sibirica Linn.- At 2500 m. Darjeeling hills. Flower purple. Rare.

CARYOPHYLLACEAE

Stellaria Linn.-

*Several species of Stelieria have been collected in Darjeeling of which **s. media* Linn., **S. sikkimensis* Hook. F. & Edw. S. monosperma Buch-Ham ex Don are commonest. *Flower* mostly white. Extends from plains to temperate hills very common.

Drymaria Linn.-

D. diandra Bl.-Creeping glandular herb, pubescent. Common in Darjeeling and Kurseong at 1400 m.

Polycarpon Linn.-

P. prostratum (Forsk) Asch (*P. loeflingiae* Benth & Hook)-Edges of ponds. Very common in rice fields petals white, leaves pseudo whorled. Stipules scarious prostrate. Common.

PORTULACACEAE

Portulaca Linn.-

P. oleracea Linn.-Everywhere common in waste ground. An annual prostrate succulent herb, common; flower yellow. Petals 5.

P. quadrifida Linn.-A small diffuse prostrate annual. Everywhere very common by road sides and in wast places. Flower yellow. Petals 4.

MALVACEAE

Malva Linn.-

**M. verticillata* Linn.-An erect annual. *Flower* pink in cluster. Cultivated.

Sida Linn.-

**S. acuta* Burm (*S. carpinifolia*)-Leaves lanceolate serrate. *Flower* yellow, Sal forests, very Common.

S. multicaulis Cav-A trailing herb. Common. *Flower* yellow. In plains of North Bengal.

**S. cordifolia* Linn.- Sal forests. *Flower* yellow. Very common.

Abelmoschus Medik (*Hibiscus* Medik)

A. pungens Roxb (*H. pungens* Roxb.)-Fairly common in grass lands especially in the riverain forests. Flower yellow.

STERCULIACEAE

Malochia Linn.-

M. corchorifolia Linn.- Herb, stellata hairy. Common in waste places. *Flower* white/pink with dense head.

TILIACEAE

Triumfetta Linn.-

T. annua Linn.-An annual herb. *Flower* orange coloured. Rare.

T. bartramia Linn.-(=*T. rhomboidea* Jacq.)-Very common along road sides and abandoned village in the plains. *Flower* yellow.

LINACEAE

Anisadenia Wall ex Meissn

A. Saxatilis Wall.-A diffuse herb. *Flower* pink. At 1400-200 m in foot hills of Himalayas.

GERANIACEAE

Geranium Linn.-

Several species are common in Darjeeling Himalayas. *Flowers* are mostly pink. *Geranium nepalense* SW is most common in temperate zone.

Oxalis Linn.-

**O. corniculatea* Linn.-*Flowers* yellow. Common everywhere.

**O. acetosella* Linn.-*Flower* white or rosy; in the hills.

Biophytum Dc.-

B. sensitivum Dc.-A small herb with a spreading crown of sensitive leaves. *Flower* yellow. Very common eveywhere on road sides and in cultivated ground.

B. reinwardtii Walp. - A graceful herb. Rare

Impatiens Linn.-

**I. balsamina* Linn.- A species of the plains; others are in the temperate hills. Impatiens have about 25 species common in Darjeeling hills. The genus is very wide spread.

SAPINDACEAE

Cardiospermum Linn.-

C. halicacabum Linn.-A climbing herb with wiry stems. Sporadic in occurrence.

FABACEAE

Vicia Linn.

V. sativa Linn-A small herb. *Flower* red/blue. Jalpaiguri plains. Sometimes cultivated.

V. angustifolia Linn-A small prostrate or climbing herb. Generally throughout the area.

V. hirsuta Koch.-A samll prostrate herb. Darjeeling hills 200 m. It is also a field weed.

Crotalaria Linn

**C. pallida* Ait (*C. striata* Dc.)-Common in waste places. Jalpaiguri plains and Kurseong.

C. spectabilis Roth. (*C. sericea* Retz.) *Flower* yellow. Jalpaiguri plains.

C. tetragona Roxb.-Sal forest of Jalpaiguri.

C. alata Buch-Ham ex Benth-1300 m. Kurseong, road side waste.

**C. albida* Heyne.-very common. *Flower* yellow. Kurseong.

**C. sessiliflora* Linn.-Along water courses or margings of swaps in the Kurseong hills.

Atylosia W. & A.

A. scarabaeoides (Linn) Benth.- Climbing herb. Stem tomentose; *Flower* yellow. Kurseong, Bagdogra . Common.

Melilotus Linn. Mill

**M. indica* Linn.-In Sevoke forests. *Flower* yellow. Common.

Uraria Desv.

U. lagopodioides Mer.-

U. rufescens (Dc) Sehind - Waste places Jalpaiguri. Sporadic. (*U.hamosa* Wall) -

U. laqopus Dc. Var. neglecta (Prain) Onasi (*U. neglecta* Prain)-A short, erect herb.-North Bengal, Duars.

Desmodium Desv.

D. renifolium (Linn) Sehind. (*D. reniforme* Dc.)-A diffuse trailing herb. Leaves reniform. North Bengal.

D. gangenmticum Dc.-Undershrub. Leaves 1-foliolate, carolla purplish-white. In Sal forests; very common.

Alysicarpous Desv.-

A. vaginalis (Linn) Dc.-A suberect weed.-Jalpaiguri plains.

Psorale Linn.-

P. corylifolia Linn.- An ecrect annual 20-60 cm. tall. Leaflet repand or toohed gland dotted. Rare.

Parochetus Buch -Ham ex. D. Don

P. communis Buch-Ham ex D. Don.- Trailing on road sides and forest paths above 1400 m. in the hills; conspicuous for its beautiful voilet flowers. Widely occurring.

Campylotropis Bunge (Lespideza Michx)

C. macrostylla (Don) Lindl (*L. eriocara* Dc. Prodr.) A common herbs of Sevoke forests.

Astragalus Linn.

A. donianus Dc. (*A. pycnorhizus* Wall.)
A sikkimensis Benth.
A. Coneretus Benth (*A. viphocarpus* Benth.)
} From 3500 m. at Phalut. Common.

ROSACEAE

Potentilla Linn.-

**P. fruticosa* Linn.-*P. fruticosa* var. rigida (Wall ex Lehm) Wolf

P. cuneata wall ex Lehm;

P. griffithii Hook.-All are comon in temperate hills of Darjeeling.

Duchesnea Smith-(*Fragaria* Linn)-

D. indica (Andr) Forke (=*F. indica* Andrews) and *D. daltoniana* J. Gas. are common in Darjeeling Hills.

SAXIFERAGACEAE

Astilbe Buch-Ham ex D. Don.

A. rivularis Buch Ham ex D. Don-An erect herb. *Flowers* yellow. Darjeeling Hills (1600-2000 m) A common herb.

HALORAGEAE

Myriophyllum Linn.-

M. tuberculatum Roxb.-Submerged water plant. *Flowers* pink. Common.

M. indicum Wild.-Submerged water plant. *Flower*-white.

MELASTOMACEAE

Sarcopyramis.-

S. nepalensis Wall (=*S. lanceolata* Wall)-Common in Senchal, Rangiroom in temperate hills. *Flower* pink. Common.

LYTHRACEAE

Rotala Linn.-

**R. rotundifolia* (Buch-Ham ex Roxb) Kochne—A weed or rice field and road side ditch. Capsule 4-valved. Very common.

R. indica (Wild) Kochne.-All over in rice field and ditches. Common.

Ammania Linn.

A. baccifera Linn.-Very common weed in marshy areas.

ONAGRACEAE

Epilobium Linn.-

E. amurense Haussk

E. teragonum Linn.-*Flower* purple. Senchal hills (1400-2000) m. Common.

E. angustifolium Linn.-*E. royleanum* Haussk E. *sikkimense* Haussk. sub-species *Ludlowianum* Ravel.-All these species occur in Senchal hills.

Ludwigia Linn. (*Jussiaea* Linn.-)

**L. adscendens* (L) Hora (*J. repens* Linn.)-A herb, creeping in mud on margins of tanks and floating on the surface of the water. In plains.

**L. octavalvis* (Jacq) Raven (*J. suffruticosa* Linn).-A herb of undershrub, in moist places. Flowers yellow. Common.

L. perennis L. (*L. prostrata* Roxb.)-A prostrate or decumbent herb, in rice fields. Common.

TRAPACEAE

Trapa Linn

T. bispinosa Roxb.-Leaves floating denticulate, incised, beneath red densely pubescent. Petole inflated. Found all over India in ponds, marshs.

BEGONIACEAE

Begonia Linn.

**B. roxburghii* (Miq) A. Dc.-A herb of Darjeeling (1300 m.) *Flowers* White. Common.

B. inflata Cl.-A herb of Darjeeling (1300-1950) *Flowers* white. Common.

B. josephi A. Dc.-A herb of Darjeeling (900-1400 m.) *Flower* rosy. Common. Many other species are common in the hills.

AIZOACEAE

Trianthema Linn.

T. portulacastrum Linn.-A diffuse, glabrous herb. *Flowers* solitary, sessile white or pinkish. In plains.

Glynus Linn. (*Mollingo* Linn.)

G. tatoides Linn. (*M. stricta* Linn.)-A sub-erect herb. All over.

G. oppisilaefolius (L.) Dc. (*M. spergula* Linn.)-A diffuse leafy herb on walls; waste places on sandy soil. *Flowers* white. All over.

UMBELLIFERAE

Hydrocotyle Linn.-

**H. nepalensis* Hook. (*H. javanica* Thumb).-A very common herb of temperate hills.

Sanicula Linn.-

S. elata Buch Ham ex D. Don. Ham. (*S. europaea* Linn.)-A herb of Kurseong (900-1800 m.)

Oenanthe Linn.-

O. bengnalensis Benth.-In moist place. Darjeeling hills. Common.

Centella Linn.

C. asiatica (L) Urb (*Hydrocotyle asiatica* Linn)-A prostrate herb; Leaf reniform. Very common.

Heracluem Linn.-

H. nepalense D. Don.-Above 3500 m. Common in waste places.

Bupleurum Linn.-

**B. candollii* Wall ex Dc.-At 3600 m. Singalila Range. Common.

Acronema Edgew *(=Pimpinella)*

A. sikkimensis C.B. Clarke and *A. wallichii* C.B. Clarke,—Occurs in upper temperate hills of Darjeeling Himalayas. Common.

ARALIACEAE

Aralia Linn.-

A. pseudoginseng Benth.- Darjeeling (1800-2100 m.) usually in shaddy situations, it is now rare.

RUBIACEAE

Hedyotis Linn.-

H. thomsoni Hook.-A small annual herbs on muddy river banks in Jalpaiguri.

H. verticillata (L) Lam. (*H. hispida* Retz.)-A diffuse herb, in plains. Very common.

H. vestita Don.-A stout villous herb. *Flower* pink-1000 m. Tista valley.

H. auriculata Linn.-Diffuse herb in wet Sal forests.

Oldenlandia Linn.-

O. affinis (R&S) DC. (*O. dichotoma* Koen.)-In North Bengal. A very slender extensively branching herb of dry places.

**O. corymbosa* Linn.-In rice field and in Tista valley. Common.

O. diffusa Roxb.-Tista valley. Sporadic.

O. biflora Linn. (*O. paniculata* Linn.)-Plains of Jalpaiguri. Very common.

Knoxia Linn.-

K. corymbosa Wild.-An erect slender annual. Plains (upto 1524 m.) *Flower* white purplish.

K. gracilis (Wall) Dc. (*Oldenlandia gracilis* Dc)-An erect thickly softly stemmed herb. In plains not common.

Dentella Forst.-

**D. repans* (Linn) Forst.-Riverbed; *Flowers* white. Dicholomurshly branched, leaves whorled. Very common.

Orphiorrhiza Linn.-

O. prostrata D. Don-*Flowers* white. Sporadic.

O. mungos Linn., *O. hirta* Hook. f.-Tista valley sporadic.

O. rosea Hook.- Middle hills. Sporadic.

Borreria.-

B. articularis (Linn) Wall (*B. hispida* Schum).-A weed, grows very thick in open grassland or plantation. Very common, all over Terai and Duars.

VALERIANACEAE

Valeriana Linn.-

V. hardwickii Wall.- Upper temperate hills (1600/1950 m.) *Flowers* reddish. Common.

V. jatamansii Jones (*V. wallichii* Dc.)- Upper temperate hills (1600-1950 m). Leaves deeply cordate. *Flowers* reddish. Common.

Nardostachys Dc.-

(*N. grandiflora* Dc. Prodr.-A herb of 10 cm. height perennial of economic importance. *Flowers* pink of blue in head of 1, 3 or 5. Upper temperate hills. Rare.

DIPSACEAE

Dipsacus Linn.-

D. asper Wall.-1950 m. in Darjeeling. *Flowers* white. Sporadic.

D. inermis Wall.-*Flowers* white in hemispherical head 6-8. Temperate hills. Sporadic.

Morina Linn.-

M. polyphylla Wall. ex DC.-A robust herb. Stem 6 m. Leaves doubly spinous-toothed. At 3500 m. Common.

M. nepalensis Don. Prodr.-Temperate hills. At 3500 m. Sporadic.

ASTERACEAE

Elephantopus Linn.-

E. scaber Linn.-A rigid, dichotomously branched, scabrid herb; in plains and hills. Sporadic.

Adenostenema Forst.-

A lovenia (L) O. Kuntze (*A. visocosum* Forest.)-An erect, rather slender annual. In grassy blanks at 1,800 m.

Ageratum Linn.-

**A. conyzoides* Linn.-Anuual herb. *Flower* blue. All over plains & foot hills. A very common and wide spread herb.

Mikania Wild.-

**M. cordata* Wild.-A climbing herb with long-petiole and opposite leaves. Very common.

Conyza Less.-

C. leucantha (D. Don) Ludlow & Roven (*C. viscidula* Wall).-At Bagdogra. Sporadic.

Blumea Dc.-

**B. aromatica* Dc. Prdor.- An aromatic herb of foot hills. *Flowers* yellow, pappus red. Very common.

B. fistulosa (Roxb) Kurz (*B. glomerata* Dc.)-An erect, rather slender much branched herb. Common in Darjeeling.

**B. laciniata* Dc.-A tall erect; *Flowers* yellow. Very common in Darjeeling.

B. densiflora Dc.-A herb. Common.

Xanthium Linn.-

**X. indicum* Voenig (*X. strumarium* Linn.)-A coarse, unarmed annual, with bur-like heads. Very common in road side burrow pits.

Siegesbickia Linn.-

S. orientalia Linn.-A glandular pubescent herb, *Flowers* yellow, Sporadic. Darjeeling.

Enhydra Lour

**E fluctuans* Lour.-A glabrous amphibious marsh herb. Very common.

Wedelia Jacq.-

W. calendulacea Less.- Sporadic. Tista valley.

W. wallichii Less.-Sporadic. Tista valley.

Bidens Linn.-

**B. bitereata* (Lour) Merr. & Sherjf (*B. pilosa* Linn).-An erect herb with adhering bared achenes. Common. At 1700 m. in temperate hills.

Tridax Linn.-

**T. procumbens* Linn.-A weak, straggling perennial herb. Very common all over plains and foot hills.

Myriactis Less.

M. nepalensis (Less) Dc. Prodr.-Temperate hills. Common.

Galinsuga Ruiz & Pavon

G. parviflora DC.-Common in Darjeeling at 1900 m.

Lactuca Linn.

L. gracilliflora Dc.-Occurs in Sureli.

Ganaphalium L.

G. affine D. Don. and *G. hypoleucum* Dc.-Both occur at Darjeeling.

Anaphilis Dc.-

A. triplinerveis (Sims) Clarke.-A very common herb.

**A. triplinerzeis* Var *intermedia* (Dc Airy Shaw.-Phalut. Common. 10 (ten) species are distinct in Darjeeling Himalayas. *A. contorta* and *A. margaritacea* are sporadic.

Senecio Linn.-

Darjeeling and Sikkim have about 25 species of which several are herbs. These species occur mostly in upper temperate forests. Common are *S. scandens* Buch-Ham. at Sonada, *S. Cappa* Buch Ham ex D.Don.-at Romam, *S. alatus* Waxi.-at Senchal.

Taraxacum Weber.

**T. officinale* Wigg.- A scapigerous herb. *Flowers* yellow. All over Darjeeling very common.

Artemisia Linn.-

A. japonica Thunb and *A. campbellii* Hook.-are two herbs of Alpine grassland of Darjeeling.

Emilia Cass.-

E. sonchifolia L (Dc) and *E. prenanthoidea* Dc. Prodr.-are two species of foot hills and plains of Darjeeling. Common.

Eclipta Linn.-

**E. prostrata* L. (E. alba Hassk.)-A very common herb. *Flowers* white. All over the plains and foot hills.

CAMPANULACEAE

Campanumoea Blume.-

C. parviflora Benth.-In Darjeeling hills upto 3000 m., *Flowers* 4-merous, Corolla small. Sporadic.

C. javanica Bl. *C. celebica* Bl., *C. inflata* Cl.-Occurs in Darjeeling hills. Sporadic.

Lobelia Linn.-

L. seylanica Linn., *L. pyramidalis* Wall., *L. rosea* Wall.-At Darjeeling. Sporadic.

Sphenoclea Gaertn.-

S. zeylancia Gaertn.-An annual erect herb. In swamps in Darjeeling.

Cephalostigma A. Dc.-

C. hirsutum Edgew.-A rather rigid herb. 12 cm. high. Terai.

Pratia Gaud.-

P. nummularia (Lam) A. Br. & Aschor.-Creeping herb at Tista valley.

P. montana (Rienw ex Bl.) Hassk.-Creeping herb at 2,000 m. in Sureli in Darjeeling.

Codonopsis Wall.-

C. benthami H.F., *C. thullictrifolia* Wall.-In Phalut, Sandakphu. Other species of Darjeeling are: *Campanula silvatica* Wall., *C. Colorata* Wall., *C. aristata* Wall., *C. fulgens* Wall., *C. camchericana* Royle.

N.B. *Sphenoclea* has been put under Sphenocleaceae by some Botanists.

PLUMBAGINACEAE

Plumbago Linn.-

P. zeylanica Linn.-A rambling herb. In Darjeeling it is wild, but usually cultivated. Tista valley.

PRIMULACEAE

Primula Linn.-

P. denticulata Smith.-At temperate hill (upto 1950 m). There are many other species.

Lysimachia Linn.

L. ramosa Wall.-A perennial weed. Darjeeling common.

Androsace Linn.-

A. umbellata (Lour) Merr (*A. saxifragaefolia* Bunge.)-It occurs in Terai; *Flower* in March.

GENTIANACEAE

Swertia Linn.-

S. delatata Clarke., **S. bimaculata* (S & Z) Clarke., *S. chirita* Ham., *S. nervosa* (G. Don) Clarke, *S. cordata* (G. Don) Clarke. Common at 2000 m. Most common of all these species is *S. bimaculata* Hk.f.

Exacum Linn.-

E. tetragonum Roxb.-An erect annual, 1 m. high. *Flower* azure blue. On river bed common.

L. indicum Thw.-In ponds. An aquatic herb.

Crawfordia Wall.-

C. speciosa Wall.,-C. affinis Wall., C. japonica.-Occurs in Sittong, Kalapokri and Tonglu. Sporadic.

Nymphoides Hill

N. hetrophyllum (Lour) Kuntz - In ponds with flouting leaves Common in plains.

HYDROPHYLIACEAE

Hydrolea Linn.-

H. zeylanica (Linn.) Vahl.-A weed of swamp; rice field, *flowers* deep bluish purple. Not common.

BORAGINEAE

Heliotropium Linn.-

H. indicum Linn.-*Flower* blue-white. In plains, sporadic.

H. strigosum Wild.-A procumbent, much branched hern. In plains. Sporadic.

CONVOLVULACEAE

Evolvulus Linn.-

**E. alsinoides* Linn.- A prostrate herb, small, diffuse, perennial flower blue-white. growing in open, grassland, common.

Merremia Dennst.-

**M. umbellata* Hallier.-occcures in plains, common.

SOLANACEAE

Solanum Linn.-

S. nigrum Linn.- Darjeeling hills 1650 m at Road side and waste places.

**S. ferox* Linn.-A stout herb. Leaves prickly. Flower white. Very common on sandy soil.

S. viarum Dunal (*S. khasianum* Cl.)-A medicinal plant and under-shrub, young part hirsate. L-lobed, *flower* Berry yellow about 2.5 cm. dia. black dotted.

Physalis Linn.

P. peruviana Linn.-A stoutish annual herb. Cultivated in gardens.

Nicotiana Linn.-

N. tabacum Linn.-Generally cultivated, Corolla rosy white.

N. plumbaginifolia Viv.-A herb of waste place. Flower white.

Nicandra Adams.-

**N. physaloides* (Linn.) Gaertn. A herb of waste places. In the plains and foot hills.

Physalis Linn.-

**P. peruviana* Linn.-In plains, cultivated.

SCROPHULARIACEAE

Linaria Juss.

L. cymbalaria (L) Mill.- A sporadically occurring herb of foot hills.

Verbascum Linn.

V. thapsus Linn.-A herb upto 30-90 cm. clothed with stellate hairs. Stem winged; *flower* in wooly spike, yellow. In hills.

Mazus Lour.-

**M. pumillus* (Burm. f) Van Steenis (*M. rugosus* Lour.)- *M. dentatus* Wall., *M. delavyi* Bl.-Occur in plains. They are small annuals with tufted radical leaves.

Lindenbergia Lahem.

**L. muraria* (Roxb) Bruhl (*L. urticifolia* Lehum).-An annual herb with brittle stem and branches, on walls and waste land everywhere. *Flower* yellow, sporadic.

L. grandiflora (Buch-Ham ex D. Don) Benth is another species of plains.

Limnophila R. Br.-

L. roxburghii G. Don.-In North Bengal. An aromatic annual. *Flower* blue purple. Leaves opposite. Sporadic.

L. chinensis Merrill (=*L. polyantha* Kurz).-A graceful, slender, annual. Leaves whorted. In North Bengal, Duars. Sporadic.

L. sissiliflora Bl.-A herb in swamps and rice fields, submerged below. *Flower* rose purple. In plains. Common.

L. erecta Benth.-Duars plains. Sporadic

L. repens (Benth) Benth (*L. conferta* Benth.)-*Flower* purple, Duars plains. Common.

Torenia Linn.-

**T. edentula* Griff.-Sureil, Darjeeling. An erect or decumbent, sparsely hairy herb. Common.

T. diffusa D. Don.-In temperate hills. Sporadic.

Lindernia Linn.-

**L. sissiliflora* Benth.-A small erect, simple or branching herb. Common.

**L. hookeri* Cl.-In Terai,. *V. scabra* Benth. in Foot hills, *V. erecta* Benth. in Tista valley have also been found to occur sporadically.

L. ruelloides (Colsm) Penn.- A prostrate, creeping herb. Common.

L. antipoda (L) Alston.-A decumbent herb, creeping. Flower violet, common. In Darjeeling.

Scoparia Linn.-

**S. dulcis* Linn.-A rigid perennial herb in waste places. Flower petals white very common.

Veronica Linn.-

V. cana Wall. ex Benth.-North Bengal, 2500 m.

**V. capitata* Benth.-A pubescent annual with perennial roots. Flower blue, common.

V. javanica Bl.-A much branched pubescent annual, 1600 m.

Striga Lour.-

S. densiflora Benth.-A herb, 15-40 cm. high. Flower white.

Curanga Juss.-

C. amara Juss.-A glabrous straggling annual. Riana, Darjeeling 2,000 m. flower red brown.

Hemiphragma Wall.-

H. heterophyllum Wall.-Flower red, common. At Darjeeling 1800/3000 m.

LENTIBULARIACEAE

Utricularia Linn.-

**H. aurea* Lour (*U. flexuosa* Vahl.)-A floading herb of road side ditches in rains, very common.

GESNERACEAE

Didymocarpus Wall.-

At least 9 species are common in Darjeeling areas. They give flowers in July and August of which common are—*D. mortoni* Cl., *D. pulchra* Cl., *D. albicalyx* Cl., *D. oblongus* Wall., *D. aurantthiacus* Cl., and *D. andersonii* Cl.

Chirita Buch-Ham ex D. Don

**C. urticaefolia* Buch-Ham ex D. Don.-A small herb, epiphytic; flower violet. Common at 2400 m.

C. kurzu Cl.-At 2,000 m. Darjeeling, not common.

Epithema Blume.-

E. carnosum Benth.-A rock herb. Leaves creuate, sinuate. Birrik, Darjeeling foot hills of Himalayas.

Rhynchotoechum Blume.-

R. ellipticum A. Dc.-An erect undershurb. Tista valley.

R. vestitum Hk. f. & Ṭ.-Darjeeling, sporadic.

PEDALIACEAE

Martynia Linn.-

**M. annua* Linn. (=*M. diandra* Glox.).-A tall, coarse herb. Capsule hooked. Very common on roadsides and near villages. North Bengal upto 1,000 m.

ACANTHACEAE

Hygrophila R. Br.

**H. polysperma* T. and.-An annual herb. Sub-gregarious, flower pale blue. In plains of Bengal. Very common.

Eranthemum Linn.-

**E. pulchellum* Andr. (*E. nervosus* (Vahl) R.Br. ex R&S)-A stout ish herb, 1-2 m. high. Common in Sal forests.

Strobilanthes Bl.

S. scaber Nees.-A herb. All other strobilanthes are shurbs. Not common. North Bengal.

Lepidagathis Willd.-

**L. incurva* Buch-Ham ex D. Don (*L. hyalina* Nees).-A diffuse herb. *Flower* white. Foot hills of Himalayas. Common.

Justicia Linn.-

J. gendarussa Linn.-A gregarious plant of shady stream bed. Jalpaiguri common.

Dicliptera Juss.

**D. roxburghiana* Nees.-A tufted herb, upto 45 cm. high. *Flower* pinkish in axillary panicle. Common in temperate hills.

Rungia Nees.-

R. himalayensis Cl.-At 1500 m. in hills.

**R. parviflora* Nees.- A diffuse herb. *Flower* blue purple. In plains and waste places.

Cardanthera Ham.

C. uliginosa Ham.-*Flower* purple rose, swamps, rice field in Duars. Common.

VERBENACEAE

Phyla Linn.

**P. nodiflora* (L) Greens.-A rostrate much branched herb, often rooting at the nodes. Plains and foot hills. Very common.

Premna L.

P. herbacea Roxb.-Flower white small. Common in grass land. In plains.

LABIATAE

Ocimum Linn.-

O. basilicum Linn.-A herb. *Flower* white. Cultivated and also grow in wild.

Acrocephalus Benth.-

A. indicus (Burm f.) ktze (*A. capitatus* (Roth.) Benth.)-A small pubescent herb. *Flowers* in dense heads pale purple in December. At Sukna.

Anisomeles R. Br.,

**A. indica* O. ktze. (*A. obovata* (R. Br.)-A strong smelling softly pubescent herb. Corolla purple in axillary whorl. *Flowers* in February, Sporadic in plains.

Anisochilus (Wall) Benth

A. polystachyus Benth.-A herb with strict, rigid stems, 1-3 m. high. Plains in Duars.

A. pallidus Wall. ex Benth.- Flower in October. Kalimpong, Mogpoo.

Eusteralis Rafin (*Dysophylla* Bl.

**D. pumila* (Grah) Rafin (*D. crassicaulis* Benth).-Gregarious, *Flower* blue purple. In swamps, Terai, North Bengal. Common in February.

E. cruciata (Benth) Panigrahi (*D. quadrifolia* (Roxb) Benth.)-*Flower* common in moist situation in February, in Terai & Tista valley.

Leucas R. Br.-

L. indica (L) R. Br. (*L. linifolia* Spreng.)-An annual herb. In all over.

L. cephalotes (Roth.) Spreng.-An annual herb. In most of the places.

L. mollissina Wall.-A slender branching herb. North Bengal.

Ajuga Linn.-

A. macrosperma Wall ex Benth.-In plains and foot hills. *Flower*-January, blue, violet.

Perilla Linn.-

P. frutescens (L.) Britton. (*P. ocimoides* Linn.).-Aromatic herbs. *Flower* white. Gregarious, also in the hills. Common.

Elscholtzoa Willd.-

E. strobillfera (Benth) Benth.-A herb of alpine hills.

E. blanda Benth.-*Flower* white, in November. At 1600 m.

Hyptis Jacq.-

**H. suaveoloens* (L) Poit.-*Flower* blue. Hyptis has atleast 7 species in temperate hills.

Pogostemon Desf.-

**P. benghalensis* (Burm. f.) C. Kuntze. (*P. plectranthoides* Dest.).-Form thickets of flower-white. Darjeeling.

Mosla Maxim

**M. dianthera* (Buch-Ham ex Roxb) Marim.-*Flower* in June. At 1600 m. Common in Tista valley.

Calamintha Moench.-

**C. umbrosa* Benth.-*Flower* in July, blue violet. Riang, Senchal at 2300 m. Very common.

Melissa Linn.-

**M. axillaris* (Benth) Bakh. (*M. parviflora* Benth).-A gregarious herb; *Flower* white, numerous. A medicinal plant at 1600 m.

Bruoella Linn.-

**B. vulgaris.-Flower*-July. Common in grassland. At 2300-3500 m.

Notochaete

N. humosa Benth.-*Flower* in December. Common in grass land in Senchal at 2500 m.

NYCTAGINEAE

Boerhavia Linn.-

**B. diffusa* Linn. (*B. repens* Linn.)-A small, diffusely branched herb. Very common in waste places in plains.

AMARANTHACEAE

Allamania R. Br.-

**A. nodiflora* R. Br.-A diffuse herb, dichotomously branched, common.

Amaranthus Linn.-

**A. spinosus* Linn.-An erect, spinescent herb. Common.

A. paniculatus Linn.-A tall robust annual. In plains, cultivated.

A. viridis Linn.- A slender annual, common. A weed of cultivated area.

A. lividus Linn (*A. blitum* Linn).-A herb common in Rongbong, Labdah-Darjeeling.

Aerva Forsk.-

**A. scandens* Wall.-An erect herb. *Flower* white, very common.

Achyranthes Linn.-

A. aquatica R. Br.-A tall, stout herb straggling in water and rooting on the margins of ponds and river banks. In plains adjoining Bihar.

**A. aspera* Linn.-A coarse weed of way sides and waste places. Very common.

**A. bidentata* Bl.-A rambling herb of temperate hills. Common.

Alternanthera Forsk.-

A. sissilis (L) DC.-A prostrate herb. Very common in waste moist places.

CHENOPODIACEAE

Chenopodium Linn.-

C. album Linn.-A tall herb 1 m. high. Cultivated and weed in cultivated ground. Common.

**C. ambrosioides* Linn.- A tall much branched herb, aromatic. Common, North Bengal.

PHYTOLACCACEAE

Phytolacca Linn.-

P. acinosa Roxb.-A succulent herb, *Flower*-7. Fruit blue black. In temperate hills. 1600-2500 m. Sporadic.

POLYGONACEAE

Polygonum Linn.-

**P. plebejum* R. Br.-A prostrate, diffusely branched weed in cultivated ground and in lawns. Common in plains.

Rumex Linn.-

**R. trisetifer* Stokes (*R. maritimus* Linn.)-An erect annual 1 m. high in marshy places. Common in plains and foot hills.

R. oricutalis Bernh.-In plains of Jalpaiguri.

**R. nepalensis* Spreng.-In temperate hills.

R. acetoella Linn.-In temperate hills.

Fagopyrum Hill

**F. asculentum* Moench.
F. dibotrys (D. Don) Hara
R. cymosum Meissa.
} Temperate Darjeeling hills.

EUPHORSIACEAE

Euphorbia Linn.-

**E. hirta* Linn. (*E. pilulifera* Linn.)-A very common annual weed of waste places, fields, and gardens. North Bengal.

Phyllanthus Linn.-

**P. urinaria* Linn.-A herb, usually, sometimes perennial. In plains of North Bengal.

P. niruri Linn.-An annual weed, North Bengal.

Crozophora Neck.-

**C. plicata* A. Juss.-A coarse annual weed, very common.

Acalypha Linn.-

A. indica Linn.-An annual weed, 20-60 cm. high, sporadic.

URTICACEAE

Urtica Linn.-

**U. ardens* Linn (U. *parviflora* Roxb).-Leaves ovate cordate wrinkled. Temperate Himalayas (1000 m.); *Flower*-in January Common.

U. dioica Linn.-Himalayas; decoction of plant used in medicine, common.

Pilea Lindl.-

**P. bracteosa* Wedd.-*P. anisophylla* Wedd., *P. approximata* Cl. & *P. glaberrima* (Bl) BL (*P. smilacifolia* Wedd.)-are important species in the temperate Himalyas.

Lecanthus Wedd.-

**L. peduncularis* (Royle) Wedd. (*L. wightii* Wedd.).-A succulent and a very common gregoriuos weed of temperate forests. Leaves opposite, has fleshy discoid receptacle. Common.

Elatostemma Forst.-

**E. rupestre* (Buch. Ham ex D. Don) Wedd., **E. sessile* Forst., **E. hookerianum* Wedd.- are common in Darjeeling & Sikkim hills.

Pouzolzia Caud.-

**P. zeylancia* (L) Benth & Brown. (*P. indica* Gaud.)-A perennial herb. Darjeeling and Sikkim foot hills. Common.

**P hirta* Hassk.-A slender decumbent or sub-erect herb. Darjeeling temperate hills. Common.

**P. sanguinea* (Bl.) Merr. (*P. viminea* Wedd.)-A herb, leaves silvery margin serrate, base 3- nerved. Leaves edible. In foot hills and temperate hills. Common.

(iv) Analytical Synopsis of Vascular Epiphytes

The study included listing of vascular epiphytes. A brief description of the species has been madE. Fidelity of the species has been mentioned. The numericals viz., 1, 2, 3 means January, February, March, and so on. Some monocotyledons which occasionally assume epiphytic habit have not been included in this section (included in section VIII).

The zone is not rich in vascular epiphytes. Most common ones are marked with asterisk (*).

RUBIACEAE

Hymenodictyon Wall.

H. flaccidum Wall.-Tree, often epiphytic, in young stage. *Flowers* 6; almost sessile, fasciculate in dropping spikes. *Fruit* 12-1. Occurs in foot hill forests upto 1,524 m. Sporadic.

Hymenopogon Wall.

**H. parasiticus* Wall.-Epiphytic small shrub of the Temperate hill forests from 1,534-2,438 m. stem curved, shoots villous. Leaves oblonceolate. *Flowers* 5-8; white with large whitish bracts. *Fruit* 10-1. Frequent.

VACCINIACEAE

Agapets. D. Don ex. G. Don.

**A. sikkimensis* Airy Shaw (*A. aurticulata* Hook)-An epiphytic shrup of the temperate hills, from 500-1,800 m. Leaves with distinct intramarginal veins often falsely whorled; 15-20 cm. long. Epicalyx double. Corolla tubular, elongate, lobes less then half its length. *Flower* 6-9, red in corymbs. Corolla red with V-shaped bar. Sporadic

**A. saligna* Hook ex Benth.-Large epiphytic shrub of the lower hill forests from 1,200-2,100, m. Stems often greatly swollen at the base. Leaves entire. *Flowers* 3-5; dark red, long with V-shaped bars.(Abundant).

A. discolor Clarke.-Epiphytic shrub of Temperate hills, 914-1,524, m. Leaves 3-8, by 1.3, cm. pedicels thickened upwards. *Flower* in corymbs. Sporadic.

**A. serpens* (Wighth.) Slermer.- Epiphytic shrub of the middle & upper temperate hills forests from 1,200-2,400 m. Densely leafy, hispid, common. *Flower* 1-5; red, with black V-shaped markings, calyx 5- angled. *Fruit* 4-6; edible. Abundant.

At least ten more species have been recorded from Eastern India.

Vaccinium Linn.

**V. vacciniaceum* (Roxb.) Sleumer -Shrub, bulbous -rooted, often epiphytic, found from 1,200-2,400 m. *Flower* and *Fruit* 4-5; both white. Frequent.

V. *dunalianum* Wight.-Shrub, often epiphytic, of the Upper Hills Forest, from 1,500-2,400, m. *Flower* 5; small pink *Fruit* 8-10; black. Sporadic

V. *nummularia* Hk.f. & Th ex Clarke.-A small epiphytic shrub with small round leaves and pink *flowers* found in the R.F. from 2,400-3,000, m. *Flowers* 5-6; 10; blue -black, with the flavour of black berris. Sporadic.

V. *retusum* (Grifl) Hk.f ex Clarke.-Large shrub, often epiphytic, of the temperate hills forest from 2,134-2,743, m. *Flowers* 5-6; pink with red stripes. Bracts pure white. *Fruit* 8-9; purple. Sporadic.

ERICACEAE

Rhododendron Linn.-

R. dalhousiae Hk.f.-A shrub of temperate hill forest 2-3 m. high, found from 1,800-2,700 m. on rocks, steep banks of land slips or epiphytic, of the leaves. *Flower* 5-6; large, white, tinged with rose on the outside. *Fruit* 10-11. Sporadic to rare.

GESNERACEAE

Aeschynanthus Jack,-

*A. *hookeri* C.B.Clarke.-Epiphytic shrub of temperate forests. *Flowers* red, 5-7. Frequent.

*A. *bracteata* Wall.-Epiphytic shrub of the temperate forests. *Flowers* 7-9. Frequent.

A. *parviflorus* (Don) Spr.(*A. ramosissima* Wall.) -Epiphytic shrub of the lower hills forests 1,829, m. *Flowers* and *Fruits* at all seasons. *Flower* 1. Frequent.

Lysionotus D. Don.

L. serrata Don.-Small epiphytic shrub of the lower and temperate hill forests. *Flowers* 8-11; blue. *Fruit* 10-12. Palmajoa. Sporadic.

Loxostigma Clarke

L. griffithii (Wight) Clarke.-A small epiphytic shrub. *Flower* (pale yellow) and *fruit* at all seasons. Birch hill. Sporadic.

VERBENACEAE

Premna Linn.-

P. interrupta Wall ex Schaner.-A shrub or small tree of foot hills from 1,500-1800, m. usually climbing, youngest shoots and inflorescence pubescent, long. *Flower* sessile, in compact nearly sessile cymose clusters supported by foliaceous lanceolate bracts, forming a simple cylindrical terminal spike, often interrupted near the base, Frequent.

**P. bracteata* Wall ex Clarke.-A climbing shrub of 600-1,500, m. Leaves glabrous, petiole 1 cm., fine reticulate veins conspicuous on the underside, spikes panicled. Frequent.

P. coriaceae Clarke -A large ornamental shrub, generally climbing, often epiphytic. Corymbs large, red or purple, petols white, calyx truncate. Sporadic in temperate hills.

ORCHIDACEAE

Orchids (epiphytic) collected in the forests have been listed under Epiphytes as components of vertical structure of forests.

(v) Analytical Synopsis of Families Genera and Species of Climbers

The study area is rich in climber, but the listing of those falling inside the quadrats and transects and adjoining areas has been made. Besides incorporating a very brief deseription, fidelity, *flowering* and *fruiting* time (indicated by numericals viz., 1,2,3 etc. meaning the months January, February, March and so on) have been mentioned. Conspicuous and frequently occurring species have been marked with asterisks(*).

RANUNCULACEAE

Clematis Linn.

C. montana Buch -Ham ex DC .-Climber of the temperate forests. *Flower* 6-7; large white. *Fruit* 9. Young parts pubescent. Sporadic.

**C. gouriana* Roxb ex DC.-Climber of the plains and foothills and upto 1,500, m. *Flower* 9; small, greenish white, clustered. *Fruit* 12-3. Branches purple, pubescent when young.Common.

C. *buchananiana* DC. - Climber of the Mountain and temperate hill forests from 1,500 - 3,000 m. *Flower* 9-12, *Fruit* 2. Sporadic.

C. grewiaefolia DC. - Climber of the mountain and temperate hill forests from 914 - 1,524 m. Very tomentose. *Flower* 11-12: *Fruit* 12-1. Rare.

Naravelia Dc.

* *N. Zeylanica* DC. - Climbing shrub, common throughout the plains and foothills forests. *Flower* 6-7 ; yellowish-white. *Fruit* 10-11. Common.

DILLENIACEAE

Tetracera Linn.-

* *T. sarmentosa* Linn. (syn. *Delima sarmentosa* Linn.)- Climber with very scabrid leaves, found all over North Bengal and foothills. Grows on dark forest floor. The stem when out gives copious potable water. Abundant

SCHIZANDRACEAE

Schizandra Michaux.-

S. grandiflora (Wall) Hook & Th .- Large woody climber of the temperate hills forest from 2,100 - 2,700 m. Common round Darjeeling. *Flower* 5-6; white, fragrant. *Fruit* 6-10; red, edible; carpels on a long receptacle. (sporadic.)

Kadsura Kaemp.-

K. heteroclita (Roxb.) Craib (=*Kadsura roxburghiana* Arn) - Large woody climber of the foothills, middle and temporate hill forests upto 2,400 m. *Fruit* 8-9, Carpels in a globose head; seed edible.

ANONACEAE

Desmos Linn.-

* *D. praecox* nov. comb. (Syn. *Unona praecox* Hk. f. & T.) - Climber found near Riang, Darjeeling, Sikkim at about 1,200 - 1,500 m. common.

* *D. chienesis* Lour. (*Unona discolor* Vahl.) - A spreading or climbing shrub, 12 to 24 cm. long of the evergreen forests and lower hills forest of Darjeeling. *Flower* 5-7. *Fruit* 11 - 3. Common.

Uvaria Linn.-

U. hamiltonii Hk.f. & T. - Large climber of the lower hills forest and plains. *Flower* 6-8; brick red. *Fruit* 8-9; carpels scarlet. Sporadic

MENISPERMACEAE

Tinospora Colebr.-

* *T. cordifolia* Miers. - Climber of the waste places of the Terai and Duars. It sends down long filiform roots from the branches of the trees over which it climbs. All deer, bovine species and elephants are particularly fond of it. Abundant.

* *T. Sihensis* (lour) Merr.(*T. Malabarica* Miers.) - A woody fleshy climber of the lower hills forest of the Darjeeling. Branchlets clothed with white hairs. *Flower* 2-5. Carpels crimson. Abundant.

Tiliacora Colebr.-

T. acuminata (Lamk) Miers (=*T. racemosa* Colebr,) - A climbing shrub. Panicle 3-7 *flowered* (male), 1-*flowered* (female), *flower* yellow. Common.

Cocculus DC.-

C. orbiculaters (L) DC. *C. mollis* Wall.)- A Climber of middle hills. Branches blackish, striate, petals notched 2,100 m. at Darjeeling. Sporadic.

Aspidocarya Hook.-

A. uvifera Hk.f. - A climber of the foot hills. Stem 15-2 cm. in diameter. *Flower* 4-5. *Fruit* 6-7; edible, tasting rather like green grapes. Branches ribbed.

Stephania Lour.-

* *S japonica* (Thunb) Miers. (*S. hernandifolia* Walp.) - Slender climber of the foot - hills. *Flower* 7, umbels capitate. *Flowers* yellow. *Fruit* 9. Common.

S. elegans Hk.f. & T. - Small wiry climber of the temperate hill forests, about 1,500 - 2.100 m. *Flower* purple or greenish, inflorescence lax and *Fruit* 9-12, red and succulent. Common.

Pericampylus Miers.-

P. glaucus Lam Merr. (= *P. incanus* Miers.) - Climbing tomentose shrub of Darjeeling, *Flower* greenish 2-4. *Fruit* 5-9; red. Sporadic.

Cissampelos Linn.-

C. pareira Linn. var. hirsuta (Buch-Ham ex Dc) Forman.- Large common climber of the plains also Terai, Duars and lower hill forest of Darjeeling and Sikkim. Common.

Parabaena Miers.-

P. sagittata Miers.-A climber of Darjeeling foot hills and over all plains. Stem yields milky juice. Leaves sagittate. *Flower* 4-5, *Fruit* 9-12. Sporadic.

Cyclea Arn.-

C. bicristata (Griff) Diels. (=*Lophophyllum bicristatum* Griff.)- Large woody climber of the foot hills forests, Darjeeling. *Flower* 4-5. Sporadic.

Pycnarrhena pleniflora Miers.- A woody climbing shrub. Sal forests of Jalpaiguri Forest Division. *Flowers* white, 3-4. *Fruit* 5-6. Sporadic.

LARDIZABALACEAE

Holboellia Wall.-

H. latifolia Wall.-Large climber of Himalayan temperate hill forests of Darjeeling from 1,200-3,000 m. *Flower* 4-5; purple fragrant. Stamens 6 carpel 3. *Fruit* 10-11; edible; seeds black. Sporadic.

CAPPARIDACEAE

Capparis Linn.-

C. sikkimensis Kurz (*C. cathcarti* Hemsl.)-Large climbing shrub of the foot-hills forests of Darjeeling about 1,200 m. *Flower* 6. Common.

C. pumila Champ., *C. zeylanica* Linn., *C. lenera* Dalz., are all straggling shrub found in North Bengal. Sporadic.

STERCULIACEAE

Buettneria Linn.-

B. grandiflora De (B. aspera Colebr.)-Strong woody climber of the foot-hills forests and the forests of Darjeeling, upto 1,200 m. *Flower* brownish 5. *Fruit* 6-8.: Capsule with stout conical bristles. Sporadic.

MALPIGHIACEAE

Hiptage Gaertn.-

H. benqalensis (Linn) Kurz (*H. madablota* Gaertn.)-Large climber of the forests of Darjeeling. *Flower* white 3-5; handsome, fragrant. *Fruit* 6-7. Sporadic.

Aspidopterys A. Juss.

A. glabriuscula A. Juss. (*A. roxburghiana* A. Juss)-A woody climber of the foot hills forests of Darjeeling. *Flower* 8-10. *Fruit* 12; winged. Sporadic.

A. nutans Hook.-Slender climber of the foot hill forests of Darjeeling upto 1,200 m. *Flower* and *fruit* 7-9. Samara, hairy. Common.

RUTACEAE

Zanthoxylum Linn

Z. hamiltonianum (Wall) Engl.-A large climbing shrub of the temperate hill forests; covered with large conical prickles. *Flower* 5 green; *Fruit* 10; edible. Sporadic.

Z. oxyphylla (Edgew.) Engler. (Edgew.)-A throny shrub, often climbing, of the temperate hill forests, upto 2,700 m. petioles armed throughout with hooked spines. *Flower* 5: lilac; large. *Fruit* 10-1; red. Common.

Toddalia Juss.-

**T. asiatica* (Linn) Lamk. (Syn. *T. aculeata* Pers.)-Large rambling or climbing shrub of the foot hills and plains with 2,000 m. Stem covered with sharp recurved prickles on broad woody bases. *Flower* 2. *Fruit* 12; yellow like a small orange. Common.

Paramignya Wight.-

P. monophylla Wight.-Large shrub, with sharp curved axillary spines. Common in the lower hills forests valleys. *Flower*-3; white. *Fruit*-4. Common.

OLACACEAE

Erythropalum

E. scandens Bl.-Occurs in foot hills of Darjeeling. *Flower* minute in dichotomous cyme. Sporadic.

E. vagum Masters.-Occurs in Darjeeling and Sikkim. Leaves with basal nerve 3-5.

ICACINACEAE

Natsiatum Buch-Ham ex Arn.

N. herpeticum Ham.-A thin wiry climbing shrub of Bamanpokri. *Flower* 11-1, greenish yellow. *Fruit* 2-4. Rare.

CELASTRACEAE

Celastrus Linn.-

C. paniculatus Willd (C. *multiflorus* Roxb)-Large climbing shrub of the forests of Darjeeling upto 1,800 m. *Flower* 7. *Fruit* 1-2. Capsule with arillous seeds. Sporadic.

C. stylosa Wall.-Large climbing shrub of Darjeeling from 1,820-2,400 m. *Flower* 4-5; greenish white. *Fruit* 1-2; yellow. Seed arillous. Abundant.

C. monosperma Roxb.-Large climbing or straggling shrub of the forests of Darjeeling. Branchlets hollow; *Flower* 5; whitish. *Fruit* 8-9.

HIPPOCRATEACEAE

Reissantia Halle (*Hippocratea* Linn.-)

R. arborea (Roxb) Hara (H. arlborea. Roxb.) -A large woody climber that climbs by hooked branches of the lower hills forests and plains. *Flower* 6. *Fruit* 2; seeds winged. Rare.

Salacia

S. roxburghii Wall.-A climbing shrub of the lower hills forests from 300-600 m. *Flower* 6-12. *Fruit* 2-5; edible. Berry, Rare.

RHAMNACEAE

Ventilago Gaertn.-

*V. *denticulata* Willd. (*V. madraspatana* Gaertn.)-All over the plains and foot hills. *Flower* numerous, greenish winged *fruit*. Abundant.

**V. calyculata* Tulsane-All over the plains and foot hills. Panicles deeply pubescent. Abundant.

Gouania Linn.-

**G. leptostachaya* DC. (syn. *G. napalensis* Wall).-Climbing shrub of the lower hills forest of Darjeeling upto 1,800 m. *Flower* 7, white. *Fruit* 9-10; three angled. (Abundant).

Berchemia Neck.-

**B. floribunda* (Wall) Bronga.-A large shrub of the forests of

Darjeeling from 1,200-1,800 m. *Flower* 5-12. *Fruit* 2-3. Abundant.

AMPELIDAE (VITACEAE)

Parthenocissus Planch

**P. himalayana* Planch. (syn. *Vitis himalayana* Brandis).-Very large climber of the temperate hill forests from 1,500-3,000 m. *Flower* 5-6. *Fruit* 9; edible. Abundant.

Cissus Linn.

**C. elongata* Roxb. (syn. *Vitis elongata* Wall.)-Large glabrous climber of wet areas upto 1,200 m. Stem flattened *Fruit* 10. *Flower* 4 merous. Common.

C. pedata. Lamk.-5-foliate. All over. Common.

**C. repanda* Vahl (syn. *Vitis repanda* W. & A.)-Large deciduous climber of the lower hills forest and all over Sal forests. *Flower*-5. *Fruit* 8-9, eaten. Abundant

Tetrastigma planch

**T. bracteolata* (Wall) Planch.-Leaves 3-foliate. *Flower* 2-4. *Fruit* 12 edible. Foot hills. Common.

T. serrulatum (Roxb) Planch.-Available at Darjeeling and Sikkim.

**T. rumicispermum* Planch. (*Vitis rumicisperma* Laws).-Very large climber of the temperate hill forests of Darjeling from 1,500-2,100 m. *Flower* 4-5. *Fruit* 10-11; ebible. Common.

T. hookeri Planch.-Darjeeling, Foot hills, sporadic.

T. dubium Planch.-Darjeeling. Foot hills, sporadic.

T. elongata Kurz.-Darjeeling. Foot hills, sporadic.

SABIACEAE

Sabia Coleb.-

S. limoniacea Wall.-A large climber of Foot hills. *Flower* yellowish. Panicles axillary or supra axillary. Sporadic.

Other species recorded are:-

S. campanulata Wall. (Sikkim Darjeeling), *S. lanceolata* Colab.,

CONNARACEAE

Connarus Linn.-

C. paniculatus Roxb.-Climbing shrub of Foot hills. *Flower* 2-4. *Fruit* 3-5. Common.

FABACEAE

Abrus Linn.-

**A. precatorius* Linn.-Slender climber of the Terai and Darjeeling.

Flower and *Fruit* 10-1, seeds scarlet with a black spot, Abundant.

A. pulchellus Wall.-Slender climber of the Terai and foot hills upto 1,200 m. *Flower* 9. *Fruit* 10-4; Seeds black (Common).

Dalbergia Linn.-

D. mimosoides Ranch.-Climber of the foot hill forests. Sporadic.

**D. volubilis* Roxb.-Large woody climber of Sal forests. *Flower* 2-4. *Fruit* 4-6. Abundant.

**D. stipulacea* Roxb.-Large climber of straggling tree of the Sal forests. *Flower* 4-8; purple. *Fruit* 10-4; 1-seeded. (Abundant).

D. rimosa Roxb.-Climbing or straggling shrub of Darjeeling upto 300 m. *Flower* 4-6 white. *Fruit* 12-1. (Rare).

Derris Lour.-

D. polystachya Benth.-Climbing shrub of the forests of Darjeeling. *Flower* 8-10; red. *Fruit* 10-3. Sporadic.

D. ferruginea Benth.-Climbing shrub of the forests of Darjeeling. *Flower* 5. *Fruit* 10. (Sporadic).

D. cuneifolia Benth.-Climbing shrub of the forests of Sikkim and Darjeeling. *Flower* 4-5; red *fruit* 6-12.

D. microptera Benth.-A large climber of the forest of Sikkim. *Flower* 4; purple. *Fruit* 8; pods flat indehiscent not winged.

D. monticola Prain.- Large handsome climber of the forests of Sikkim and Darjeeling from 1,800-2,100 m.

Millettia Wt & Arn.-

M. extensa (Benth) Baker (*M. auriculata* Baker)-A large woody climber of the forests of the Sikkim and Darjeeling. *Flower* 9-3; white. *Fruit* 3-6. (Abundant).

M. cinerea Benth.-Leaves large climber of the forests of Sikkim and Darjeeling. *Flower* 2-7; petals white with purple tips. *Fruit* 9-1. Common.

**D. pachycarpa* Benth.-Large climber of the forests of Darjeeling. *Flower* 4-5; pale pink. *Fruit* 11. Abundant.

Butea Roxb.-

**B. parviflora* Roxb. (*Spatholobus roxburgic* Benth)-Huge climber of the forests of Darjeeling. Leaves 3-foliate. *Flower* 8-10; white. *Fruit* 10-2. (Abundant).

B. *butei formis* (*voigt*) Grierson & Long (*B. minor* Ham.)-Large shrub of climber of the forests of Darjeeling and Sikkim. Leaves very large, silky tomentose. *Flower*-1.

Mucuna Adans.

**M. nigricans* (Lour) Stend (*M. imbricata* DC.)-A climber of riverain forests. *Fruit* 4-12; covered with short stinging hairs and winged along both sutures. Seeds 2-3. (Abundant).

M. monosperma DC.-A huge climber of the Sal forests. *Flower* 1-2; purple. *Fruit* 6-7; long falt, velvety covered with dense short stinging hairs and containing one seed. (Common).

**M. macrocarpa* Wall.-An enormous climber of Sal and foot hill forests. *Flower* 5; dirty yellow-white. *Fruit* 8; very long, rather twisted and soolen round the seeds. (Abundant).

**M. prurita* Hook.-A dangerous weed in Sal plantations. *Flower* 6-7; purple. *Fruit* 1: covered with golden brown stinging hairs, which cause great irritation. (Abundant).

CAESALPINIACEAE

Bauhinia Linn.-

**B. vahlii* W & A.-A gigantic and a very common climber of the foot hill forests. *Flower* 4-6 cream. *Fruit* 1-2; pod flat, bursting open with a land report; seeds edible. (Abundant).

B. scandans 1 var. *horsfieldii* (Miq) Oharhi (*B.anguiana* Roxb.)-A large climber of the foot hill forests. Sporadic.

Caesalpinia Linn.-

C. bonduc (Linn) Roxb. (*C. crista* Linn.)-Large prickly straggling shrub of the banks of streams and waste places and all over plains and foot hills. *Flower* yellow. Pods oblique 3-5 cm. (Common).

C. microphylla Ham.-A climber of Jalpaiguri in plains. *Flower* 8; *fruit* 3. Sporadic.

C. cucullata Roxb. (*M. cucullatum* W & A)-Large straggling climber of the forests all over upto 1,524 m. *Flower* 9-11; orange yellow. *Fruit* 2; flat 1-seeded, red. Common.

MIMOSACEAE

Entada Adams.

**E. pursetha* DC.-A very big climber of the Sal and foot hill forests. *Flower* pale yellow, 3-6. *Fruit* 10-12; pod often 0.6 m. or more long with very big seeds. Abundant.

Acacia Willd.

A. torta (Roxb) Craib (*A. caesia* Willd.)-A large prickly climber of all over plains and foot hills forests. *Flower* 4-8 yellow. *Fruit* 8-3, 10-15 cm.. long and 3 cm. wide. (Sporadic).

A. pennata Willd.-Very large climbing shrub of various forests. *Flower* 4 white or pale yellow. *Fruit* 10-12. Abundant.

A. Sinuata (Lour) Merr (*A. concinna* DC.)-Large, very prickly, climbing shrub of the forests. *Flower* 3 yellow. *Fruit* 1-3. Sporadic.

Mimosa Linn.

**M. himalayana* Gamble.-Common in plains and foot hills. *Flower* 9-10. *Fruit* 12-2. Abundant.

COMBRETACEAE

Combretum Linn.

C. flagrocarpum Clarke.-Climbing shrub of the foot hill forests. *Flower* 4-7. *Fruit* 11-12; red-brown. Sporadic.

C. wallichii DC.-Climbing or straggling shrub of the forests of waste places and old cultivation. Common.

**C. roybarghii* Spring.-Large climbing shrub of foot hills forests. *Flower* 13-3. *Fruit* 3-6. Abundant.

PASSIFLORACEAE

Passiflora Linn.-

P. geminiflora D. Don.-A glabrous climber with angular branches of hills from 1,200 to 1,800 m. *Flower* 7-9. *Fruit* 10, 1.3 cm. across, globose, purplish. Sporadic.

P. foetida Linn.-A climbing shrub, foetid smelling when bruised. *Flower* 3 cm. greenish white, coronal process filiform, white and purple. Sporadic.

Modecea Ham.

M. trilobata Roxb.-A large climber of foot hills. *Flowers* uni-sexual, peduncle axillary 15-25 cm. long, many flowered cyme. Staminodes 5. *Flower* 7. *Fruit* 8. Common.

CUCURBITACEAE

Hodgsonia Hk.f.

**H. macrocarpa* (Bl.) Conqn (*H. heterochita* Roxb.)-An extensively woody climber of Sal and wet mixed forests with extremely bitter bark. *Flower* 2-3, *fruit* 11. *Fruit* 10-25 cm. red brown. Common.

Trichosanthes Linn.

T. Wallichiana (Serir) Wight.-Available in the Sal forest. *Fruit* bright red with orange stripe 5-8 cm. Sporadic.

T. tricaspidata Lour (T. *bracteata* Lamk.)-An extensive climber of plains and foot hills. *Fruit* 3-5 cm. globose, red with orange streaks, ultimately deep red. *Flower* 7-9. *Fruit* 10-11. Common.

Gymnopetalum Arn.

G. cochinchinense Kurz.-A slender scandent or spreading herb. *Flower* bright, scarlet ornamental. Sporadic.

Biswarea Cogn.

B. tonglensis Cogniaux.-A climber of temperate hills. Common near Tonglu, Rangiroom, Senchal and Ghoom. *Fruit* oblong, trigonous, pilose, many seeded. Common.

Mukia Arn.

M. maderaspetana (L) Roem (*M. scabrella* Arn).-Common scabrid climber of Sal forests. Leaves cordate-palmately. Common.

Thladiantha Bungue

T. Cordifolia (Bl) Cong

T. calcarta colebr.-Available in the forests upto 1,200 m. Sporadic.

Edgaria Cl.

**E. darjeelingensis* C.B. Clarke.-A very common climber in Senchal areas; *Flower*-large yellow-8, *Fruit* a dry fibrous capsule. Common at 1,500 to 2,400 m.

Solena Laur (*Melothria* Linn).-

S. heterophylla Law (*M. heterophylla* Cog.)- Occurs in the foot hills forest of Darjeeling.

ARALIACEAE

Pentapanax Seem.

**P. racemosus* Seem.-Very large climber of the forests of the Darjeeling and Jalpaiguri plain. Easily recognised by its particularly racemose inflorescence that form cluster. (Abundant).

Schefflera Ham.

S. bengalensis Gamble.- Large straggling shrub or climber of the lower hill forest from 900-1,500 m. *Flower* 12-1. *Fruit* 2-3. Common.

S. venulosum (W&A) Ham-A soft climber. Leaflets 3-7. *Flower* umbellate rachis purplish in a branched panicle. Common.

CAPRIFOLIACEAE

Lonicera Linn.-There are about eight common species in the Darjeeling hills. All of them are sporadic.

Leycesteria Wall.-There are five species common to Darjeeling hills. All of them are sporadic.

RUBIACEAE

Uncaria Schreb.-

U. macrophylla Wall.-Large climber of foothills. *Flowers* 10. Sporadic.

U. pilosa Roxb.-Large straggling shrub of Darjeeling upto 1,500 m. *Flower* 4, Yellow; *Fruits* 10-12. Common.

U. sessilifructus Roabge-Climbing shrub of Darjeeling upto 1,200 m. *Flower*-11. *Fruit*-12-3. Sporadic.

Paederia **Linn.**

Paederia foetida Linn.-A small climbing shrub from plains to 1,800 m. It has a characteristics foetid smell when bruised. *Flower* 9-10 red-fringed. *Fruit* 12-2. Leaves and *fruit* eaten. (Abundant).

Galium Linn.

**Galium* Linn.-There are more than 10 species in the temperate and sub-tropical forests. Leaves 3-nerved in whorl. *Flowers* numerous. *Fruit* either smooth of hooked. (Abundant).

Rubia Linn.

**Rubia* Linn.-There are 3-4 common species in the temperate hills. Herbaceous climber Leaves in whorl. *Flowers* 5-merous. Common.

COMPOSITAE (ASTERACEAE)

Vernonia Schreb.-

V. scandens DC; Prodr.- A rambling woody climber of the lower hills forest at 1,200 m. *Flower* 2-3. *Fruit* 3-4. Sporadic.

Mikania

**M. cordata* (Burn.) Robinson.-A recent invader and has been a menace to the plantations and forests. *Flower* in corymbose 4-flowered heads, white. (Abundant).

MYRSINEAE

Embelia Burm.-

E. ribes Burm.-A climber of the temperate hills forests and foot hills forest upto 1,300 m. *Flower* and *Fruit* 11-4. (Sporadic).

E. floribunda Wall.-A large climber of foot hills and middle hills. Inflorescence paniculate. Sporadic.

E. vestita Roxb.-(=E. *nagushia* Don.).-Big climber of the temperate hills forest. *Flower* 5-9. Inflorescence simple racemose. Sporadic.

E. gamblei Kurz.-Climber of the temperate hills forest above 1,800-2,400 m. *Flower* 5-8, *Fruit* 8-12. Sporadic.

OLEACEAE

Jasminum Linn.-

J. amplexicaule Ham.-Scandent shrub of the foot hills forest and temperate hills of forests from 600-1,300 m. *Flower* 9-11; *Fruit*-4. Common.

J. multiflorum (Brum.) Ardr. (*J pubescens* Wild.) - Small much branched shrub of the temperate hills forest and foot hills forest. *Flower* 9-11; white, scented *Fruit*-4. (Common).

J. napalense Spr (*J. glandulosum* Wall).-Climbing shrub of the foot hills forest and temperate hills forest. *Flowers* 7: white, scented. *Fruit* 1. Sporadic.

J. caudatum Wall.-Large climber of the foot hills and temperate, hills forests, 1,300 m. *Flower* 11, *Fruit* 6. Sporadic.

J. scandens Vahl. - Large climber of the temperate hills forests, *Flower* 11-1; with a heavy sweet scent. *Fruit* 2-3. Sporadic.

J. dispermum Wall-An extensive climber of the temperate hills forest from 1,700-2,400 m. *Flower* 4-5 white or lilac, sweet-scented. *Fruit* 12-3. Sporadic.

Myxopyrum Bl.-

M. smilacifolium Blume. E.-Climbing shrub of the temperate hills forest with 3-nerved leaves. *Fruit* 1. Sporadic.

APOCYNACEAE

Vallaris Burn. f.-

**V solanacea* (Roth) O. Ktze.-Extensive woody climber of the foot hills and riverain tracts. *Flower* and *Fruit* 2-4. *Flowers* large, white sweet-scented. Abundant.

Beaumontia Wall.-

**B. grandiflora* (Roxb) Wall.-Large climber of the forest of foot hills and Sal Forests. *Flower* 12-5; large trumpet-shaped white, very handsome. Common.

Chonemorpha G. Don.-

**C. fragrans* (Moon) Aiston (*C. macrophylla* G. Don)-Large climber of the forest of foot hills and Sal forests. *Flower* 5-6; white large. *Fruit* 1-2. (Common).

C. griffithii Hk. f.-Large climber of the forests of foot hills. It differs from the proceeding in having 5 distinct sepals and smaller leaves. *Flower* 5-6, pale yellow.

Aganosma G. Don.-

A. caryophyllata G. Don.-Rare in the forests of Darjeeling and Jalpaiguri. Sporadic.

Ichnocarpus Bl.

**I. frutescens* Br.-Slender woody climber of the forest of Darjeeling and Jalpaiguri and all over Sal and miscellaneous forests. Abundant.

Rhychodia Benth.-

R. wallichii Hk.f.-Large woody climber, with tubercles on the branches. *Flower* 5; white with red centre, fragrant. Sporadic.

Trachelospermum Lemaire.-

T. lucidum (D. Don) Schum (*T. fragrans* Hk.)-Large climber of the forest Darjeeling. *Flower* 4-6, white. *Fruit* 8; long, slender, glabrous. Cyme long peduncled. Sporadic.

T. axillare Hk.f.-A large climber of the forests of Darjeeling. Corolla cylindric. *Fruits* long, curved pubescent. Cyme short peduncled. Sporadic.

Micrechites Miguel.-

M. elliptica Hk.f.-Large climber of the forests of Darjeeling. *Flowers* small, yellow. *Fruit* 8-10. Sporadic.

ASCLEPIADACEAE

Toxocarpus Wright & Arn.-

**T. himalensis* Falc.-Slender woody climber of the forest of Darjeeling and Sikkim. *Flower* 5-6. Young parts rusty. Follicles rusty. Sporadic.

Gymnema R. Br.-

G. tingens W&A-An extensive climber. Cyme many flowered; branches densely puberulous. *Flower* 1.3 cm. across pale yellow. 5. *Fruit* 12. (Common).

Telosma Coville.-

T. pallida (Roxb). Craib.-A large twinning shrub. *Flower* yellowish umbellate. *Flower* 6-8. Sporadic.

Marsdenia Br.-

**M. tenacissima* W & A.-Large climber of the lower hills forest. Holds a strong silky fibre. *Flower* 5-6; yellow. Bark of old stem corky, deeply farrowed. *Flower* greenish yellow. Common.

M. roylei Wight.-Large climbing shrub. *Flower* pale orange, corolla fleshy. Available at foot hills. *Flower* 5. *Fruit* 9. Common.

Treutlera Hook.f.-

T. insignis Hk.f.-A stout climbing shrub in the upper hills forest from 1,830-3,000 m. tubercled. *Flower* 8-10; purplish, handsome. Sporadic.

Tylophora R. Br.-

T. longifolia Wight Twinning undershrub. *Flower* umbelliform. Calyx 5-partite glandular, corolla deeply 1-lobed. Coronal processes tubercled. Pollinia Oblong. Several species. Common.

Dregea E. meyer.-

**D. volubilis* Benth.-In Sal forests and riverain forests. *Flower* green. Corolla dark green. Corona scales stellately spreading, cuspidate; pollinia oblong. (Abundant).

Raphistemma Wall.-

R. pulchellum Wall.-Twinning shrub. *Flower* large white, long pedicelled, corolla 5 glandular, corona legulate. Sporadic.

Cryptolepis Br.-

C. buchanani Roem.- A large twining shrub. *Flowers* pale greenish, yellow in short axillary panicled cymes. (Abundant).

C. elegans Wall.-Slender climbing shrub of the forests of all over the States, Darjeeling and Sikkim. *Flower* 6. *Fruit* 11-15. Common.

BORAGINACEAE

Tournefortia Linn.-

T. hookeri C.B.C.-Climber of the foot hill forests. *Flower* and *Fruits* at most seasons. *Fruit* greenish yellow. Style short 2-lobed. Common.

CONVOLVULACEAE

Erycibe Roxb.-

E. laevigata Wall.-A straggling shrub of the forests from 1,500-1,800 m. Cymes short, axillary towards the end of branches and terminal but scanty, panicled; *Flower* 5-6; yellowish white. *Fruit* 12-3. Sporadic.

Rivea Chois.-

**R. roxburghii* Prain.- Climbing shrub of the Duars and Terai. *Flower* 7-8 white. *Fruit* 12. Peduncles usually 7-flowered. Common.

Argyreia Lour. (=Lettsomia) Lour.-

**A. wallichii* Choisy var. coriacea.-Large climber of the Terai and lower hills and upto 1,200 m. *Flower* 10; white tinged with pink. *Fruit* 12. Common.

**A. hookeri* Clarke. *Puttokbok,* Linn.-Large climber of the lower hill forests and upto 1,200 m. Terai and Duars. *Flower* 6-7; red. *Fruit* 9-10. Common.

**A. argentea* Choisy.-Large climber of the lower hill forests Terai and Duars. Common.

**A. capitata* (Vahl) Choisy. (*L. strigosa* Roxb.)-Climber of the banks of streams and waste places in the terai, also of the lower hill forests. *Fruit* clothes with golden hairs, Common.

A. sikkimensis (Cl.) Var. oost. (=*L. sikkimensis* Clarke)-Large climber of the Lower hill forests. *Flower*-8. Common.

A. roxburghii Chois.-Corolla purplish (Darjeeling and Jalpaiguri). Common.

A. splendens S.W. Foothills of Darjeeling, Jalpaiguri and Sikkim.

Ipomoea Linn.-

**I. pestigridis* Linn.-*Flowers* capitate; bracts persistent Stem hirsute. Corolla pink. In the forests of Darjeeling and Jalpaiguri. *I. kurzii* Prain-(Darjeeling). & *I. vitifolia* SW. (Darjeeling). One other species.

Porana Burm.-

P. racemosa Roxb.-A slender climber. *Flower* white bracts pinkish, 10-11, *Fruit* 12. Rare.

**P. paniculata* Roxb.-Large handsome climber of the forests of Darjeeling and Jalpaiguri. *Flower* 11-12 white. *Fruit* 1-3. Abundant.

P. grandiflora Wall.-Darjeeling and Jalpaiguri. *Flower* crimson.

ACANTHACEAE

Thunbergia Retz.-

**T. grandiflora* Roxb.-An extensive climber of the plains of Jalpaiguri and Darjeeling District. *Flower* 8; blue or white. Abundant.

**T. coccinea* Wall.-Large climber of Darjeeling hills from 610-2,130 m. *Flowers* in fascicle at the nodes of lax elonggated pendent racemes; *Flower* 8, orange red, *Fruit* 12. Sporadic.

VERBENACEAE

Holmaskioldia Retz.-

**H. sanguinea* Retz.-A common climbing shrub all over Sal and Foot hills forests. *Flowers* very conspicuous, flame coloured. Calyx bow shaped. *Flower* 10-1. *Fruit*-3. Common.

AMARANTACEAE

Deeringia R. Br.-

**D. amaranteoides* (Lamk.) Meer.-Large climbing shrub of the forests of Darjeeling and Jalpaiguri district. *Flower* 8-9. Pale yellowish green. *Fruit* 1; red. Succulent. Berry. Common.

Stilbanthus Hook.-

A large climber of temperate forests. *Flower* 9-10. Covers tall trees with handsome white *flower*, whitish leaves. Branches 4-angled silky. Rare.

POLYGONACEAE

Polygonum Linn.-

P. chinense Linn.-Rambling or erect shrub, about 1.5 m. high of the lower hill forest. *Flower* 9-12; white. Frequent.

P. molle Don.-Large trailing shrub, very common all over the Darjeeling Hills, sending its branches even over big trees. The stems grow to about .3 m. in girth, the shoots are eaten. *Flower* 6-7. *Fruit* 11-12. There are many such species of Polygonum in temperate hills.

ARISTOLOCHIACEAE

Aristolochia Linn.-

A. tagala Cham. & Schlect.-Twining shrub of Duars and lower hills forests, upto 1,200 m. *Flower* 5-6, *Fruit* 10-1. Tista valley, Riang. Sporadic.

A. saccata Walll.-A lofty, woody climber upto 2,000 m. *Flower* 3; purple, generally borne on the main stem, often in clusters at the root. *Fruit* 7-8. Sporadic.

A. platanifolia Duchart.-Large climber of the Lower hill and upto 2,000 m. *Flower* 3: pink with yellow inside, chiefly on the woody stem. Sporadic.

PIPERACEAE

Piper Linn.-

P. attenuatum Ham.-A rambling shrub of the Lower hill Forest. *Flower* and *Fruit* 12-3. Sporadic.

P. peepuloides Miq.-Slender shrub of the forests of Darjeeling and Jalpaiguri foot hills and Sal forests. *Flower* 10-11. *Fruit* 2-3. Common.

P. thomsoni Kh.f.-Climber of the forests of Darjeeling and Jalpaiguri district foot hills and Sal Forest. *Flower* 3-5. *Fruit* 6. Abundant.

P. nepalense Miq.-Small shrub of the Lower hills Forest. *Flower*-8 and *Fruit* 10. Sporadic.

MONOCOTYLIDONS
LILIACEAE

Asparagus Linn.-

**A. racemosus* Wild. P.A.-Climbing prickly under shrub in the all over plains. The underground roots are eaten. Common.

Smilax Linn.-

S. ferox Wall.-914-2,743 m. All over foot hills in Darjeeling and Jalpaiguri. *Flower* 10. Common.

S. perfolita Lour.-A stout shorny climber in the forests of Darjeeling upto 1,829 m. Stem 2.5 cm. diam. Leaves 3-nerved. Very big leaved. *Flowers* 3. Berry .5 cm., red. Common.

S. macrophylla Roxb.-A thorny of the lower forests. Leaves 5-7 nerved blade 2-7 cm. diam. *Flowers* 3. Umbel 2-3. Berries large. Abundant.

S. ocreata DC.-Large thorny climber of the lower hills forests. *Flowers* 3, *Fruits* 8.

ARACEAE

Pothos Linn.-Ovary 3-celled.

**P. scandens* Linn.-Common epiphytic climber on the large trees in Darjeeling and Jalpaiguri districts. Leaves 3-10 cm. Abundant.

P. vriesianus Schoot.-305-1,219 m. Peduncle large, erect; bracts large.

**P. cathcarti* Schoot.-610-1,219 m. *Flowers* 5, epiphytic climber of the forests of Jalpaiguri and Darjeeling. Leaves bigger than other two species. Leaves 8-15 cm. Abundant.

Scindapsus Schoot.-

**S. officinalis* (Roxb.) Schoot.-Common in the Sal forests, of Terai and Duars. Large epiphytic climbing shrub in the lower hill forests. Leaves ovate oblong cordate of oblong lanceolate; petiole with distinct joint. (Abundant).

Rhaphidophora Schoot.-

R. decursiva (Roxb) Scholl (*R. axiimia* Schoot).-610-1,524 m. Leaves large distictous, oblique pinnatisect. Leaves 30-60 cm. petiole thick. Spathe orange yellow. Sporadic.

**R. glauca* (Wall) Schoot.-Common all over moist forests of Darjeeling, Jalpaiguri and Sikkim. Leaves pinnetisect, petiole furrowed spathe yellow. Abundant.

R. poepla Schoot.-Sikkim, Darjeeling, Jalpaiguri. Rare.

DIOSCOREACEAE

Dioscorea. Linn.-

D. anguina Roxb.-Leaves simple cordate, tomentose, opposite and alternate, long petioled. A large climber of the Duars. Common.

D. pentaphylla L. Leaves 5. Duars Sal forests, Common.

D. deltoides Wall. Leaves hastate, Duars. *Flower* 8, Sporadic.

D. bulbifera var. *saliva* L. Leaves alternate, ovate cordate. Common, axillary bulbil wanted. Sporadic.

D. glabra Roxb. 1200 metres ft. *Flower* 9. Sporadic.

GYMNOSPERM
GNETACEAE

Gnetum Roxb.-

G. scandans Roxb.-A large woody climber. Stem flattened and twisted. *Flower* catkins (Looks like caterpillar) at the axils of cup shaped. *Fruit* orange coloured, edible.

Occurs in Darjeeling, Jalpaiguri, (Sporadic).

G. montanum Nark.-Darjeeling and Jalpaiguri.

(vi) Analytical Synopsis of Grasses and Sedges

A list of grasses of sedges collected and identified has been produced. It provides only the zone of occurrence of species and the habitat.

The text deals with the ecology and distribution of the species in detail. It has not been intended to compile a flora of the study zone. Yet widely occurring and conspicuous species have been recorded. It also gives the frequency of occurrence of various species. Species marked astoriaks(*) are very common.

LIST OF GRASSES OF STUDY AREA

1. **Acroceras munroanum* (Bal Hern-Sporadic all over; a prostrate herb.
2. *Agropyron longiaristatum* (Boiss) Boiss.-Alpine dry areas.
3. *A. semicostatum* Ness ex Steud.-Temperate Himalayas tufted grass.
4. *A. thoroldianum* Oliver.-Alpine dry area.
5. *Agrostis inaquiglumis* Grisb.-Alpine hills tufted.
6. *A. filipes* Hook.f.-Temperate Himalayas.
7. *A. micrantha* Steud.-Gregarious in alpine pasture.
8. *A. pilosula* Trin.-Sporadic.
9. *A. pilosula* Trin. var ciliata (Trin) Bor.-Wide distribution in temperate forests.
10. *A. sikkimensis* Bor.-Temperate Darjeeling hills; gregarious.
11. *A. triaristata* Hook.f.-Temperate Darjeeling hills; gregarious.
12. *A. zenkeri* Trin.-Sporadic, Temperate Darjeeling hills.
13. **Alloteropsis cimicina* (Linn) Stapf.-Wet areas in plains.
14. *Alopecurus* species.-Flat leaved grass of middle hills all states.
15. *Andropogon pumilus* Roxb.-Fodder grass of dry locality of W.B. & Bihar.
16. **Apluda mutica* Linn.-Common in foot hills, a fodder grass.
17. **Apocopis paleacea* (Trin.) Hochst.- All over temperate lower hills.
18. **Arundinella bengalensis* (Spreng) Druce Syn.-Gregarious in plains and foot hills.
19. **A. decampedalis* (O. Ktze.).-Gregarious in high temperate Himalayas. Nanowsky.
20. *A. hookeri* Munro ex King.-Gregarious on dry soil in plains and hills.
21. *A. intricata* Hughes.-Widely distributed in all hills; gregarious.
22. *A. nepalensis* Trin.-Sporadic in the Himalayas.
23. *A. setosa* Trin.-Very sporadic in Darjeeling hills.
24. *Arthraxon* hookeri (Hack).-Gregarious. Singalila Range common.

25. *A. lancifolius* (Trin) Hochst. -Sporadic temperate hills; all over.
26. *A. nudus* (Steud) Hochst.-Sporadic temperate hills; all over.
27. *A. sikkimensis* Bor.-Semi-gregarious in upper temperate hills.
28. *Arundo donax* L.-A red grass of moist savannah and hills.
29. *Avena fatua* L.-A sporadic grass.
30. **Axonopus* compressus (Swartz) P. Beauv.-A excellent pasture grass.
31. *Avena subspikata* Clairv.-Recorded from Sandakphu.
32. *Brachiaria distachya* (L.) Stapf.-A grass of waste place.
33. *B. kurzii* (Hook.f.) *A. Camus* Syn. Panicum kurzii Hook.f.-A rare grass.
34. *B. miliiformis* A. Chase.-An excellent fodder grass.
35. *B. remota* (Retz.) Haines.-Rare; temperate Himalayas.
36. **B. reptans* (L) Gardener et. Hubb. (*Urochloa reptans* Stapf.) (*Panicum prostatum* Lamk.) - Gregarious grass of waste place.
37. *B. setigera* (Retz.) C.E. Hubb. (Syn. *Urochloa setgara* Stapf.).-A leafy sporadic grass.
38. *Brachypodium sylvaticum* (Huds) P. Beauv.-Upper temperate Himalayas; sporadic.
39. *Bromu's himalaicus* Stapf.-Upper temperate Himalayas; sporadic.
40. *B. japonicus* Thumb.-Upper temperate Himalayas; sporadic.
41. **B. tectorum* L.- Gregarious in temperate Himalayas.
42. *B. unioloides* H.B.K.-Sporadic in temperate Himalayas, fodder.
43. *B. ramosus* Huds.-Sporadic, temperate Himalayas in shade.
44. *Calamogrostis* emodensis Griseb.-High altitude, temperate Himalayas.
45. **Capillipedium* assimile (Steud) *A. camus*.-Fodder of foot hills of Jhum areas.
46. *C. parviflorum* (R.Br.) Stapf.-Foot hills. dry location, common fodder.
47. *Catabrosa sikkimensis* Stapf.-Sporadic inhigh temperate hills.
48. **Centotheca* lappacea (Wur) Desv.A forest grass; excellent fodder.
49. *Chionachne koenighii* (Spreng).-Thw. (Syn. *Polytoca barbata* Stapf. -Grasses of wet place, sporadic, bad fodder.
50. **Chrysopogon aciculatus* (Tetz.) Trin (Syn. *Andropogon acicatatus* Retz.)-A fodder grass of wet place.
51. *C. fulvus* (Spreng) Chiov.-Plain not common, fodder.
52. *C. gryllus* (L.) Trin.-Temperate hills, tufted grass, not common.
53. *Coelorachis striata A. camus*.-Not common.
54. *Coix lachryma* Jobi L.-A tall fodder grass; also cultivated.
55. *C. gigantea* Roxb. var. aquatica.-In most locations.

56. *Cymbopogon martinii* Wats.-Very common in foot hills.
57. *C. hookeri* Stapf.-Very sporadic.
58. *C. jwarancusa* (Sones) Schull.-On sandy river bank.
59. *C. microtheca* (Hook.f.) A. Camus.-On bouldary river bed; sporadic.
60. *C. flexuosus* (Nees) Wats.-Sporadic but common.
61. **C. pendulus* (Ness ex. Steud) Wats.-On sandy river bank.
62. **Cynodon* dactylon Pers.-All over, upto 2000 m.
63. **Cyrtococcum accrescens* (Trin) Stapf. Syn. Panicum trigonum Retz.-A gregarious moisture loving grass.
64. **C. oxyphyllum* (Steud) Stapf. A shade and moisture loving forest grass.
65. *C. pattens. A Camus*-A widely occurring moisture loving grass. Sporadic.
66. *Dactyloctenium aegyptium* (L) P. Beauv. Syn. *Eleusine aegyptica* Desf.-A fodder grass; sporadic.
67. *Dactylis glomerata* L.-Gregarious high altitude pasture grass.
68. *Danthonia cachmeriana* Jaub et Spach.-High altitude temperate hills; sporadic.
69. *D. cachmyriana* var. minor. A sporadic grass.
70. *D. jaccuemontii* Bor.- Sporadic in high temperate Himalayas.
71. *D. preslii* (Kunth).-Sporadic in high Himalayas.
72. *D. schreideri* Pilger.-Sporadic in Himalayas.
73. **Deschampsia caespitosa* (Linn.). P. Beauv.-Widely distributed, gregarious in upper temperate and alpine hills.
74. **Deyeuxia scabrescens* (Griseb) Munro.-Alpine, upper temperate gregarious grass of Himalayas.
75. *D. pulchella* (Griseb) Hook. f.-Upper temperate gregarious grass.
76. *Dichanthium annulatum* A. Camus (Syn. *Andropogon annulatus* Forsk.)-Sporadic.
77. *Diplachne fusca* (L.) P. Beauv.-Sporadic grass of plains, wide spread.
78. **Demostachya* bipinnata (L.) Stapf.-A tough gregarious; bed fodder.
79. **Digitaria adscendens* Henrard. (Syn. *D. sanguinalis* Scop.)-Tufted grass. Common all over.
80. **D. granularis* (Trin) Henr. (*D. pedicellaris* Prain).-Tufted grass. North Bengal Plains.
81. *D. cruciata* (Ness) A. Camus.-Very common in foot hills. Fodder grass.
82. **D. longiflora* Pers. (*D. tenuiflora* Beav.)-Creeping grass. Common all over.
83. *D. preslii* (Kunth) Henr.-Common. Creeping fodder grass.
84. *D. setigera* Ŕoem et. Schult.-Common. Fodder grass.

85. **D. sanguinalis* (Linn) Scop-All over common.
86. **Echinochloa colonum* (Link.).-A fodder grass of wet place.
87. **E. crusgalli* P. Beauv. (*Panicum crusgalli)* Linn.-A very common fodder grass.
88. *E. stagnina* Beauv.-A rare grass.
89. *Eleusine indica* Gaertn. Sporadic a good fodder grass.
90. **E. coracana* Gaertn.-Cultivated in temperate hills all over.
91. **Elymus nutans* Griseb.-Upper temperate Himalayas hills.
92. *E. dasystachys* Trin.-Upper temperate Himalyan hills.
93. *Elytrophorus spicatus* A. Camus. Syn. E. articulatus Beauv.-Sporadic in plains; sometimes in the hills.
94. *Eragrostis coarctata* Stapf.-Sporadic soft grass, not good fodder.
95. *E. diarrhena* (Schult.) Steud.-Sporadic soft grass, not good fodder.
96. **E. japonica* (Thunb.) Trin (Syn. *E. interrupta* Beauv.).-Sporadic soft grass, not good fodder.
97. *E. *gangetica* (Roxb.) Steud. (Syn. *E. elgantula* Stapf.)
98. **E. nigra* Nees ex. Steud.
99. **E. tremula* Hochst ex. Steud.
100. **E. tenella* Roem et Schult.
101. *E. pilosa* (L.) Beauv.
102. **E. uniloides* Nees ex Steud. Syn. *E. ambilis* Wight.

(98–102: Sporadic soft grass, not good fodder.)

103. **Erianthus arundinaceus* Retz.-Perhaps it is *S. begalenisi* of N. Bengal.
104. **E. fulvus* Nees Ex Steud.-Sporadic; temperate hills all over.
105. **E. longiosetosus* Anderss-Sporadic; all over lower hills.
106. **E. ravennae* Beauv.-A tall grass of North Bengal.
107. *E. sikkimensis* Hook. f.-Sporadic; temperate Himalayas in Sikkim.
108. **Eriochloa procera* (Retz.) C.E. Hubb. (Syn. *E. ramosa* O. Kuntze.)-A fodder grass of paddy field.
109. *Eulalia fastigiata* (Nees) Stapf. Stout fodder grass of Jalpaiguri, Assam.
110. *E. fimbriata* (Hack) O. Kuntze.-A decorative and fodder of grass.
111. *E. mollis* (Griseb) O. Kuntze.-Semigregarious in foot hills, fodder.
112. *E. quadrinervis* (Hack) O. Kuntze.-A bad fodder grass all over the eastern states.
113. *E. trispicata* (Schult) Henr.-All over lower temperate hills.
114. *Festuca cumminsii* Stapf.-Sporadic in high temperate hills.
115. *F. ovina* L.-Sporadic in high temperate hills.
116. *F. polycolea* Stapf.-Sporadic in high temperate hills.
117. *F. rubra* L.-Sporadic in high temperate hills.
118. *F. undata* Stapf.-Sporadic in high temperate hills.
119. *F. valesiaca Schelich* ex Gan.-Alpine grass; very high altitude.
120. *Garnotia stricta* Brenan.-Gregarious grass of all the hills, fodder.

121. *Glyceria tonglens* C. B. Clarke-Sporadic in all th states in temperate hills.
122. **Hackelochloa granularis* (L.) O. Kuntze.-All over plains of N. Bengal.
123. *H. porifera* (Hack) Rhind.-Foot hills all over.
124. **Hemarthria compressa* (L.f.) R. Br.-Sporadic in waste places.
125. *Helictrotrichon virescens* (Nees) Henr.-Sporadic in temperate Himalayas.
126. *Heriochola flexuosa* Hook.f.-Sporadic in temperate Himalayas.
127. *Holcus lanatus* L.-Wide, sporadic in temperate Himalayas.
128. *H. molis* L.-Cultivated; wild sporadic in temperate Himalayas.
129. *Hordeum vulgare* L.-A cultivated grass.
130. **Hymenachne pseudointerrupta* C. Muell. (Syn. *H. amplexicaulis*).-A fodder grass of swamps and bills.
131. **Hygroryza aristata* (Retz.) Nees.-A floating gregarious fodder grass; all over the states.
132. **Imperata cylindrica* (L.) P. Beauv. Syn. *I. arundinacea* Cyrill. Very common and conspicuous all over; fodder grass.
133. *Isachne albens* Trin.-All over sporadic.
134. **I. miliacea* Roth.-All over in swamp fodder grass.
135. *I. globosa* (Thurb) C. Ktze. Br.-In rice field. Fodder.
136. *I. himalaica* Hook. f.-In temperate Himalayas.
137. *I. sikkimensis*-Bor.-Sporadic in temperate high hills.
138. *Ischaemum indicum* (Houtt) Merr.-Sporadic; bad fodder.
139. **I. rugosum* Salisb.-All over; common, fodder grass.
140. *I. semisagittatum* Roxb.-Sporadic in Bengal and Assam.
141. **Iseilema laxum* Hack.-Locally gregarious, very good fodder grass.
142. *Koeleria cristata* (L.) Pers.-Alpine grass; sporadic.
143. **Leersia hexandra* SW.-Floating and mars land grass, fodder.
144. **Leptochloa chinensis* Nees.-Sporadic.
145. *L. panicea* (Retz.) Ohwl.-Sporadic.
146. *Lolium perenne* L.-A good grazing grass, high altitude in Himalayas.
147. **Microstegium ciliatum* A. Camus.-Fodder grass. Gregarious of bhabar track.
148. *M. nudum* A. Camus. Syn. *Pollinia nuda* Trin.-Sporadic.
149. *Microchloa indica* Beauv. *M. setacea* Br.-Sporadic; drier areas.
150. *Misanthus nepalensis* (Trin.) Hack.-A sporadic and ornamental grass in temperate open areas.
151. *M. nudipes* (Griseb) Hack.-A handsome sporadic grass.
152. **Mnesithea laevis* (Retz.) Kunth.-Sporadic and open place.
153. **Muehlenbergia huegelli* Trin.-Temperate hills. gregarious in partial shade.

154. ***Narenga porphyrocoma*** (Hance) Bor-Very common in Sal forest; elsewhere in North Bengal and Assam common.
155. *Neraudia arundinacea* (Linn) Menr. *(N. madagascariens* Hook.f.)-Sporadic.
156. **Neyraudia reynaudiana* (Kunth) Keng.-Vey common and gregarious in North Bengal and Assam plains and foot hills.
157. **Oplismenus burmanii* P. Beauv.-A shade loving fodder grass of forests.
158. **O. compositus* (L) P. Beauv.-Gregarious, forests shade loving fodder.
159. Ophiurus *exaltatus* (L) Kuntze.-Sporadic in West Bengal and Assam.
160. *O. megaphyllus* Stapf.-Sporadic, W. Bengal, Assam, Khasi and Naga hills.
161. *Oropetium thomaeum* (Linn.f.) Trin.-Sporadic in dry areas of W. Bengal and Bihar.
162. *Oryza minuta* F.S. Presl.-Tidal forests grass.
163. *O. rufipogon* Griff.-Forests grass in Terai plains.
164. *O. sativa* L.-Gregarious grass.
165. *Oryzopsis acquiglumis* Duthie-A temperate hills robust grass; all states.
166. *Panicum occidentale* Retz. Rare but very useful grass.
167. *P. auritum* Presl.-A forest grass, sporadic.
168. **P Walense* Muz.-On moist soil.
169. **P. brevifolium* L.-A shade loving fodder grass.
170. **P. maximum* Jacq.-An excellent fodder grass.
171. *P. miliaceum* L.-A leafy fodder grass of wet places.
172. **P. montanum* Roxb.-Sporadic.
173. **P. paludosum* Roxb.-A gregarious grass of water margin.
174. **P. psilopodium* Trin.-Sporadic; In moist sites.
175. *P. repens* L.-Sporadic in moist sites.
176. **P. sarmentosum* Roxb. Sporadic.
177. **P. trypheron* Schult. Sporadic.
178. **Paspalum conjugatum* Berg.-A fodder grass of open land.
179. **P. dilttatum* Poir.-A sporadic grass.
180. **P. distichum* Linn. *(P. vaginatum* SW.) *P. paspaloides* (Michx) Seribn. —A fodder grass of hills, sporadic.
181. *Perotis indica* (L.) O. Kuntze.-A creeping grass of open land, sporadic.
182. **Paspalidium flavidum.*-A Camus. (*Panicum flavidum* Retz.)-A widely creeping, leafy fodder grass.
183. **P. punctatum* A. Camus. (*Panicum punctatum* Burm).-A common good leafy fodder grass. Cultivated.

184. *Penisetum pedicellatum* Trin.-Cultivated.
185. *P. setosum* Swartz. (Brum) A. Camus.-A temperate forest grass.
186. *Phalaris arunidinacea* L.-Sporadic in temperate hills.
187. *P. minor* Retz.-Sporadic in alpine Himalayas.
188. *Phleum alpinum* L.-Gregarious in low-lying areas, all over.
189. **Phragmites karka* (Retz.) Trin.-Gregarious in alpine grazing ground.
190. **Poa alpina* L.-Very common in 1600 m to 2200 m.
191. **P. annua* L.-Sporadic in alpine areas.
192. **P. khasiana* Stapf.-Sporadic in alpine areas.
193. *P. ludens* R.R. Stewart.-Sporadic in shade in alpine areas.
194. *P. nemoralis* L.-An excellent pasture grass in alpine areas.
195. *Panicum flexuosum* Retz.-Sporadic.
196. *P. pseudopratensis* Hook.-Sporadic, fodder grass of alpine areas.
197. **Poa pratensis* L.-An excellent pasture grass in alpine areas.
198. *P. sikkimensis* Bor.-Sporadic.
199. *P. trivialis* L.-Sporadic, fodder grass of alpine areas.
200. *P. tibetica* Munro ex Stapf.- Excellent pasture grass in alpine areas.
201. *Polytoca barbata* Br.-A sporadic grass.
202. *Polypogon monspeliensis* Desf.-Semi-gregarious in upper temperate, grazed.
203. **Poganatherum panicum* (Lamk) Hack.
204. **P. crinitum Kunth.*-

} Occurs in foot hills, bhabar drier areas on rocks and in shades vertical cutting; gregarious.

205. **Rottboellia exaltata* L.f.-A sporadic grass of open grass land, fodder.
206. *Saccharum officianarum* L.-Cultivated.
207. **S. procerum* Roxb.
208. **S. spontaneum* L.

} Gregarious on river bank, a good fodder grass, best colonizer grass.

209. **S. ravennae* (Linn) Murrag.-A gigantic tifted grass.
210. **Sacciolepis indica* A Chase. (*Panicum indicum L.*)-Gregarious in swamp; otherwise sporadic.
211. *S. interrupta* (Wildd.) Stapf. (*P. interruptum* Wild.)-In Swamp, a good fodder grass.
212. **S. myosuroides* (R.Br.) A. Camus. (*Panicum myosuroides* R.Br.)-A good fodder in damp locality.
213. **Schizachyrium brevifolium* (SW) Nees.-Grows in open, sporadic.
214. *S. exile* Stapf.-Grows in open, sporadic.
215. *Sclerostachya fusca* (Roxb.)-Robust gregarious grass.

216. *Sehima nervosum* Stapf.-Not gregarious in E. India, good fodder.
217. **Setaria glauca* Deauv. Syn. *S. intescens*.-Sporadic in open forest floor.
218. **Setaria palmaeflia* Stapf.-Very common forest grass of shade.
219. *S. veticillata* Beauv.-Sporadic.
220. *Sorghum halepense* (L) Pers.-Sporadic in open and wet places.
221. **Sporobolus diander* Beauv.-True shade, and moisture loving forest grass.
222. *S. fertilis* Steud.-Sporadic in temperate Himalayas.
223. *S. pallidefusca* (Schumach) Staptf et Rubb.-Sporadic.
224. *S. viridis* Beauv.-Sporadic.
225. *Stipa mongolica* Truez extrin,-Alpine dry grass land.
226. *S. purpurea* Griseb.-Upper temperate, Himalayas and Nagaland.
227. *S. roylei* (Nees).-Mez.-
228. **Themeda strigosa* (Han) Hains.-A tall perennial grass all over the state.
229. **Themeda arundinacea* (Roxb) Pidley.-Tall (5m) grass plains and hills.
230. **T. caudata* (Nees) A. Camus.-A tall grass of foot hills. Semi-gregarious.
231. **T. hookeri* (Griseb) A camus.-A hill grass; sporadic.
232. *T. quadrivalvis* (L) O. Kuntze.-A sporadic grass; wide distribution.
233. *T. triandra* Forsk.-Gregarious species of N. Bengal and Assam.
234. **Thysanolaena* maxima (Roxb) O. Kuntze.-All over plains and foot hills.
235. *Tripogon filiformis* Nees ex Steud.-Sporadic in temperate hills.
236. *Trisetum spicatum* Richt.-Gregarious, fodder of upper temperate and alpine Himalayas.
237. *T. flavescens* (L) Beauv.-Sporadic.
238. *Triticum aestivum* (L)-Cultivated all over.
239. *Urochloa panicoides* (*Panicum panicoides* Beav.)-Rare Sal forests.
240. *Vossia cuspidata* Griff.-Floats on water in plains.
241. **Vetiveria zizanioides* (L) Wash.-All over plains at the margin of marses, paddy field.
242. *Zoysia matrella* (L) Merr.-Marshy land; sea side.

CYPERACEAE OF STUDY AREA

Kobresia

**K. seticulmis* Boeck. 3800 m. Singalila Forest.

K. hookeri var dioica Boeck. 3658 m. Singalila Forest.

K. angusta Clarke. 3658 m. Singalila forest.

Scleria Borg.-

S. tessellata Wild.-Plains to 2438 m. Terai and Duars. Slender Jalpaiguri forests.

S. hebecarpa Nees.-Plains to 1524 m. Pankhabari areas.

**S. elata* Boeck.-Beaves sheath winged. Plains to 610 m. Common in Jalpaiguri and Buxa forests. Robust and hairy. Also at 2600 m.

**S. elata* var. latior.-Plains to 1066 m. Jalpaiguri Forest.

Eriophorum

E. comosum Wall.-A robust sedge of marshy areas.

Lipocarpha R. Br.-

L. chinensis Kern. Riang in foot hills.

Rhynchospora Vahl.-

R. wallichiana Kunth.-Plains to 1524 m.

R. sikkimensis Clarke. 2438 m. in upper hills.

Scirpus Linn.-

S. supinux L. var. uninodis Clarke.-Darjeeling hills.

**S. erectus* Poir.-Upto 914 m. Common in Jalpaiguri and Darjeeling, Forest Division.

**S. articulatus* L. Common in Jalpaiguri and Darjeeling Forest Division. In marshes.

S. comosus Wall (*Eriophorum comosum* Wall.)-Darjeeling hills 2500 m.

**S. squarroses* L. Common in Jalpaiguri and Darjeeling Forest Division. All over.

Fimbristylis Vahl.-

**F. quinqueangularis* Kunth.-Darjeeling and Jalpaiguri Forests.

**F. barbata* Benth. (*Bulbostylis barbata* Kenth)-All over valley and Kalimpong and Darjeeling hills. Bristles on stem.

F. schaenoides Vahl.-25 cm. Common in Darjeeling hills.

**F.diphylla* Vahl. Common; 1 spiklet.-Darjeeling and Jalpaiguri Forests.

F. junciformis Kunth.-30 cm. Darjeeling and Jalpaiguri Forests.

F. fuscinux Clarke.-Terai.

F. filifolia Boeck.-Terai.

F. aestivalis Vahl.-Temperate hills.

Eleocharis R. Br.-

E. plantaginea R. Br.-Singalila Forests.

E. fistulosa Schult.-Singalila Forests.

E. capilata R.Br.-Singalila Forests.

E. atropurpurea Kunth.-All over.

E. congesta D. Don.-Palmajoa, Rimbick.

Kyllinga

K. triceps Rottb.-All over.

Mariscus Vahl.-

**M. cyperinus* Vahl. var bengalensis Clarke.-All over plains and foot hills.

**M. microcephalus* Presl.-Plains. Leaves spongy; stout, stem 7 m. Jalpaiguri Forests.

Pycreus Beauv.-

P. sulcinux Clarke.-North Bengal.

P. angulatus-1300 m. North Bengal.

P. nitens Nees.-Tufted stem. Common in Jalpaiguri Forests.

**P. globosus* Reichb. (*P. capillaris* Nees).-Common in Jalpaiguri Forest Division.

Cyperus Linn.-

C. cephalotes Vahl.-Plains, North Bengal.

C. platystylis R. Br.-Plains and Darjeeling Hills.

**C. cuspidatus* H.B. & B.-Plains and Darjeeling hills; all over hills.

**C. haspan* Linn.-Plains of Jalpaiguri, Buxa.

C. tenuispica Stendel. (*C. flavidus* Retz.)-All over plains.

C. niveus Retz.-Plains and hills.

C. diffusus Vahl.-Plain of Jalpaiguri Buxa and Darjeeling.

C. iria Linn.- Plains of Jalpaiguri and hills of Darjeeling.

**C. distans* Linn.-Plains to 910 m. Buxa, Jalpaiguri, Darjeeling.

**C. nutans* Vahl.-Plains and Darjeeling hills.

**C. brevifolius* Hask. (*Kyllinga brevifolia* Rottb.)-Common all over.

C. kyllinga endlicher (*Kyllinga monocephala* Rottb.)-Common all over.

**C. aristatus* Rottb.-2430 m. All over.

C. tegetiformis Roxb.-All over.

C. corymbosus Rottb.-Plains all over.

**C. tegetum* Roxb.- Plains to 1800 m. All over.

C. zollingeri Stemd.-Plains to 1624 m.

C. radiatus Vahl.- Plains.

C. exaltatus Retz.-Plains.

C. sesquiflorus Mattf.-Rongpoo.

C. silletensis Nees. ex Wight.-Bongpoo.

Carex Linn.-

**C. nubigena* D. Don.-2000-3000 m.

C. longiceps D. Don.-
C. teres Boott.-
C. sikkimensis Clarke.-
C. fucata Boott.-
] Upper temperate and alpine hills.

C. rigida Gooden-Singalila Range.

C. rara Boott.-2000 m.

C. cruciata Wahl.-1500 m. Tonglu and Singalila forests.

C. vesiculosa Boott.-1200 m.

C. stramentitia Boott.- In Sal forests.

C. polycephala Boott. - 7000 - 10,000 m - Upper temperate hills.

C. decora Boott.- 9000-12,000 m.- Upper temperate hills.

C. inaequalis Boott.-2500-3000 m.
C. pulchra Boott.-10,000-14000 m.-
C. filicina Nees var minor Boott.-2000m
] Upper temperate hills.

C. baccans Nees.-1000 m.-Lower temperate foot hills.

C. curva Boott.-
C. alpina Sw.-
C. duthiei Clarke.-
] Alpine hills.

C. cruenta Nees.-3300 m.

C. finitima Bott.-Palmajoa, Phalut, 2000 m.
C. lurida Clarke - 2500 m.-
C. inanis Kunth.-2000 m. Tonglu & Singalila Range.
] Upper temperate hills.

C. insignis Boott.-Darjeeling 2000 m. Takdah.

C. rochenbruni Franch. = *C. remota* L.-Singalila Forests, Akahills.

C. capillacea Boot.-Darjeeling hills.

C. indica L.-Terai. Common.

C. olivacea Boott.-Terai.

(vii) Analytical Synopsis of Fern and Fern Allies

During the study special attention was paid to identify fern and fern allies species and study their ecology. A preliminary record of ferns of the study area was made by the author and recorded in Indian Forester Vol. 90 No.2, 1964.

Given proper habitat conditions, ferns tend to grow pure and gregarious. But some ferns are sporadic in their occurrence. Most of the species are susceptible to fire and grazing damage. Species falling inside the quadrats, transects and adjoining areas have been recorded along with other ground flora and with epiphytes.

The present synopsis gives a consolidated lis of important species and their frequency of occurrence. Since it is intended to give a list of ferns no description has been given. Most common and conspicuous ones are marked with asterisks(*).

EQUISETACEAE

Equisetun Linn.-

E. debile Roxb. Ex Vaucher-Frequent in foot hills and less in plains along water channels. Its size is bigger than *E. diffusum.*

E. ramosissimum Desf.-var-*altissimum*-Frequent in the Tea Estates around Darjeeling between 1500-2100 m. altitude. The plant attains a height of about 1.5 m. and the main stem often becomes 10 cm. in diameter.

E. diffusum D. Don.-It is very frequently found along road sides between 1520-2130 m. Grows to a height of 35 m.

SELAGINELLACEAE

Selaginella Spring.

**S. vaginata* Spring-Grows at 762 m. to 2134 m. at Kurseong, Pankhabari and at lower altitude at Tista valley.

S. involvens (SW) Spr.-Common around Darjeeling at 1829 m. on rocks in the forests. Sometimes gives metallic lustre.

S. pennata (Don.) Spr.-Occurs in Great Rangit valley.

**S. monospora* Spring.-Very common on forest floor between 914-1829 m. at Darjeeling. Leaves in some pinkish grey lustre. Common in Lebong.

S. chrysocaulos (Hook & Grey) Spr.-Common near Senchal lake 2134 m.

S. tenuifolia Spr.-Rare. Found in Sureil and Darjeeling.

S. bisulcata Spr.-Occurs in Darjeeling (Sureil). Rare.

OPHIOLGLOSSACEAE

Japanobotrychium

J. lanuginosum Nishida (=*B. lanuginosum* Wall.)-Abundant between 1372-1981 m. on road side in Darjeeling and Sikkim.

Botrychium-SW.

B. lunaria (L) Swartz. Darjeeling, Sikkim; Rare.

Sceptridium.-

S. daucifolium Lyon var. *dauchifolium (=B. daucifolium* Wall).- In temperate hills between 1524-2438 m; Frequent.

Ophioglossum Linn.-

O. petiolatum Hook.-Common along Senchal Lake. Small sized herb.

O. valgatum Linn.-Common in Senchal.

Helminthostachys Kaulf.

H. zeylanica Hk.-Darjeeling hills at road side waste places.

LYCOPODIACEAE

Lycopodium Linn.-

L. sikkimense Herter.-Reported from Darjeeling, Senchal.

**L. serratum* Thunberg.-Common in Senchal forest (2438 m.)

L. setaceum Hamilt.-On tree trunks at 1524 m. in Lebong and Darjeeling forests.

L. squarrosum Forst.-An epiphyte. The plants hung from trees. All over hills.

L. phlegmaria Linn.-It is a rare pendulous epiphyte at 1524 m. Can be seen in Darjeeling.

L. cernuum Linn.-Darjeeling rare. A pretty club-moss.

**L. clavatum* Linn.-Common around Darjeeling at 2134 m.

FERNS (FILICINAE)
MARATTIACEAE

Angiopteris Hoofs.

A. evecta (Forst.) Hoffm.-In plains and foot hills at shady and damp localities. Sporadic (not rare).

OSMUNDACEAE

Osmunda Linn.-

O. claytoniana Linn.-High temperate hills at 2800 m. Frond simple pinnate; sterile pinnae entire.

SCHIZAEACEAE

Lygodium SW.-

**L. flexuosum* (L) SW.-In plains and hills. Frequent.

**L. microphyllum* R. Br.-Abundant.

**L. japonicum* Thbg) SW.-Wide distribution, frequent.

ADIANTACEAE

Adiantum Linn.-

A. pedatum Linn.-In shaded damp places, foot hills.

A. venustum Don. Prod.-In shaded damp place, foot hills.

A. capillus-veneris Linn.-Darjeeling on rocks. Frond ovate tri-quadripinnate.

A. caudatum Linn.-In plains and foot hills and on rocks. Fronds linear, oblong. It has walking habit by the terminal bud.

A. cuneatum Langed.-Darjeeling on rocks.

A. incisum Forms.-Common in foot hills.

A. lunulatum Burm.-Very common in moist situation upto 1300 m.

VITTARIACEAE

Vittaria J. Smith.-

**V. elongata* SW. All over foot hills. An epiphyte, pendulous.

V. sikkimensis Kuhn.-In temperate hills, Darjeeling, on shaded rocks.

V. ophiopogonoides Ching.-Very abundant between 1500-3000 m.

V. flexuosa Fee.-Sporadic at 2600 m. in Darjeeling.

ANTROPHYCEAE

Antrophyum Kaulfuss.-

**A. coriaceum* (Don.) Wall.-Common in foot hills in Darjeeling.

A. reticulatum (Forst) Klf.-Foot hills in moist shaded place; grows as epiphytes. Sporadic.

CRYPTOGRAMMACEAE

Onychium Kaulf.-

O. auratum Kaulf.-In Tirap and Darjeeling hills. Frond a span to a foot and a half long, ovate-lanceolate acuminate.

O. siliculosum (Desv.) C. Chr.-Common in foot hills in Darjeeling and Sikkim.

O. lucidum Spr.-Lebong, Darjeeling.

SINOPTERIDACEAE

Cheilanthes SW.-

C. tenuifolia (Burm) SW.-Occurs in foot hills Darjeeling. Sporadic.

C. farinosa (Forak) Kaulf.-Occurs in foot hills of Darjeeling, Sikkim. Common on rocks. Sporadic.

C. albomarginata C.B. Clarke.-Common in Darjeeling.

Notholaena R. Br.-

N. marantae (L) R. Br.-A high altitude fern of Sikkim and Bhutan.

CONIOGRAMMACEAE

Coniogramme Fee.-

C. fraxinea (Don) Diels.-Temperate hills in Darjeeling. Sporadic.

Pityrogramma Link.-

P. calomelanos (L) Link.-It has been found to occur extensively in Sikkim and Darjeeling in foot hills.

PARKERIACEAE

Ceratopteris Brongniart.-

C. thalictroides (L) Brown.-Very common in plains in wet and swampy areas.

PTERIDACEAE

Pteris Linn.-

P. wallichiana Ag.-In temperate hills of Darjeeling and Sikkim. Frequent.

P. biaurita Linn.-Sikkim and Darjeeling temperate hills; Frequent.

P. quadriaurita Ritz.-In upper temperate hills.

P. pellucida Presl.-Foot hills of Himalayas.

P. cretica Linn.-In foot hills.

P. blumena Ag.-Lebong, Darjeeling.

HYMENOPHYLLACEAE

Mecodium Presl.-

M. javanicum (Spr.) Copel.-Occur in foot hills.

M. exsertum (Wall ex. Hook.) Copel.-Occurs in Lebong, Darjeeling. Frond oblong, elongata, acuminate.

M. polyanthos (SW) Copel.-Occurs in Sikkim, Bhutan and Meghalaya. Main rachis have narrow wing; lower pinnate rhomboid, Kameng.

Trichomanes-

T. bipunctatum (Poir)-Occurs in temperate Himalayas hills.

Vandenbochia copelard

V. radicans Copel.-Lebong, Darjeeling; common.

V. pyxidifera Copel-Moist area in temperate hills.

Hymenophyllum Smith.-

**H. simosianum* Hock.-Epiphytes, common on trees in temperate hills.

DENNSTAEDTIACEAE

Dannstaedtia Bernh.-

**D. scabra* (Wall).-Moore. Very common in temperate hills.

Microlepia Presl.-

M. speluncae (L) Moore.- A widely occurring fern.

M. trapiziformis Kuhn.-Lebong.

**M. strigosa* (Thbg.) Presl.-Common between 600 to 1500 m. in Lebong, Darjeeling.

M. pilosula (Wall) Presl.-Lebong.

M. khasiana Presl.-Lebong, Darjeeling.

HYPOLEPIDACEAE

Pteridium Scopoli.-

**P. aquilinum* (L) Kuhn.-Very common at temperate hill.

Hypolepis Berhn.-

H. punctata (Thbg.) Mett.-Very common in foot hills.

LINDSAYACEAE

Lindsaya Dryander.-

**L. cultrata* (Wild.) SW.-Very common all over from 1000-1800 m.

Sphenomeris Maxon.-

**S. chinensis* (Linn). Maxon.-Very common in middle temperate hills of Darjeeling.

DAVALLIACEAE

Araiostegia Copel.-

**A. pulchra* (Don) Copel.-Very common on tree trunk.

Davallia J. Smith.-

D. bullata Wall.-Very common; epiphyte between 1000-1800 m.

D. polypodioides Don.-Common in Darjeeling.

OLEANDRACEAE

Oleandra Cavanilles.-

**O. wallichii* (Hook) Presl.-Common on rocks in temperate forests all over.

GLEICHENIACEAE

Gleichenia Smith.-

**G. glauca* (Thbg.) Hook.-An exceedingly common fern in hill slopes all over temperate hills of Darjeeling.

Dicranopteris Bernh.-

**D. linearis* (Burm.).-Very common in temperate hills all over. The ultimate branches bear a pair of forked pinnate.

CYATHEACEAE

Alsophila A. Br.-

A. glabra (Br.) Hook.-A tree fern occurs in shady moist ravines in the hills of Kalimpong.

A. glauca (Bl.) J. Smith.-Available at 1500 m. all over. A lofty tree fern. Pinnae 1m. long. Veins 3 branched all over Darjeeling.

A. ornata Schott.-Available at 1500 m. Recorded from Darjeeling. Rachis purple brown.

A. latebrosa (Wall) Hook.-A lofty tree fern; stipe dark grown in Kalimpong hills at 1100-1500 m.

Cyathea Smith.-

C. spinulosa Wallich.-Commonest tree fern between 1200-2000 m. in Darjeeling hills.

WOODSIACEAE

Woodsia R. Br.-

W. elongata Hook.-Very common on rock.

ASPIDIACEAE

Dryopteris Adanson.-There are more than 20 species recorded from Darjeeling. All are high altitude ferns.

Bolbitis Schott.-

B. heteroclita (Presl.) Ching.-Extremely common on forest floor.

Polystichum Roth.-Several species are common in the temperate hills.

ATHYRIACEAE

Cystopteris Bernh.-

C. sikkimensis R.C. Ching.-Sporadic at 2500 m.

C. tenuisecta (Bl.) Metta.-Available in temperate hills. Extremely common between 1500-2500 m.

Athyrium Roth.-Darjeeling and adjoining hills have more than 25 species.

A. attenuatum (Cl.) Tagawe, *A. rupicola* (Hope) C. Chr. *A. flabellulatum* (Cl.) Tard-Blot, *A. filixfoemina* (L) Roth, *A. pectinatum* (Wall) Presl., *A. Puncticaule* (B1.) Moore, *A. japonicum* (Thbg.) Copel,. *A. drepanopterum* (Kze) A. Bra., *A. fimbristegium* Copel., *A. nigragipes* (Bl). Moore,. *A. procerum* (Wall) Mild., *A. solenopleris* (Kerze) Moore are sporadic in Darjeeling District.

A. anisopterum Chrich very common in Darjeeling.

Dryoathyrium Ching.-

**D. boryanum* (Wild) Ching.-Very common at 1400 m.

Diplazium DE.-There are more than 15 species in Darjeeling and Sikkim plains and hills.

D. australe (R.Br.) S.S. Bir., *D. spectabile* (Wall. ex Mett.) Bir, *D. polypodioides* Bl., *D. latifolium* (Don) Moore.-N. Cachar hills, *D. maximum* (Don) C. Chr., **D. esculentum* (Retz) SW.-Kameng, all these are common and large sized fern (small tree fern) in the plains and foot hills. *D. stoliczkae* var *hirsutipes* Bedd.

THELYPTERIDACEAE

Thelypteris Schmidel.-It is a large genus with more than 18 species in Sikkim and Darjeeling.

**T. uliginosa* (Kunze) Ching.-Common in Rangit and Tista valley in Darjeeling and Sikkim.
T. brunnea Ching.-Common at 1500 m.
T. repens (Hope) Ching.-Darjeeling hills common.
T. sub-villosa Ching.-Very common.

Cyclosorus Link.-
**C. molliusculus* (Wall.) Ching.-Extremely common in Sal forests in Darjeeling District.
C. dentatus (Forsk) Ching.-In plains upto 1000 m.
C. parasiticus (L) Farwell.,-Occurs in Darjeeling.
C. papillco (Hore) Ching-In upper temperate hills.

Abacopteris Fee.-
**A. multilineata* (Wall) Ching.-Extensive in foot hills. Sal Forests. and on road side. *A. rubicunds* (V.A.V.R.) Hallb-Kameng, *A. europhylla* (Wall) Ching.-Kameng.

ASPLENIACEAE

Asplenium Linn.-There are more than 18 species in this genus in the area and all are epiphytes.

A. falcatum Lamk., *A. normale* Don., (Very common), *A. tenuifolium* Don., *A. nitidum* SW. are very common. *A. nidus* Linn., is the forests 'Bird next' fern which is very common in all types of tropical moist.
**A. laciniatum* Don. is very common.

BLECHNACEAE

Blechnum Linn.-
*B. orientale Linn.-Common in Darjeeling.

Woodwardia Smith.-

*W. *radicans* (L) Smith.-Common on rocks at 2000 m. in temperate hills.

Stenochbena.-

*S. *palustris* (Burn) Bedd.-Extremely common in Wet and moist forests of plains and foot hills.

LOXOGRAMMACEAE

Loxogramme (Bl.) Presl.-

L. chinensis Ching., *L. involuta* (Don) Presl. & *L. lanceolata* Presl. Grow in the temperate hills of Darjeeling. *L. involuta* (Don) Presl. *L. lanceolata* (Sn.) Pr.-These are all common epiphytes.

POLYPODIACEAE

Pyrrosia Mirbel.-

*P. *adnascens* (SW) Ching.-A common polymorphous fern of Darjeeling, Assam, Kameng and Sikkim. *P. mollis* (Kze) Ching., *P. stigmosa* (SW) Ching., *P. lanceolata* (L) Farwell are common ferns. *P. lanceolata* forms colonies. *P. mannii* (Gies-Ching and *P. flocculosa* (Don) ching occur in Darjeeling and Sikkim.

Pteridium acquilinum (L) Kuha-A sporadic fern in temperate Darjeeling hills (Lebong).

Polypodium Linn.-

*P. *amoenum* Wall. is the commonest epiphytic fern in the genus. *P. lachnopus* Wall, *P. argutum* Wall.-These occurs in Darjeeling hills.

Lepisorus (J. Smith) Ching.-

L. excavatus (Bony) Ching, *L. scolopendrius* (Don) Mehra & *L. nidus* (Hook) Ching are common epiphytes of Darjeeling.

Pseudodrynaria C. Christensen.-

*P. *coronans* (Wall) Ching.-Common in hills of Darjeeling. Fronds not dimorphic, long.

Drynaria (Bory) J. Smith.-

*D. *quercifolia* (L) Smith. A very common and abundantly growing epiphyte of plains and foot hills.

D. propinqua (Wall) J. Smith.-Tista valley.

Microsorium Link.-

*M. *membranaceum* (Don) Ching., *M. zippelii* (Bl.) Ching,. *M. punctatum* (Linn) Copel are commonest species in Darjeeling. These are wide spread.

(viii) Analytical Synopsis of Monocotyledon Flora

A list of monocot plants collected and identified has been produced. A very brief description of the species their fidelity and the zone of occurrence and have been provided. Grasses and Sedges collected and identified from the study area have been listed separately (in Seclion VI). They study was not intended for compiling a flora of the area, rather it was intended to give a quantitative assessment of various components of vegetation to some extent. Conspicuous and widely occurring species have been recorded and marked (*) asterisk.

MUSACEAE

Musa Linn.-

**M. thomsoni* King. Mss.- In sheltered spots in the foot hills; Seeds few, soft, black; pulp soft, sweet. Common.

Occurrence:- Foot hills, Sukna, Sevoke.

M. hookeri King. Mss.- In the middle hills from 1372-1829 m. Seeds few, large, pulp hard, sweetish. Sporadic.

Occurrene:- Latpanchan, Monpoo.

**M. sapiantum* Linn. *dubia* King Mss.- In the foot hills, stem 3-4 m. high. *Fruits* smaller and with more pulp than in *M. hookeri* and more angular and with less pulp than in *M. thomsoni*. Common.

Occurrence:- Sukna.

M. babisiana Colla (=*M. Paradisiaca* Linn., *M. sapientum* Linn).- Stem tall upto 4 m. cylindrical. Bracts boat shaped or ovate subtending many *flowers* in 2 rows, scorpioid cymes. Bracts deciduous. A tall herb. Sporadic.

Occurrence:- Foot hills, cultivated.

ZINGIBERACEAE

Hedychium Koenig:-

H. coronarium Koenig.- A herb with sweet scented *flower*. *Flowers* are generally white to yellow; may be red and white tinged red flush in Aug. Sporadic.

Occurrence:- Middle to lower hills, Tindharia.

H. spicatum Sm.- A herb with glabrous leaves, epiphytes also. *Flowers* white. Sporadic to common.

Occurrence:- Kurseong 1400 m.

H. spicatum Var. *acuminatum* Wall.- Recorded from Darjeeling (2438 m.) *Flower* 11. Darjeeling.

H. spicatum var *accuminatum* (Rose) Wall.- A herb. leaves oblong. *Flowers* small. *Flowers* 7.

Occurrence:- Foot hills to 2700 m.

H. thyrsiforme Sm.- A tall leafy herb. Leaves 10 cm. wide spike with very dense pale yellow *flower*. *Flowers* 2.

Occurrence:- From 1000 to 2500 m.

H. coccineum Buch - Ham.- A tall leafy herb. *Flowers* 8. *Flowers* red.

Occurrence:- Foot hills.

**H. gracile* Roxb. var. glaucescens:- Spike long. *Flowers* 8. *Flowers* small.

Occurrence:- Foot hills.

Globba Linn.-

**G. bulbifera* Roxb.- Spike lex flowered, lower bracts bulbuliferous. Another winged. Common.

Occurrence:- Sal forests and upto 1600 m. hills.

**G. racemosa* Smith.- A herb about 60 cm. high. *Flowers* small bright yellow. *Flowers* 6. Bulbils 0, Panicle 30 cm. Common.

Occurrence:- Foot hills upto 2000 m.

G. clarkei Baker.- *Flowers* yellow finged brown. Bulbils many at the upper portion of the inflorescence.

Occurrence:- Foot hills.

**G. hookeri* Clarke, ex Baker- *Flowers* - 7. Leaves finely hairy below he mid vein. Widely occurring.

Occurrence:- Foot hills upto 2000 m.

G. multiflora Wall, ex- Baker.- *Flowers* - 8. A sporadically occurring herb

Occurrence:- Upto 2000 m.

Kaempferia Linn.-

**K. angustifolia* Rosc. - A stemless herb. *Flowers* white. With red tip, sessile spike. Common.

Occurrence:- Temperate forests 1800 m.

K. rotunda Linn.- A stemless herb. *Flowers* reddish. Abundant.

Occurrence:- At 2000 m. in temperate forests.

Curcuma Linn.-

C. angustifolia Roxb.- A small herb; spike 30 cm. Corolla tube white or yellow. Staminode and labellum bright yellow. Frequent.

Occurrence:- Plains of Sal forests.

**C. aromatica* Salisb,- A Large herb, cultivated. *Flowers* 5-7. Common in waste places and Sal forests. Wild and cultivated. Abundant.

Occurrence;- Plain of Sal forests.

**C. zedoaria* (Bergius) Rosecoe.- Large herb, leaves have purple tinge up to the middle. *Flowers* yellow with white inside. Fl.3. Abundant.

Occurrence:- All over plain forests.

C. domestica Val (=*C. longa* L.)- Leaves short petioled, *flowers* pale yellow. Tubers bright yellow inside. Cultivated.

Cautleya Royle.-

**C. gracilis* (Smith) Dandy (=*C. lutea* Royle).- *Flowers* bright yellow, bracts red. Frequent.

Occurrence:- Temperate hills 1800-2100 m.

**C. spicata* (Smith) Bak.- Leaves broader, sessile. Bracts of *flower* red, labellum wide, yellow. Frequent.

Occurrence:- At 2100 m.

C. cathcarti Bak.- A robust herb, bracts green shorter than clayx. Sporadic.

Occurrence:- Rangesoom.

Hemiorchis sulp Kurz.-

H. pantlangi King.- Corolla and rachis bright yellow; *flowers* in raceme; lateral staminodes yellow with brown margin.

Occurrence:- foot hills.

Alpinia Linn.-

*A. *malaccenensis* (Burn.f.) Rose.- A tall herb 3 m. *Flower* in simple racemes, corolla segments white; labellum ovate emerginate, centre red yellow. Margin incurved. Capsule globose, yellow. Abundant.

Occurrence:- In lowlying areas in places.

**A. allughas* (Retz) Roscoe.- A tall (1.8 m.) leafy herb. Corolla segments greenish white. Capsule globose black. *Flower* - 3. Abundant.

Occurrence:- In low lying areas in plains.

Amomum Linn.-

A. dealbatum Roxb.- Leaves large, pubescent below. Capsule red 2.5 cm. with winged crenulate rubs, globose, Corolla white.

Occurrence:- Upper temperate hills.

A. subulatum Roxb.- Stem 1 m., Leaves glabrous. Capsule red brown, globose, crisped, winged.

Occurrence:- Sal forests in Tondu Block.

A. kingii Bak.- A stout stemmed herb; bract place brown. Corolla tube white, tinged yellow.

Occurrence:- Foot hills.

A. cardamon Linn. (=*A. costatum* Benth = *A. cardamomum* Roxb). - *Fruit* leathery ribbed capsule. Seeds aromatic. Beng - Bara alach.

Costos . Linn.-

**C. speciosus*(Koenig) Smith.- A herb or shrub upto 2.5 m. high; *flower* white in terminal globose head like conical spike; *fruit* capsular red seeds arillate; bracts ending into throny species. Frequent. *Flowers* 6-10.

Occurrence:- Plain Sal forests and foot hills.

Roscoea Smith.-

**R. purpurea* Smith. - A herb with leafy stem, corolla purple or white. Abundant.

Occurrence:- Temperate hills 2200 m.

MARANTACEAE.

Schumannianthus (Clinogyne)

S. dichotomus (Roxb.) Gagnep (*C. dichotoma* Salisb).- A shrub, cane like; stem polished green. Seeds arillate, *Fruit* capsule. Rare.

Occurrence:- In road side burrow pit.

Phrynium Wild.-

P. pubineve B1. (*P. capitatum* Wild).- Common in shady marshes in plains. Gregarious

IRIDACEAE

Iris Linn.-

I. clarkei Bak.- A herb with stout stem. *Flower* bright or lilac. *Flowers*-7. Sporadic.

Occurrence:- Upper temperate hills.

I. nepaulensis Dc. Don.- Leaves linear. Spathes 1-3 flowered. Crost of style arms toothed. *Flowers* lilac.-Sporadic.

Occurrence:- Upper temperate hills.

AMARYLIDACEAE

Crinum Linn.-

C. asiaticum L.- Leaves large wide (12 cm. or 50 cm.). Umbel many flowered (more than 15 cm.). Perianth tube 8 cm.; segments linear as long as tube. Sporadic.

Occurrence:- Sal forests.

C. wattii Bak.- Scape flaccid, umbel 6-8 flowered, pedicel short. Sporadic.

Occurrence:- Sal forests.

Pancratium Linn.-

P. verecundum Ait. - A bulbous herb of Duars. In Sal Forests. *Flower* white, not common. It differs from *Crinum* in that the filaments here are united towards their base by petaloid membrane. Sporadic.

Occurrence:- Sal forests.

Curculigo Gaertn.-

**C. orchioides* Gaerth.- In Sal forests. *Flower* yellow; *fruit*-11. Common.

Hypoxis Linn.-

H. aurea Lam.- Leaves upto 30 cm. long, narrowly linear, scape - 2 flowered. *Flowers* yellow. *Flowers* - 6. Frequent.

Occurrence:- Sal forests.

Molineria Colla -

**M. recurvata* (Dryand) Nel.- *Flowers* in sub-specate spike. *Flowers* - 5. A shout herb. Common.

Occurrence:- Temperate forests from 1500 m.

DIOSCOREACEAE.

Dioscorea Linn.-

D. anguina Roxb.- A large climber. Leaves simple cordate, tomentose, opposite and alternate, long petioled. Common.

Occurrence:- Sal forests.

D. alata L.- Large climber, quadrangular, spinous towards base; leaves linear cordate. Cultivated; Tubers long. Sporadic.

Occurrence:- Sal and foot hill forests.

D. hamiltoni Hook.f.- Leaves simple cordate, ovate, branches 4-angled or slightly winged. Twisted stems. Frequent.

Occurrence:- Sal and foot hill forests.

**D. daemona* Roxb.- A large climber, 3-leaves, big *fruit*. *Flowers*-7. Frequent.

Occurrence:- Sal and foot hill forests.

**D. pentaphylla* Linn.- Leaves 5, pinnately nerved. Frequent.

Occurrence:- Sal and foot hill forests.

D. deltoidea Wall.- Leaves hastate, *flowers*-8. Sporadic.

Occurrence:- Sal and foot hills forests.

**D. oppositifolia* L.- Leaves opposite, margins cartilaginous. Common.

Occurrence:- Sal and foot hills forests.

LILIACEAE.

Gloriosa Linn.-

G. superba Linn.- A climbing herb with yellow orange tepals, sporadic.

Occurrence:-Sal forests (Wet type).

Chlorophytum Ker-Gawl.-

C. arundinaceum Baker.- Scape taller (20-50 cm.) than leaves, Anothers longer than filaments. *Flowers* white. Common.

Occurrence: Sal forests.

C. nepalense (Lindle) Baker (*C. undulatum* Wall.)- Scape naked, paniculately branched. Racemes with *flowers* remotely scattered, *flowers* drooping. *Flowers* - 8. Sporadic.

Occurrence:- Temperate hills Senchal.

Lilium Linn.-

L. roseum Wall.- (3000 m.), *L. wallichianum* Schultz (1700 m.), *L. nepalense* D.Don (1800 m.)- Darjeeling.

Dracaena Linn.-

D. angustifolia (Rumph) Roxb. (*D.ensiflora* Wall.) - A tree with persistent leaf scars on the stem (2-3 cm.), *Flowers* white. Frequent.

Occurrence:- Sal forests (Wet type).

Asparagas Linn.-

**A. racemosus* Wild.- A tall climbing shrub with fleshy fasciculated roots cladodes 2-6 in cluster. Stem scandent, strongly spinous. *Flowers* in racemes. Abundant.

Occurrence:- Sal forests and dry type forests.

Mondo (=Ophiopogon)

M. Clarkei Hook.f. *M. wallichianus* (Kunth) Baily *M. dracaenoides*(Baker) Forwell.	Temperate hills at 1800 m. at Palmajoa.

Peliosanthes.

P. griffithii Baker - Tadan at 1700 m.

P. macrophylla Wall - Sal forest in Sukna and Bagdogra.

**Paris*

P. polyphylla Smith- Var Brachystemon Franchet - Temperate hills - Common, *Flower* in May.

Tupistra

T. clarkei Hook - A herb at 1600 m.

Tovaria =(Smilacina).- Spike dense flowered.

T. fuscu Baker.- At 2000 m.

Streptopus

S. simplex Don Prodr.- A herb with ovate, cordate leaves.

Polygonatum

P. punctatum Royle
P. cirrohifolium Royle } Herb from Tiger hills.

Smilax Linn.-

**S. zeylanica* Linn (=*S. macrophylla* Roxb.
**S. ferox* Wall - in plains
**S. perfolira* Lour (S. Prolifera Roxb.) in plains } Widely occurring in Sal forests.

S. wightii A.D.C. - 100 m. at Pesoke.

**S. glaucophylla* Khotzsch
S. ocreata A.D.C.
S. rigida Wall. } 2000 m. from Senchal Rimbick.

PONTEDERIACEAE

Eichornia Kunth.-

E. crassipes (Mart) Solms.- (Water hyacinth). Common in ponds and rivers introduced from America. The posterior inner perianth segment yellow rhomboid spot with bluish. Sporadic.

Occurrence:- Plains of swampy areas.

Monochoria Presl.-

**M. vaginalis* (Burn f) Presl ex Kunth.- Marsh plant. *Flower* blue. Inflorescence spike like. Common.

Occurrence:- Plains of swampy area.

M. hastata (L) Solms-(*M. hastaefolia* Presl.) - *Flower* long-peducelled and bright blue with one large blue stamen and 5 smaller yellow ones. Common.

Occurrence:- Plains of swampy area.

COMMELINACEAE

Aclisia (Pollia) Thamb.-

**A. serzogoaensis* E.May. *(Pollia sorzogoaensis)*.- 1219 m. *Flower* 12. A herb with subsessile leaves, persistent bract. White on pink *flowers*. Fruit blue. Frequent.

Occurrence:- Wet, swampy area in plains.

**A. subumbellata* Bruckner (*P. subumbellata* Clarke).- 914 m. *Flower* 8. Pancle sessile depressed. Sepals round. Frequent.

Occurrence:- Wet swampy area in plains.

Cyanotis Don.-

C. vaga (Lour) Don
C. cristata (Linn) Don } Screeping stem at 2300 m.

Anis cholotype Hassk (Forrestia Rich)

A. hookeri (Hassk) Hara (**F. hookeri* Hassk.-).- Stem stout erect, capsule oblong larger than sepals. Common.

Occurrence:- Plain sal forests of moist situation.

A. glabrata* Hassk (F. glabrata* Hassk.-) Stout erect herb 30 cm. - 1.5 m. Leaves 10-15 cm., acuminate cuspidate. Sheath 3.0 - 5.0 cm. very damp situation. Common.

Occurrence:- plains of sal forests.

Streptolirion Edgew.-

**S. volubile* Edgeworth.- A climbing undershrub with long petioled leaves and distinct leaf-sheath. Grows in shade, gregarious, *Flowers* in May.

Occurrence:- Temperate forests 2000 m. Raugiram.

Murdannia Royle.-

M. scapiflora (Roxb.) Royle.- *Flowers* on leafless scape in cymose panicles, bracts large sheathing. Tufted herb. Sporadic.

Occurrence:- Temperate 2100 m.

M. divergens (Clarke) Bruckner.- Tall stout herb, linear leaves, Petals blue and seeds yellow. Sporadic.

M. blumei (Hassk) Brcnam (*M. hamiltonianum* (Wall) Bruckner)- Much branched weed of the marshy ground. Petals white or pink. Branches usually creeping. Sporadic.

Occurrence:- Temperate forests.

Commelina Linn.-

**C. sikkimensis* Clarke.- Leaves lanceolate, stems slinder and creeping. Sapthes stalked, ovate lanceolate. Common.

Occurrence:- Tista valley upto 1000 m.

**C. apendiculata* Clarke.- Diffusely branched tall herb. Leaves sessile, linear, narrow lanceolate glabrous. Common.

Occurrence:- Sevoke

C. diffusa B wim f (*C. nudiflora* Linn.) - A creeping herb, leaves lanceolate the two interior petals obovate with long claws, dark blue, the exterior one sabsessile, orbicular, pale blue or white. 914 m. Common.

Occurrence:- Tista valley.

Floscopa Lour.-

**F. scandens* Lour.- *Flowers* 7. A semi prostate herb with distinct leaf sheath; *flowers* white or pink, seeds transversely wrinkled. Common.

Occurrence:- Plains of sal forests.

JUNCACEAE

Luzula D.C.

L. plumosa. F.Meyer.- Appendage of seed large, curved look like. Capsule oblong apiculate.

Occurrence:- Upper temperate hills.

Other species are - *L. effusa* Buch., *L. chinensis* N.E. Brown., *L. Campestris* D.C., *L. leschenaultii* (=*J. prismatocarpus*-R.Br.)

Occurrence:- In upper temperate hills.

Juncus Linn.-

J. bufonius Linn.- *Flowers* pale green Upto 3000 m. Rare.

Occurrence:- Marshes.

PALMAE

Licuala Thunb.-

L. peltata Roxb.- A gregarious shrub 5 m. high L.H.F. and ascerding to 1800 m. *Flowers* tomentose, large. *Fruit* red, 1.3 cm. long. Rare.

Occurrence:- Tista valley.

Livistona R.Br.-

L. jenkinsiana Griff.- A fan palm upto 9 m. high. Not common.

Occurrence:- Tista valley.

Trachycarpus H. coendl.-

T. martianus H. Wendl.- A tall fan palm of 1800-2400 m. Rare. *Flowers* yellow. Drupe yellow then glossy blue.

Occurrence:- Temperate hills 2000 m.

Plectocomia Mart et Blume Schult.-

**P. himalayana* Griff.- A scandent cane of 1300-2000 m. Very common. *Fruit* 1.3 cm. in diameter. Scales shining fimbriate. Frequent.

Occurrence:- All over temperate hills.

Daemonorops Bl.-

**D. jenkinsianus* Mart Dangre Bet.- Commonest of all areas. Stems very long, stout and scandent. Used for making baskets. Petiole rachis and flagellum with many hooked spines or 3-5 fid clavs. When cut yields white juice. Abundant.

Occurrence:- In plains of sal forests.

Calamus Linn.-

**C. erectus* Roxb var. Schizospathus Rhong, L.- An erect densely tufted Rattan. Common.

**C. flagellum* Griff.- Leaf sheath flagellate with very unequal spines. Common.

Pinanga Bl.-

**P. gracilis* Blume. - Found in Sal and miscellaneous forests in deep shade. Usually gregarious. Fruiting spadix 30 cm; *fruit* in 3 rows, 1.3 cm. in diameter, red. Common.

Occurrence:- Sal forests.

Phoenix Linn.-

P. humilis Royle. var *typica* - Found in the steep rocky slopes of sub-Himalayan tract. Stem upto about 3.0 m. high. *Fruit* 2-5 fruiting stalk very long. *Fruit* orange to black.

Occurrence:- Tista valley.

P. rupicola, T. Andrews - Found in the foot hills in the Tista and Mahanda valleys. It generally grows on rocks and especially or very steep cliffs on both sides of the Tista above Sevoke. *Fruit* is shining yellow.

P. acaulis Roxb.- Found in the Terai and foot hills. Common Sal forests in the Tista valley. *Flowers* cold. Season *fruit* 4-5 edible. *Fruit* bright red to blue black.

Caryota Linn.-

C. urens Linn. Rangbhang. P. The Indian Sago palm, palm attaining 18.3 m. Common in the foot hills and occasionally upto 1219 m. Sometimes planted in gardens. Stem 61.0 cm. in diameters, annular, smooth, grey, and shining.

Occurrence:- Foot and middle hills.

Wallichia Roxb.

W. densiflora Mart.- A stemless palm, common in the L. and M.H.F. upto 1219 m., especially on rocks under dense forest. *Flowers* 5-7 *Fruit* 3-4.

Occurrence:- Middle hills.

PANDANACEAE

Pandanus Linn.-

**P. furcatus* Roxb.- A small tree with numerous, thick, soft, aerial roots from the lower part of the stem. Found commonly all over sub-Himalayan tract. Common.

Occurrence:- Foot hills.

P. minor Ham.- A dwart shrub with densely matted stem and roots. Common in the plains. Stamens connate, filaments in palmate clusters. Carpels free. *Fruit* drupe, red. Common.

Occurrence:- All over.

TYPHACEAE.

Typha Linn.-

**T. elephantina* Roxb.- It is a tall plant of marshes. Very common in Terai and Duars on chars of rivers. *Flowers* monoecious. Male *flower* with 1-5 stamens. Female *flowers* consists of a single carpel.

Occurrence:- Over low lying areas.

ARACEAE

Pothos Linn.-

**P. sandens* L.- *Flowers* - 3. Common epiphytic climber of the foot hills and plains. Spadix with a pedicel above the spathe which is usually deflexed. Abundant.

Occurrence:- All over plains and foot hills.

**P. cathcarti* Schoot.- Epiphytic climber of the L.H.F. Leaves bigger than other two species. Pedicel erect. Fl.-5. Berries scarlet. Abundant.

Occurrence:- All over plains and foot hills.

P. vriesianus Schoot.- Spadix large erect, inclined, bracts large climbing shrub.

Occurrence:- Extend upto 1400 m.

Scindapsus Schott.-

**S. officinalis* (Roxb.) Schott.- Sal forests. A large epiphytic climbing shrub and spathe green, ultimately yellow within, terminating in long acumen, deciduous. Common.

Occurrence:- Plains and foot hills.

Rapidophora Schott.-

**R. hookeri* Schott.- A lofty epiphytic climber. Leaves entire membraneous, cuspidately caudate acuminate, base unequal. Petiole furrowed upto joint at petiole. Beak of spathe slender.

Occurrence:- Foot hills.

**R. glauca* (Wall) Schott.- Lofty climber of Sal forests of Terai and Duars and sub-Himalayan tract in damp situations. Petiole 25 cm. Spadix pale yellow.

Occurrence:- Sal forests and foot hills.

R. deeursiba (Roxb) Schott (*R. eximia* Schott.-) Leaf pinnatisect, long 30-60 cm. Spathe long, orange yellow.

Occurrence:- Foot hills.

Pistia Linn.-

**P. stratiotes* Linn. var *cuneata* Engl.- Leaves triangular, connate crisped, floating herb.

Occurrence:- Common in ponds.

Lasia Lour.-

**L. spinosa* Thwaites (=*L. heterophylla* Schott.) - A very spiny stout herb. Shathe deep purple, twisted, convulute, thick and rather brittle. Common in Marshes.

Occurrence:- In forest swamps all over.

Amorphophallus Blume.-

A. companulatus Blume.- A very stout herb. Peteole mottled dark and light green very stout and rough with small tubercles. Leaves large pinnatisect and warted corms. Spathe companulate.

Occurrence:- Sal forests.

Gonatanthus Klotzsch.-

**G. pumillus* (D. Don) Engler (=*G. sarmentosus* Klotz)-*Flowers* 7. Berries yellow. Leaves peltate, ovate cordate. Abundant.

Occurrence:- Foot hills, middle hills.

G. ornatus Schott.- Leaves narrow oblong, lanceolate, base cordate, coppery purple between the nerves. Abundant.

Occurrence:- Middle hills.

Alocasia Schott.-

A. cucullata (Lour) Schott.- A herb, leaves slightly peltate, lower nerves of lamina digitate, cordate, wide petiole very long. Rhizomes 1-2 feet. Spathe narrow, boat shaped, fleasly appendage short.

Occurrence:- Rangit valley.

A. falax Schott.- A herb, leaves broadly ovate, orbicular, ovate-sagittate, spathe tubes oblong, the lanceolate, limb cuspidate, spathe yellow.

Occurrence:- Foot hills.

Arisaema Mont.-

A. qriffilhir schott var *verrucosum* Schott (*A. verrucosum* Schott.) At 2400 - 3000 m. Limb of spathe broad, obcordate, spadix purple.

Occurrence:- Tonglue

A. nepenthoides (Wall) Mart.- Limb of spathe dilated into 2 recurved auricles. Leaves 2, leaflets 5, sessile.

Occurrence:- At 3000 m., Tonglu.

A. jacquemontii Bl.- Leaves 2, digitate, leaflets 5-7 sessile. Spathe limb lanceolate, margin recurved. Appendage with subcylindric tall.

Occurrence:- At 3200 m. in Phalut.

A. concinnum Schott. var *alienatum* Engler.- 1524 m. Spathe lanceolate to lanceolate; limb tailed.

A. tortuosum Schott.-Leaves 2-3 digitate, pedate, leaflets 5-18. Limb of spathe pale green or purplish, boat shaped.

Occurrence:- At 1900 m.

A. speciosum (Wall) Mont.- var *mirabile* 305-3658 m. Tube of spathe striped purplish red. Leaf 1, leaflets petiolate.

Occurrence:- At 2100 m. in hills.

Typhonium Schott.-

T. trilobatum (L) Schott.- Spathe green or inside red-purple, appendage bright red stipitate and muricate. Common.

Occurrence:- In marshes, in plains.

LEMNACEAE

Lemna Linn.-

L. perpusilla Torry (=*L. paucicostata* Hegelm.) Floating stem asymmeric. Seed erect.

Occurrence:- Common in plains.

**L. polyrrhiza* Kurz.- Stem herbaceous, roots many.

Occurrence:- Common in plains.

Wolffia Hook. f.-

W. glabosa (Roxb) Hantog & Mas (*W. arrhiza* Wimm.-) Smallest known monocot on water floating. Stem subglobose.

Occurrence:- Common.

W. microscopica Kurz.- Extremely minute, floating annual, stem flat above, conical below.

Occurrence:- Common.

ALISMACEAE

Alisma Linn.-

A. plantago Linn.- Leaves 5-7 nerved, linear, lanceolate; achenes membranerous, compressed. Petals pink with yellow cloud.

Saggitaria Linn.-

**S. sagittifolia* Linn.- In marshes, leaves hastate or sagittate. *Flowers* white with usually purple centre.

Occurrence:- Common in plains

Limnophyton Miq.-

L. obtusefolium Miq.- Aquatic or marshy herbs with milky juice and erect sagittate leaves. *Flower* white.

Occurrence:- Common in plains.

NAJADACEAE

Aponogeton Thunb.-

A. crispum Thunb.- Root stock stoloniferous, leaves linear with compressed petiole. *Flowers* crowded white.

Occurrence:- Common in plains, in water.

Potamogeton Linn.-

P. indicus Roxb.- *Flowers* densely packed. Sepals with long claud. Floating leaves large.

Occurrence:- Common in water.

Najas Linn.-

N. kurziana Rendle.- Leaves minutely dented, numerous; sheath shortly auricled.

Occurrence:- In water common.

ERIOCAULACEAE

Eriocaulon Linn.-

E. capillusnaiadis Hook. f., *E. xeranthemum* Mart., *E. lazulifolium* Mart. *E. alpestre*, *E. nepalense*, are a few species that occur in swamps and marshy areas. These are grass-like annuals or perennials.

15

Floristic and Ecological Findings

A. FLORAL DIVERSITY

The quantitative figures based on the analysis of various data have been analysed in this diversity studies. While determining the specific diversity, samples were taken by units from a population arranged in log-series distribution. In all random samples of any size from one population the ratio of nl: the number of groups with one unit to x (a number less than unity), the

$$\text{specific diversity} = \frac{n1}{x}$$

The ratio of the number of individuals (N) to the number of species (S) is given by

$$\frac{N}{S} = \frac{X}{(1-X)\,(\log_e \overline{1-x})}$$

The value of x having been found out, the specific diversity is determined as follows:-

$$\text{Specific diversity} = \frac{N\,(1-X)}{x}$$

The generic diversity is determined in the same lines as specific diversity, with number of species (S), replacing the number of individuals (N) and the number of genera (G) replacing the number of species (S) in the aforesaid formula for specific diversity.

$$\text{Generic diversity} = \frac{S\,(1-X)}{x} \quad \text{where S = number of Species,}$$

x is value obtained for each of the different combinations of S/G.

It has been found mathematically that if in community the species are spread over a large number of genera, the generic diversity is high and low, if only a few genera are represented.

The generic coefficient is given by the following formula:

$$= 100 \times \frac{\text{Total number of genera}}{\text{Total number of species}}$$

In the case of log series distribution which many genera and species distributions follow, the relation between the average number of species (S) per genus (G) in a sample form population with diversity (d).

In other words, it is dependent on the ratio between the number of genera represented in the sample (i.e.) on the size of sample) and the richness of population. If 'd' remains constant i.e. if a series of random samples are taken with same number of genera from two different communities the coefficient is simply is a measure of diversity.

But actually in field many factors interfere with normal or natural distribution of vegetation. In this particular vegetation zone—

(i) exploitation of economically important plants started more than a century back.

(ii) *Shorea robusta* forests had their origin and distribution in fire protection.

(iii) Strict fire protection measure rendered the soil moist resulting in invasion of evergreen species in the *Shorea* forests.

(iv) Many fire tender species suffered from surface fire.

(v) Man-made forests replaced the original natural forests over large areas.

(vi) Grazing, lopping, flood and fire had tremendous effect on the forests.

As such, any mathmetical finding should be evaluated in consideration of the biotic and climatic factors that each of the study zones has suffered.

Although pure *Shorea* forests form bulk of the plains forests, maximum diversity is evidenced in the wet-mixed forests (dry-mixed types are very local). But in the typical dry mixed forests both specific and generic diversities are very high. One significant indication is that in mixed forests the species have more chances of survival than the pure forests. In pure *Shorea* forests the absence of regeneration, fungal disease in wet *Shorea* forest, poor coppicing power etc. indicate bleak future unless this species is artificially raised over extensive areas. But the mixed forests have maximum wild life species. Most the trees in the mixed forests are fruit bearing. The ground vegetation in the dry-mixed forest is not as rich as those of wet-mixed Forests. Gregarious occurrence of a few grasses are significant for browsing by wild life population. An analysis of ground vegetation showed that *Shorea* forests (along with scattered *Shorea*, pure *Shorea* and wet-mixed) have 112 species against 72 in Riverain Forest, 72 in Dry-Mixed forest and 55 in plantations (man-made forests). Monoculture in the man-made forests is detrimental to wild-life species as seen in the less diversity of species.

Generic and Specific Diversity Calculations in the foot hills

Baman Pokri, Khairbani and Chamta reserves show maximum diversity of species and genera. These are typical forest areas and are much disturbed owing to biotic interference. Nevertheless these zones with big paches of dry-mixed forests are sometimes, ideal habitats for wild life. Maximum number of bird species, both arboreal and ground and inter-layer, and animal species especially deer, bison and elephant herd are found in such forests. But these forests are gradually being converted and replaced by valuable timber species in pure formations.

Generic and Specific Diversity Calculations of Tree Vegetation in the Kalimpong hills

The vegetation of Kalimpong hills has a strong diversity and spread over from foot hills upto 1800 metres. This is not so in the Darjeeling hills where the diversity is maximum at 1000 metres which indicates richness of vegetation (tree) in the foot hills. In these zones there is abundance of arboreal avifauna and animal species.

Floristic Composition of Ground Vegetation in the Hills

The hills were classified into three zones viz., lower hill zone upto 1000 metres, Middle hill zone between 1001 and 1980 metres and upper hill zone between 1951 and 2920 metres. Methods of study are as follows:-

Lower hill zone:-
Total number of species (S) recorded = 60
Total number of genera (G) recorded = 52

$$\frac{S}{G} = 1.15$$

Corresponing 'X' from logarithmic series- .298
Therefore, generic diversity

$$= \frac{60 \times .702}{.298} = \frac{42.120}{.298}$$

The following table gives the frequency distribution and frequency percentages of different species:-

Frequency % (x)	Number of species (f)	fx	fx^2
10	13	130	1300
20	12	240	4800
30	7	210	6300
40	11	440	18600
50	6	300	15000
60	7	420	25200
70	2	140	9800
80	1	80	6400
90	1	90	8100
100	0	0	0

Mean and Variance have been calculated. The value of variance has been calculated at 91.42.

Similarly,

For middle hill forests (genera 102, species 114)

Generic diversity - 288.82
Mean - 31.40
Variance - 530.70

For upper hills (species 105, Genera 94)

Generic diversity - 266.02
Mean - 35.33
Variance - 69.60

Generic and Specific Abundance

(a) Genera—Among the dicots, there are 28 families with only 1 genus, 25 families with 2 genera, 11 families with 3, 10 families with 4, 16 families with 5 and 15 families with 8 genera.

Among the monocots, 7 families have 1 genus each.

(b) Species—Among the dicots, there are 26 families with 1 genus, 15 families with 2 genera and 12 families with 4 genera.

Among the monocots, 3 families have 1 genus each.

Only 12 families had good number of generic abundance among the dicots (5 families among the moncots); 19 families had good number of specific abundance among the dicots (8 families among the monocots).

A critical observation on the generic and specific abundance and count of individuals showed that about 50% of the angiosperm families had either 1 or 2 genera and few species among them are gregarious. This gives a rather gloomy picture of the inspite of high diversity in the wet mixed forest.

(c) Classification of herbs, shrubs, trees and canopy classes.

A synthetic analysis of vegetation shows that 50% of the entire Dictyledons and Monocotylendons is herb; 23 % of the herbs (1251) species in the plains. There are 540 tree species and 628 species of shrubs, most of the species of the latter are found in the plains; large climbers have 111 and climbing shrubs have 105 species. Furthermore, 205 species of ferns, 11 species of Gymnosperm and about 300 species of Lichens, mosses ad hepatics make the study extremely interesting for diversity study of plants and animal.

Shorea robusta form about 50.0%, 20.0%, 6.0% and 3.0% of the species in pure *Shorea*, scattered *Shorea*, wet mixed and dry mixed types of forests respectively; Lauraceous species form about 10.0%, 13.0%, 25.0% and 13.0% of the species in the aforesaid four types of forests respectively; Meliaceous species form about 3.0%, 12.0%, 20.0% and 14.0% of the species in the aforesaid four types of forests respectively. Canopy analysis of the tree species of the aforesaid four types of forests shows that top storey, middle storey and lower storey has about 26.0%, 50.0% and 24.0% of the tree species; in this canopy classification wet mixed, scattered *Shorea* and pure *Shorea* forests represent 132, 114, 102 and 77 numbers of species respectively.

These data indicate that herb vegetation in the hills, shrub in the plains and middle storey species have high diversity of species and that the wet-mixed forest is the most important types of forests from the point of view of species diversity which invariably form most suitable environment for wild life species.

Calculation of Specific Diversity:

It was not possible to count the number of individuals in many cases of ground vegetation as the number of many species are extremely high.

The specific and generic diversity at different altitudes were calculated by a special mathematical method and the findings are as follows:-

Region	*Altitude*	*Specific diversity*	*Generic diversity*
Lower hills	upto 1000 metres	29.82	141.34
Middle hills	between 1001 to 1950 metres	58.42	288.82
Upper hills	between 1951 to 2920 metres	51.07	266.02

Once again middle hills zone i.e. from 1001 metres to 1950 metres altitude show maximum diversity in the ground vegetation. Taking the hill forests as a whole the zone of vegetation from plains upto 1950 metres is most viable. As such foot hill tree vegetation and lower middle hill ground vegetation together with ground vegetation of the upper hills form the most important biologically viable zone.

Richness of Vegetation:

As compared to adjoining geographical areas the vegetation of the study area is luxuriant. It is the smallest area amongst rich floral zones (except Bihar and Orissa) till now described. Diversity is maximum among the monocotyledons and pteriodophytes.

A survey of dicotyledons in undivided Assam (indicluding Sylhet, Meghalaya, Arunachal, Mizoram, Nagaland and Manipur) which over more than 10 times the study area shows the occurrence of only 820 genera and 2355 species against 774 and 1920 respectively in North Bengal. It is extremely significant that the study area including Sikkim which has about 4000 species of plants is one of the richest and most interesting floral and faunal zone of the world. The abundance of avifauna (about 600 species), mammals (about 90 species), fish (about 125 species), innumerable species of butterflies etc. is definitely due to wide diversity of vegetation.

Correlation between Generic and Specific Diversity:

Computations show that in an ecosystem the generic diversity maintains a positive correlation with the specific diversity. In the present study high generic diversity has been found to be associated with high specific diversity value. The study of specific diversity is of prime importance for evaluation of particular community or ecosystem. A low specific diversity gives a warning signal for existence of a species which may be due to biotic factor, environmental pollution, lack of regeneration and reproductive capacity, over exploitation etc.

There are many families with monotypic genus and very few scattered individuals. Many medicinal plants, decorative plants, economically important plants have become rare owing to over-exploitation.

(a) Generic Diversity in Dicotyledon Families:

The following families have very high generic diversity (figures in bracket):

Menispermaceae (11.59), Sterculiaceae (10.73), Meliaceae(12.94), Sapindaceae (10.06), Anacardiaceae (15.28), Asteraceae (28.89), Apocynaceae (23.05), Asclepiadaceae (21.75), Scrophyulariaceae (16.74), Acanthaceae (30.29), Labiatae (25.67) Euphorbiaceae (18.17).

(b) Generic Diversity in Monocotyledon Families:

The following families have very high diversity (figures in bracket).

Orchidaceae (34.12), Liliaceae (18.00) Palmae (10.04), Araceae (13.33). Gramineae (18.01).

(c) Diversity and faunal abundance:

Most of the families having high diversity (also many families not mentioned above) are fruit bearing and also fodder yielding and obviously attract many species of frugivorous birds (Hill mynah), Horn bills, Barbets, Parakeets, Pigeon etc. and gramminvorous animals (Bison, Rhinoceros, Pigs, Thar, Serow, various deer, Sambar, Nilgai, Wild Buffalo).

In the foot hill and adjacent areas there is heavier concentration of birds and animals which include squirrels, civets, mongooses, weasels, martins, monkeys etc.

Observations in the foot hills regions (200 to 1400 metres) show that maximum concentration of bird Babblers 92%, Laughing thrushes 82%, Warblers 52%, Red start 37%, wood peckers 100% and similar percentage of Hornbills, King fishers, Fairy blue bird, Mynas, and lesser percentage of Barbets, Minivets, Cuckoo shrikes, Bulbuls, Nuthathes, Sunbirds etc.) and 80% of animal species (big and small) occurs in this region and in wet-mixed forests.

But diversity alone cannot be responsible for aboundance of animal. Higher productivity (in dry weight) has been found correlated with animal species diversity. High productivity is derived from stem, Leaves, flowers and fruits. But caloric concentration which is of prime importance for animal species, may or may not go parallel with high productivity. Caloric production is directly invloved in animal nutrition than in dry-weight production. Pure crop forests and plantations of few species have poor diveristy due to poor productivity. More animals depend on leaves and fruits and as such the more the leave production the more animal may be expected.

A classification of ground flora, based on statistical analysis, show that *Shorea* forests, Riverain forests, dry-mixed forests and plantations have 112, 72, 72 and 55 significant species respectively. Grasses, shrubs and herbs form a small precentage of ground flora in plantations. Plantations of selected few species, are therefore, not rich in diversity and unsuitable habitat for wild-life.

The present study gives a clue to the selection of rich floral zone. The study of parallelism in the vegetational diversity increase in animal population/species should be of great economic importance. A very low generic and specific diversity warrants adoption of special protective measure.

B. CLASSIFICATION OF FORESTS WITH SERAL, EDAPHIC AND BIO-EDAPHIC TYPES:

Seral, Edaphic and Bio-edaphic Forests:

During field survey and collection of data, special types of vegetation occupying limited areas but distributed widely in plains and hills forests were observed. The composition of such vegetation was found to be widely differing from the general vegetation types; it was found to consist of a few endemic species.

While analysing the synthetic data in earlier chapters the occurrence of several such forests have been indicated and some such Edaphic and Bio-edaphic forests are as follows:-

I. Seral forest:

Soil conditions effect the growth, development and succession of forest crop, and various forest associations are intimately related to the soil conditions. The edaphic forests are very much distinctive, and do not cover wide areas in North Bengal except the riverain tracts (Maps. 4, 5, 6 and 7). There are several pockets of vegetation in *Shorea robusta* forests which reflect complex soil and micro climatic conditions. Such areas have been isolated and discussed. Similar types of areas have been isolated in the foot, middle and upper hills forests where the vegetation is the result of complex soil conditions, and their compositions have been analysed in detail.

The surface soil of riverain forests is porous, hot and dry during summer, but has an ample water at a moderate depth. The soil near the rivers is practically sandy; away from the river bed soil changes into sandy loam to deep loam. The vegetation also changes as the soil condition changes. On the sandy soil *Dalbergia sissoo* and *Acacia catechu* come up and as the soil conditions change several types of crop dominate and form various associated viz., Bombax-Albizzia, Toona-Gmelina, Trewia-Bischofia, Terminalia--Wrightia and also pure Bauhinia. Such sandy river beds which are not annually inundated by flood water are invaded by various types of coarse grasses which when get fire burnt hold the ground fast and Savannah type grassland grow.

Successional association in the Riverain Forests

Pure Acacia catechu Consocies:

Place of occurrence - There are innumerable small and big rivers, the sandy and bouldery beds of which bear those pure forests. These forests occur in small patches but over extensive areas.

Associates - Pure patches in fresh deposit of silt and in boulder beds. The crop is very dense. Tamarix diolica invade the silt deposit. Such forsts do not have any understorey vegetation except grass.

Pure Dalbergia sissoo consocies:

Place of occurrence - Same as above.

Associates - Pure patches occur on sand and silt deposit. Soft grasses like *Paspalum, Panicum, Eragrostis, Parotis, Oplismenus, Paspalidium,* Eleusine species come in. They are followed by coarse grasssed like *Saccharum, spontaneum* and *S. procerum* over extensive areas. *Tamarix dioica* and *Exacum tetragonum* come in sporadically in such forests. There is no understorey.

Dalbergia-Acacia associes:

Place of occurrence - Same as above but on comparatively older silt deposit. This association occurs over extensive areas in various proportions.

Associates - Common associates are *Erythrina stricta, Erhetia auminata, Hymenodicytyon excelsum, Premna latifolia, Albizzia stipulata, Toona ciliata, Trema orientalis, Bridelia stipulata, Ichnocarpus frutescens, Litsaea* species, *Hedyotis scandens, Paederia foetida, Trewia nudiflora and Anthocephalus cadamba.* Undergrowth of *Clerodendrum viscosum, Coffea bengalensis, Croton caudatus, Zizyphus* and other species are common.

Bombax-Albizzia associes:

Place of occurrence - Same as above; the forests of Diana, Torsa, Mahananda, Rydak, Sankos, Jaldacca, have extensive such forests.

Associates - Common associates are *Bombax ceiba, Albizzia procera, A. stipulata, Hymenodictyon excelsum, Trewia nudiflora, Dalbergia stipulacea, Craetava unilocularis* and *Tetrameles* nudiflora. There are very many species of coarse grasses. Gradually more of climbers and evergreen species invade. Moist pockets have colonies of *Biscofia javanica and Dillenia pentagyna.*

Toona-Gmelina associes:

Places of occurrence - Riverain forests as above, but on a good mixture of sand silt. More the silt and moisture on soil the better the forests. These species are rather interspersed between former and *Shorea* soil.

Associates - *Toona clliata, Gmelina arborea, Hymenodictyon excelsum, Trewia nudiflora, Dalbergia stipulacea, Wrightia tomentosa, Biscofia javanica* and *Dillenia indica.* Owing to excessive grazing and exploitation of timber, natural regeneration cannot establish. Lopping is also extensive in such forests. *Mikania cordata* is invading the forest.

Trewia-Biscofia associes:

Place of occurrence - Same as above. Areas are often flooded during rains.

Associates - *Trewia nudiflora, Biscofia javanica, Dillenia indica, Toona ciliata, Gmelina arborea,* and *morus leavigata;* other associates are as above.

Bauhinia consocies:

Place occurrence - Diana, Rydak and Pana river beds.

Associates - *Bauhinia purpurea, B. variegata* are the principal species.

Wrightia-Holarrhena associes-

Place of occurrence - Rydak and Pana forests.

Associates - *Wrightia tomentosa* and *Holarrhena antidysenterica* are predominant species. Other species are *Grewia disperma, Caseria graveolans, Terminalia belerica* and *Toona ciliata.*

Macaranga denticulata consocies:

Place of occurrence - In the river bed where there is deposit of silt and where the area is above flood level (in the rivers Rydak, Diana and Murti.)

Associates - The crop is practically pure. Regeneration of evergreen species starts on such forests.

Wet Riverain Grassland associes.

Place of occurrence - In Gorumara, Jaldapara, Balapara and Barobhisa blocks which are often flooded during rains. The stiff soil dries in winter. There are no trees. Fire is common.

Species - Principal species are *Phragmites karka, Arundo donax, Narenga porphyrocoma, Saccharum* species, *Erianthus* species. Details of the species in such forests may be found in the chapter on grassland types.

All the types and associations described above are edaphic under Tropical Dry-Deciduous and Tropical Moist Deciduous Forest.

II. Edaphic forests:

(i) Tropical Fresh Water Swamp

In the forests of Dalkajhar (Bagdogra Range), Moraghat and other Ranges where small jhoras originate, the lands surrounding the jhoras are swampy and bear a special type of vegetation. The following are the conspicuous species of such localities.

Tree-*Syzygium* species, *Jambosa formosa* (stilt root). *Carallia integerrima* (only in Dalkajhar), *Elaeocarpus* species (stilt root and buttressed) *Myristica* species (buttressed).

These species are characterised by the presence of "Knee roots" similar to those of Mangrove forests and have some of the genera common to those "Myristica Swamps" of Westernghat Ranges of South India.

Diplazium esculentum is a graceful species similar to tree fern and is typical of such swamps and grows in profusion along with *Pandanus furcatas* and *Lygodium* species and *Jambosa formosa*. Several species of canes such as *Calamus tenuis*, *C. latifolius* and *Calamus guruba* occur in such swamps.

In the marginal lands sporadic occurrence of tree species such as *Gynocardia odorata, Aporosa dioica, Saurauia* species, *Casearea vareca, Quercus spicata, Machilus gamblei* makes the forest physiognomy conspicuous.

Shrubs and Climbers - Among the climbers most conspicuous are Hoya *(Asclepiadaceae)* species, *Stenoclaena palustris* (fern). Common shrubs are *Ardisia nerifolia, Fagraea obovata,* (Myrsinee) *Phajus wallichii, P. blumei* var. *pulchara* (Orchids), *Psychotria fluva (Rubiaceae) Callicarpa longfolia (Verbenaceae) Aslenium nidus* epiphytic fern).

Ground flora - Most common is *Phyrenium capitatum (Marantaceae), Pinanga gracilis (Palmae), Justicia gendarussa (Acanthaceae), Pleris biaurita (ferns), Lasia heterophylla (Ar-*

aceae), Monochoria hastifolia (Araceae), Syzygium balsamea (Myrtaceae), Nephrodium cochileatum (fern).

Apart from these, some grasses occur in various percentages. They are:-

Narenga porphyrocoma, Saccharum species, *Arundo donax, Imperata arundinacea, Phragmites karka* etc.

Swamy lowlands are often found covered with gregarious growth of dwarf herbs that turn entire landscape blue and blue-violet; these species are *Limnophia, Sessiliflora* and *Dysophylla crassicaulis* var *pumila.*

(ii) *Syzygium-Jambosa* Swamp

Place of occurrence - In the wet mixed forsts of Moraghat, Lower Tondu, Dalkajhar, Damanpur and Rajabhatkhawa Ranges.

Species and Associates - *Syzygium cuminii* and *Jambosa praecox* are the dominant associations. The trees are conspicuous by their whitish bark. In the stream beds espcially those bouldery, *Jambosa formosa* forms a pure corp with major parts or their roots exposed. In such stream beds colonies of *Phyreniun capitatum* and *Gendarussa* vulgaris grow. Soil is heavy and water logged through the year.

(iii) *Pinanga-Dracaena* community

Place of occurrence - On the moist damp shaded forest floor of different Ranges either in moist *Shorea* or Wet-Mixed Forests.

Species and Associates - Pure and mixed colonies of *Pinanga gracilis* and *Dracaena ensifolia* can be found in the forests over limited areas. The colourful fruits of *Pinanga* is edible by birds and animals; leaves form a good fodder for elephants.

(iv) *Salix tetrasperma* Swamp

Places of occurrence - Low lying swamps and road side burrow pits of the plains forests.

Species and Associates - The crop in practically pure and regeneration is profuse. Sometimes *Trewia nudiflora* and *Calumus tenuis* are found in such areas.

(v) Cane brakes

Places of occurrence - Scattered all over the wet-mixed forests, stream banks of *Shorea forests,* moist foot hills forests and moist temperate forests.

Species - Canes form impenetrable thickets and very characteristic vegetation of the evergreen, semi-evergreen and wet type forests and extend in the hills upto 2100 metres. The soil is permanently wet, usually clayey and rich in humus. In fresh water swamps they are common. Most common and widely distributed species are *Daemonorops jenkensianus* and *Calamus latifolius* in the forests of Rajabhatkhwa, Rydak, Dalkajahar and Moraghat. Both these canes form about 70% of the cane output of plains forests. Other species of plains forests are *Calamus flagellum C. guruba, C. leptospadix, C. acanthouspathus, C. inermis* and *C. tenuis, Calamus erectus* is found in the foot hills but is sporadic. *Plectocomia himalayana* occurs in the temperate hills and is a graceful cane.

All the canes are useful and economically important. The fruits are edible both by men and animals. The young shoots are eaten by the forests labours. Regeneration is poor and owing to over exploitation and forest fire the future of all *Calamus* sp. seems to be bleak.

(vi) Bamboo brakes of Temperate and Tropical Forests

Place of occurrence - In the moist and shady forests of Tista valley, foot hills forests, shifting cultivated and fire affected forests of Kalimpong hills.

Species and Associates. - *Dendrocalamus hamiltoni* is the commonest bamboo of *Shorea* forest of foot hills. Though it occurs as a under storey crop of *Shorea* its best growth and development is on moist thick loamy soil of Tista Valley and in several blocks of Kalimpong hills. *Bambusa nutuns* and *B.*

pallida are rare in the forests. In the tropical belt of forests in the foot hills (Latpanchar, Lethi, Neam, Mana blocks) *Pseudostachium ploymorphum* and *Cephalostachium capitatum* form extensive thickets at moist and shady places. Where soil is deep loam *Dendrocalamus patellaris, D. hookeri* and *D. sikkimensis* have a very limited distribution as in Kalimpong hill; so also are the species of *Cephalostachium latifolium* and *C. fuchianum.* Most wide-spread bamboos of middle and upper hills (1800 to 3200 metres) which are results of fire are *Arundinaria maling, Chimonobambosa griffithiana,* and *Thambocalamus aristatus.* Other bamboos of limited occurrence are *Semiarundinaria pantlingii,* and *Chimonobambusa hokkeriana.*

(vii) Swamp Forests and marshes

Place of occurrence—Such areas are scattered all over the plains forests occupying very limited areas.

Floristic composition—*Ranuculus sceleratus, Polycarpon loeflingiae, Rotala* species, *Airdsia nerifolia, Aeschynomene aspera, A. indica* and several grasses (*Saccharum* species, *Erianthus* species, *Vetriveria zizanioides, Leersia hexandra, Coix* species, *Isachne miliacea, Elytrophorous spicatus,* and *Polytoca bracteata*).

In running water—*Eriocaulon rivulare, Homonia riparia, Rhabdia* species, *Monochoria hastifolia.*

In stagnant water—*Jussieau repens, Traps bispinosa, Limnanthemum cristatum, Pista stratiotes, Fagaria* indica and *Juncus prismatocarpus.*

In rice field—*Limnophila sessiliflora, Cyanotis cristata, Aneilema liniolatum, Dysophylla crassicaulis, Ammania baccifera, Vandellia* species, *Mollugo* species, *Blumea* species, *Cynodon dactylon, Alternenthera sessilis, Lippia nodiflora, Mnesithea perforata, Hemarthia compressa, Eragrosris* species, *Ranges parviflora, Solanum indiucm, S. torvum, Aerua lanata, Sida* species, *Vernonia cinera, Cyperustrotundus, Urena lobata* and *Elephantopus scaber.*

In Sultunga bill—*Hydrilla verticillata, Lagerosiphon roxhurghii, Ottelia alismoides, Eichornia crassipes, Monocharia hastifolia, Aponogeton monstachyon, Najas kurziana, Potamogeton indicus, P. javanicus, P. crispus, P. mucronatus, Eriocaulon capillusnaidis, Alisma plantago, Sagittaria sagittifoila, Butomopsis lanceolata* and *Vallisneria spiralis.*

III. Special Edaphic Forests at Micro Sites:

In the forests of plains and hills there are some sites which are restricted to small areas and have distinct types of vegetation completely different from the adjoining forests or of entire region both in species association and distribution of species. Special features of such sites are the presence of some species which are endemic to those sites only, and are not found anywhere in the districts. Thirteen such sites have been identified and presented in this type as detailed hereunder:

IV. Bio-edaphic Forests

Typical bio-edaphic forests are absent from North Bengal. There are, however, some isolated forests of Tea Estates and a big *Shorea* forest viz., Baikunthapur forest on sandy loam and bouldery slopes which are results of years of maltreatment by men. Many such Tea Estate forests have been encroached upon recently (map 4) leaving behind remnants of such forests.

Although the rainfall is high during the monsoon months, the soil gets absolutely dry during the months from November to April. This condition followed by extensive grazing, felling and lopping brings in a condition where no evergreen type of vegetation can grow and, therefore, the soil is exposed to erosion. Shrubby layer of vegetation and ground flora is very light.

(a) Depleted Tea Estates Forests

Butea monosperma is a typical species of such forests. There are very scattered *Dendrocalmus hamiltonii, Lannea coromendelica, Premna* species, *Ficus hispida, Cordia myxa, Hollarrhena antidysenterica, Wrightia tomentosa, Bombax ceiba, Albizzia procera* and *Stereospermum chelonoides. Butea* grows on clayey soil. *Croton caudatuls* form isolated bushes along with *Clerodendrum viscosum, Eupatorium odoratum, Vallaris solanacia, Cassia tora,* and *Zizyphus* species.

Overgrazed forests of Rehti, Diana, Dima, Raimatanga and Rydak have a tendency to revert to such bio-edaphic types by elimination of many mesophytic species (retrogression). Different vegetative layers are as follows:-

Forest type	*Top Storey*	*Middle storey*	*Shrub layer and ground vegetation*
Dry-mixed forests and Riverain forests/degenerated into bio-edaphic type.	*Bombax ceiba* *Butea monosperma* *Stereospermum* sp. *Albizzia* species	*Butea monosperma* *Holarrhena* sp. *Wrightia* sp. *Bauhinia* sp.	*Zizyphus* species *Groton* species *Vallaris* species *Erianthus* species *Themeda* species

Edaphic types in the plains Shorea forests of Darjeeling and Jalpaiguri districts

Area	*Characteristic type*	*Composition overwood*
1	*2*	*3*
(i) Bania, South Moraghat and Poro blocks.	A special type of wet mixed forest on damp, deep loam soil cut up by streams, pure patches of *Shorea Lophopetalum fimbriatum* are highly conspicuous not found elsewhere so prominently.	Very tall *Quercus spicata, Castanopsis tribuloides, Trewia nudiflora, Machilus* species, *Shorea rubusta Sterculia villosa, Stereospermum tetragonum, Michelia champaca Artocarpus chaplasa, Tetrameles nudiflora, Gemlina arborea, Amoora wallichii and Lophopetalum fimbriatum.*
(ii) Panbari, Buxa, and Santrabari blocks of Borojhar Reserve.	A special type of *Shorea robusta-Michelia champaca* forests where these two species occur in the top storey in pure mixture	1. Shorea robusta scattered overMichelia champaca in top storey with excellent growth. 2. Other species are *Amoora wallichii, Schima wallichii, Premna bengalensis, Castanopsis*

1	2	3
		tribuloides, and *Toona ciliata*.
(iii) Bengdubi block in Dalkajhar forests.	A special type of wet mixed forest where there are so many assemblages of ever-green species in such a small area. *Shorea robusta-Ternstroemia gymnanthera-Podocarpus neriifolia* association is very conspicuous.	1. *Shorea robusta, Ternstroemia gymnanthera, Podocarpus neriifolia, Homalium Zeylanicum, Endospermum chinense, Diplospora singularis, Lephopetalum fimbriatum.* 2. *Schima wallichii, Syzygium cuminii, Michelia oblonga M. champsca, Ailanthus grandis, Elaeocarpus varunua, Wallichii and Sterculia alata.*

Edaphic types in the plains Shorea forests of Darjeeling and Jalpaiguri districts

Underwood	*Ground and regeneration*	*Climate, rock and soil*
4	*5*	*6*
Saraca indica, Aphanamixis polystachia, Dysoxylum binectariferum, Toona ciliata, Syzygium species, *Knema* species, *Alphonsea ventricosa, Bauhinia purpurea, Aesculus assumica, Dillenia indica.*	*Machilus, Polyalthia semiarum Sterculia villosa, Artocarpus chaplasa.*	Rainfall 3750 mm deep alluvium, damp; cut up by streams.
Actinodaphne obovata, Meliosma simplicifolia, Machilus species, *Dysoxylumbinectariferum, Meliosma pinnata, Sauraia roxburghii, Pygeum acuminatum, Polyalthia semiarum, Knema longifolia.*	*Michelia champaca, Artocarpus chaplasha, Grewia* species.	Rainfall 3750 mm. Deep clayey loam, moist water drained by perennial stream.
Turphinia pomifera, Actinodaphne	*Coffea bengalensis,*	Rainfall 300 mm.

4	5	6
obovata, Dillenia pentagyna Ilex godajam, Aporosa dioica Acronychia laurifolia, Aphanamixis polystachia, Magnolia pterocarpa, Eurya japonica, Mallotus philippinensis, Casserea vareca, Macaranga denticulata, Careya arborea, Saurauia roxburghii.	*Clerodendrum viscosum Piper* species *Calamus* species *Adiantum* species, *Cyclosorus* species, *Microstegium ciliatum, Oplismenus comositus, Panicum* species, *Setaria* species.	Soil sandy loam with thick layer of humus.

Eidphic types in the plains Shorea forests of Darjeeling and Jalapaiguri Districts

Area	*Characteristic type*	*Compsition overwood*
1	2	3
(iv) Lataguri block In lower Tondu Range.	A typical forest having an association of *Shorea robusta* and *Michelia montana*. *M. montana* does not occur in any of the wet mixed forests elsewhere.	1. *Shorea robusta, Michelia montana.* 2. *Premna bengalensis, Michelia* species, *Dillenia pentagyna, Lophopetalum fimbriatum, Elaecarpus aristatus, Artocarpus chaplasha* and *Castanopsis tribuloides.*
	Edaphic types in the Bhabar forest adjoining foot hills	
(v) Adma forest in Borojhar Reserve.	A special type of dry mixed forest not comparable to seral type described elsewhere. It is conspicuous by the occurrence of *Firmiana colorata* in the top storey and *Walsura tabulata* and	1. *Firmiana colorata, Tetrameles nudiflora, Lagerstroemia parviflora, Cinnamomum cecidodaphne, Chikrasia tabularis, Michelia champaca, Canarium sikkimense, Amoora Wallichii.*

1	2	3
	climbers in the lower storeys.	2. *Gmelina arborea, Cassia fistula, Stereospermum tetragonum, Sterculia villosa, Duabanga grandiflora,Toona ciliata* and *Bombax cieba.*
(vi) Bhutanghat forests in Borojhar Reserve.	A special type of dry mixed forests where *Sapium insigne* dominates in the upper storey with *Dalbergia sissoo Acacia catechu* and *Terminalia belerica* in the top storey. *Sapium insigne* is not found anywhere in the whole of North Bengal.	1. *Sapium insigne, Dalbergia sissoo, Terminalia belerica, Lagerstroemia parviflora.* 2. *Emblica officinalis, Sterculia villosa, Bombax ceiba, Cassia fistula, Premna bengalensis, Erythrina indica, Steroesper-mum tetragonum.*

Edaphic types in the plains Shorea forests of Darjeeling and Jalpaiguri districts

Underwood	*Ground and regeneration*	*Climate, rock and soil*
4	5	6
Litsaea species, *Turpinia pomifera, Dysoxylum benecteriferum, Meliosma simplicifolia, Aporosa dioica.*	*Calamus* species, *Oplismenus* species, *Setaria* species.	Rainfall 300 metres Soil moist loam with thick humus.
Edaphic types in the Bhabar Shorea forests adjoining foot hills		
Walsura tabulata, Chisocheton paniculatus, Holarrhena antidysenterica, Litsaea polyantha, L. sebifera, Adenanthera Pavonina Premna integrifolia, Macaranga pustulata, Syzygium operaculatum and *Hymenodictyon excelsum.*	*Coffea Bangalensis Millettia aurculata, Mucuna pruriens, Dalbergia stipulacea, Ichnocarpus frutescens* and many other species. Regeneration of *Aphanamixis, Mallotus, Firmiana, Pterospermum,*	Altitude 300 metres. Rocks-Sand stone and Gneiss Soil-Sandy to sandy loam-slope1 in 10. Rainfall 3700 metres.

4	5	6
	Microstegium, *Leea*, *Urena*, *Croton* and *Daedelocanthus* species.	
Sapindus detergens, *Alstonia scholaris* *Oroxylum indicum* *Albizzia* species, *Trewia nudiflora*.	*Microstergium ciliatum*, *Urena lobata*, *Saccharum* species, *Lagerstronemia parviflora*, *Clerodendrum viscosum*, *Leea crispa*.	Rain fall 3700 metres. Well drained sandy loam with a subsoil of sand and gravel. Undulating ground.

Eiaphic types in the Plateau Forests of the foot hills

Area	*Characteristic type*	*Composition Overwood*
1	2	3
(vii) Khumani and Mal forests in Jaldacca Reserve.	A rare evergreen type in the wet mixed plateau forests where *Mesua-Phoebe* predominate.	1. Almost 60% of the crop is composed of *Mesua ferrea;* other associates are *Phoebe attenuata* and *Syzygium* species. 2. *Amoora wallichii, Michelia montana, Endospermum chinense, Mangifera sylvatica, Artocarpus chaplasa.*
	Edaphic types in the hill forest of North Bengal	
(viii) Birrick block of Tista valley	*Shorea-Terminalia* association with *Dendrocalamus hamiltonii* in the understorey sprinklings of *Ulmus lancaefolia* makes the	1. *Shorea robusta, Terminalia tomentosa, Schima wallichii, Duabanga grandiflora, Toona ciliata, Lagerstroemia*

1	2	3
	type distinct.	*parviflora.* 2. *Sterculia villosa, Chikrasia tabularis, Tonna microcarpa, Syzygium cumini, Canarium sikkimense.*
(ix) Latpanchar forests at 1000 metres altitude	Pure *Betula cylindrostachys* consociation Pure *Ostodes paniculata* formation.	1. *Betula cylindrostachys.* 2. *Castopsis tribuloides, Terminalia myriocarpa, Allanthus grandis, Albizzia marginata, Chukrasia tabularis, also some of the species of the genera Quercus, Schima, Spondius* and *Mechilus.*

Underwood	Ground and regeneration	Climate, rock and soil
4	5	6
	Edaphic types in the Plateau Forests of the Foot Hills	
Dysoxylum species, *Mesua ferrea, Phoebe attenuata, Cinnamomum obtuseflium, Syzygium formosum, Litsaea Panamonja, Meliosma simplicifolia.*	*Mesua ferrea* regenerating over the undergrowth of *Alpinia allughus, Oplismenus compositus, Panicum* species, *Pilea* species, *Curcuma* species, *Coffea bengalnesis.*	Rainfall 3500 metres Altitude 300 metres Soil sandy loam deep and rich over lying sandstone and boulders.
	Edaphic types in the hill forests of North Bengal	
Dendrocalamus hamiltonii, Ilex godajam, Aporosa dioica, Phoebe lanceolata, Sarcosperma arborea, *Trichillia connarioides, Castanopsis* indica,	*Boehhmeria macrophylla, Gerardiana heterophylla Desmodium tilaefolium, Setaria palmaefolia, Piper* species, *Abacopteris multiliniata*, Coffea	300-600 metres altitude Soil is sandy loam Rock is Gneiss.

4	5	6
Garuga pinnata.	*bengalensis, Piper* species *Desmodium floribundum, Piper peepuloides.*	
Ostodes paniculata Wrightia tomentosa, Syzyguim kurzii, S. tetragonum, Symplocos racemosa, Cinnamomum obtusefolium.	*Boehmeria marcophylla, Gerardiana heterophylla, Hovenia dulcis, Maesa indica.*	3500 metres rainfall Soil stand stone, sandy loam Rock is Gneiss.

Ediaphic Types in the hill forest of Darjeeling district

Area	*Characteristic type*	*Composition Overwood*
1	2	3
(x) Mana forests of Kurseong Hills.	Pure *Machilus* and *Engelhardtia* in the top storey. *Ostodes and Macaranga* in the second storey. There are sprinklings-*Phoebe attenuata, Fraxinus paxima* in the top storey.	1. *Machilus edulis, M. gammiena* and *Engel-hardita spicata* and *Cinnamomum impressivernum* in the top storey along with species of lower frequency as follows:- *Syzygium kurzii, Phoebe attenuata, Litsaea khasyana, L. kingii, Glochidion thompsonii, Fraxinus paximia, Spondius axillaris.*
(xi) Lopchu forest in the Rangit river valley.	Gigantic *Podocarpus nerifolia* forest in the temperate hills. A rare species in the hills and plains. It has limited distribution	*Podocarpus nerifolia, Castanopsis indica Sloanea dasycarpus, Meliosma wallichii, Engelhardtia spicata.*

1	2	3
	in the *Shorea* forests of Dalkajhar and Moraghat (wet mixed).	
(xii) Rimbic forests of Singalila Range.	Pure *Quercus lamellosa* formation.	*Quercus lamellosa* and sprinklings of *Michelia doltsopa* and *Acer campbellii.*
(xiii) Ramam forests of Singalila Range.	Pure *Tsuga dumosa* formation.	*Tsuga domosa.*

Edaphic types in the hill Forest of Darjeeling district

Underwood	*Ground and regeneration*	*Climate, rock and soil*
4	5	6
Ostodes paniculata, Macaranga pustulata, M. indica, Leucoceptrum canum, Evodia fraxinifolia, Mallotus neplaensis.	*Massaeda macrophylla, Cephalostchyum capitatum, Polygonum chinense, Elatostoma hookerianum, Pilea nepalense, P. thomsonii.*	Rainfall 3500 metres Gneiss, Sandstone and Sandyloam slope 1 and 2.
Symplocos theifoila, Prunus nepalenis, Maaramga Pustulata, Hoveina dulcis.	Seedlings of *Podocarpus* Several specis of Ferns and grasses Described Elsewhere).	Deep humus Rainfall 2000 metres. Very humid, Sandy loam soil.
Symplocos theifolia Eurya species *Saurauia* species.	*Vivurnum cordifolium, V. grandiflorum, V. mullaha, V. erubescens, Arundinaria maling, Polygonum molle, Rubus* species.	Rainfall 2000 metres Slope 1 in 20 Soil-Sandy loam Rock (Darjeeling).

4	5	6
Rhododendron arboreum, *Brassiopsis alpinum*, *Symplocos phyllocalyx* *Acer pectinatum*, *Daphne cannabina*, *Viburnum* species, *Rubus* species, *Strobilanthus* species.	*Arundianaria maling*	Rainfall 2000 metres Slope 1 in 20 Soil Shallow sand rock-very prominent.

(b) Baikunthapur-Shorea Forests

These forests extend over an area of about 80 sq. km. The soil in the upper 30-40 cm. is sandy which is laid over 110-120 cm. sandy loam soil. There is no humus on forest floor because of fires and grazing.

Since 1955 these forests came under the management of Government from private individuals and strict fire protection was enforced.

Investigation reveals that the forests appear to be the result of poor and improper management in the past, resulting in over felling of nature trees, complete absence of cultural operation and lack of protection. Fire had its annual depreciation reducing the forest in the present poor state. The species being *Shorea* no amount of biotic interference could take advantage of the potentialities. There are sporadic regeneration of pole crop, but most of them are crooked owing to repeated burning. Miscellaneus species have very localised distribution, and the mixture is in the lower storey and the *Shorea* crop is practically pure. There are blanks which have been invaded by *Imperata cylindrica.* The crop is third quality but the density is very high. Fire damaged the root system and the seedlings used to die back. After lot of competition poor crooked saplings came out of root stock. Extensive areas were under *Imperata cylindrica* and *Narenga porphyrocama.*

Along the streams original vegetation escaped biotic depredation where *Calamus, Syzygium* and *Aphanamixis* are conspicuous. Over storey species consists of *Gamelina arborea, Terminalia belerica, Lagerstroemia parviflora, Sterculia villosa, Schima wallichii, Ternminalia tomentosa, Adina cordifolia, Wrightia tomentosa, Machilus* species, *Stereopermum* species, *Hymenodictyon excelsum, Premna* species, *Ficus* species and *Toona ciliata.*

Middle storey species are *Careya arborea, Lannea coromendelica, Syzygium cumini, Syzygium operculatum, Garuga pinnata, Dillenia pentagyna, Bridelia* species, *Gasearea* species, *Wendlandia* species, *Zizyphus* species, *Emblica officinalis,* and *Holarrhena antidysenterica.*

Wet Mixed Forests—This type of forests exists over a limited area. Species of the genera *Machilus, Litsaea, Syzygium,* occur with scattered *Sterculia villosa, Toona ciliata, Premna* species and *Lannea coromendelica* in them.

Creepers and climbers-Following are the creepers that are found in the forests:-

Piper hamiltonii, Ichnocarpus frutescens, Smilax species and *Dioscorea* species.

Analysis of Ground Flora

Laltong Block

Analysis was made in quadrats and it was found that 8 square metre quadrat correctly represents the vegetation with 15 species and 34 individuals. In a fire affected area 1 sq. metre quadrat showed 8 species and 63 individuals (one *Dioscorea* species had 31 individuals). Quadrats of 6 sq. metres showed three species. *Boreria hispida* and *Biophytum sensitivum* are two most widespread plants of this forest. Common species are *Largerstoemia parviflora, Dillenia pentagyna, Careya arborea, Wrightia tomentosa, Aporosa dioica, Litsaea* species, *Holarrhena antidysenterica, Mezoneuron cuculatum, Coffea bengalensis, Clerodendrum viscosum, Acacia pernata, Ichnocarpus frutescens, Mikania cordata, Dalbergia lanceolaria, Oplismenus compositus, Kempferia rotunda, Wrightia tomentosa, Cyrtococcum oxyphyllum, Bridelia retusa, Ixora* species, *Grewia* species, *Randia* species, *Ardisia solanacea, Uvaria* species, *Asparagus* species, *Selaginella species, Ageratum conzoides, Eriocaulon* species and regeneration of *Syzygium operculatum.*

Sarugarah Block

Common ground flora are *Asparagus* species, *Clerodendrum viscosum, Piper* species, *Leea aspera, Rungea parviflora, Dalbergia stipulacea, Phyllanthus* species, *Flemingia* species, *Oplismenus compositus, Urena lobata, Osbeckia* species and *Lygodium flexuosum.*

Regeneration of tree species are of *Aphanamixis polystachia, Grewia* species *Litsaea* species, *Syzygium operculatum, Sterculia villosa, Garuga pinnata, Syzygium* species, *Terminalia belerica, Lagerstroemia parviflora, Aporosa dioica, Wrightia tomentosa, Millettia auriculata, Bauhinia* species, *Meliosma simplicifolia* and *Ficus cunea.*

Laltong Block-7th Mile

There is practically no difference of the ground flora from Laltong Block.

An area in Sevoke Block very close to Laltong showed extensive growth of *Murraya koeneigii* over large areas which shows some marked difference in the type of soil (more of humus).

Ambari Block

Analysis of quadrats of 1 sq. metre showed 10 species and 114 individuals of which the leading species was *Biophytum*, quadrats of 2 sq. metres showed 14 species, 3 sq. metres showed 19 species and 6 sq. metres showed 26 species.

Common species are *Clerodendrum viscosum, Imperata cylindrica, Ichnocarpus frutescens, Dioscorea* species, *Bridelia* species, *Spatholobus roxburghii, Leea crispa, Dillenia pentagyna, Lygodium, flexuosum, Boreria hispida* and *Dalbergia stipulacea.*

In Simulguri and Lalitabari blocks 8 sq. metres quadrat showed 36 species as follows:-

1 sq. metre .... 12 species, 2 sq. metro .... 19 species
3 sq. metres .... 21 species, 4 sq. metres .... 25 species
5 sq. metres .... 30 species, 6 sq. metres .... 32 species
7 sq. metres .... 34 species, 8 sq. metres .... 36 metres
9 sq. metres .... 36 metres, 10 sq. metres .... 37 species.

Only six species of tree are found in any of the aforesaid Blocks which shows that only the resistant and hardy species could survive in the struggle for existence.

(c) Bio-Edaphic Forests of Kalimpong hills

Gambel's "Convolvulus Mictum" is type of forests that have invaded the forests destroyed by fire or felled for cultivation. Such forests do not exist now although remnant still exists.

Common species are *Croton caudatus, Capparis multiflora, Dalbergia stipulacea, Derris* species, *Zizgphus apetala, Munronia wallichii, Argyrea* species, *Porana species, Pueraria sikkimensis, Smilax* species, *Mucuna* species. The thickets are impenetrable.

Tree species that invade the area *Trema orientalis, Premna mucronata,* Bridelia species. etc.

The forests of Chel regions in the past used to be burnt, almost regularly. This led to the invasion of many species of bamboos. On the fire burnt soil invaded *Careya herbacea, Olax nana, Premna herbacea* and *Grewia sapida.*

Forests that were felled for cultivation but not cultivated or planted are invaded by herbaceous and then by semi-herbaceous species and then by herbaceous creeper. *Blumea balsamifera* is a common invader. Gradually *Callicarpa arborea, Zizyphus jujubas, Callicarpa macrophylla, Buddleia asiatica, Urena lobata* and a few other climbers invade.

In the hills such areas are invaded by *Artemsia vulgaris, Maesa chisia, Croton caudatus, Ficus* species etc. In the middle hills Alnus nepalensis invade many types of eroded soil. Species of the genera viz., *Artemsesia, Legerstum, Viburnum, Priotropis* and *Leycesteria* also invade such areas.

C. CLASSIFICATION OF PLANT COMMUNITIES INTO TYPES

Ecologists have classified plant communities in various ways. Schimper (1898) classified the plant communities according to climatic zones such as grasslands and deserts. Warming (1909) based his classification on water relations such as hydrophytes, mesophytes and xerophytes,. Clements (1916), however, evloved a complete system of classification in which he said that vegetation occurs in climax as well as in developmental or seral stages. In the present classification terms used by clements have been retained by the author.

The different climax types on plant formation are main natural vegetation types. Such climax vegetation is not uniform but made up of a number of well defined smaller units such as associations (a climax unit with two or more dominent forms), 'Associes' (a developmental equivalent to the association and is a seral or successional stage), 'Consociation' (a climax community with a single dominant), 'Consocies' (a developmental community with only one dominant species). In the present classification use of the terms 'Faciation', 'Facies', 'Location', 'Locies', 'Society', 'Socieo' etc. was not made as elaborate classification was not found necessary. The author thinks that every species or a group of species, has greater or lesser indicator value (similar to Mont pellier-zurich school led by Favillard and Braun Blanquet) and his descriptions, graphs and diagrams indicate some relationship among various elements of vegetation.

Champion (1936) made complete classification of vegetation of Indian Forests which was revised by Champion and Seth (1968). This classification was based on climatic features (Temperature & Rainfall primarily). No classification other than this has been tried in this study area. Puri (1960) commented on Champion's Northern & Southern variation of major types which he thought was due to biotic or ecological features and that vegetation changes occur in east-west direction; but Champion & Seth (1968) do not accept the argument. Puri considered the importance of edaphic and biotic factors; the present analysis of the author upholds this suggestion in grouping several associations and consociations under edaphic and bio-edaphic types. Even Bourne (1936) was of the opinion that several types differentiated in various parts of India as climatic and edaphic climates were only seral in status. He suggested poly-climax theory, but Champion and Seth uphold Clement's (1928) monoclimax theory. In the present analysis the author did not consider it necessary, otherwise, to change the nomenclature used by Champion and Seth. Classifications which are based on (i) Physiognomy and structure compositions as proposed by Beard (1944) and Forsberg (1958), (ii) Climate as done by Schimper (1898) and Thornthwaite (1933), and (iii) Vegetational and climatic factors as done by Brutt-Davy

(1938) and Swain (1938) do not fit in at all with the classification of vegetation of the study area. Moreover, such classification would be very complex and would not serve any useful purpose. As such Champion and Seth's grouping into the following principal zones on temperature was found most appropriate.

Tentative temperature-vegetation zones of India

Temperature vegetation zone	*Mean annual temperature in summer*	*Mean annual temperature in winter*	*The nature of winter*
I. Tropical	Over 24° C	Over 18° C	None, no frost.
II. Sub-tropical	17° C to 24° C	10° C to 18° C	Definite but not servers; frost rare.
III. Temperate	7° C to 17° C	-1° C to 10° C	Pronounced, with frost & some snow.
IV. Alpine	Under 7° C	Under -1° C	Severe, with much snow.

Depending on this moisture conditions the authors (Champion and Seth) classified the vegetation of the observed field under various major and sub-groups and types as follow:

Group	*Sub-group*	*Types*
1. Tropical wet evergreen forests	1. (a) Northern tropical semi-evergreen forests.	1. (a) Eastern sub-motane semi-evergreen forests including edaphic and seral types.
	(b) North India tropical moist deciduous forests.	(b) Eastern Hima-layan Sal, Bhabar Sal, Terai Sal etc.
2. Tropical dry deciduous forests.	2. (a) Northern tropical dry deciduous forests.	2. (a) Northern dry mixed deciduous forests including seral types.
	(b) Northern tropical broad leaved forests	(b) Sub-tropical wet hill forests

Group	*Sub-group*	*Types*
3. Montane wet temperate forests.	3. (a) Northern montane wet temperate forests.	3. (a) Eastern Himalayan wet temperate forests (b) Moist temperate forests
4. Sub-Alpine forests.	-------------------	4. Eastern Himalayan sub-Alpine Birch-Fir forests.
5. Moist Alpine scrub	-------------------	5. Birch - Rhododendron forests.

The present classification is based on a different approach (synthetic survey). The basis of such classification has been described under the Chapter design and methodology and elsewhere. It is on such synthetic survey the vegetation has been classified under different ecological groups.

Bulk of the plains forest are Tropical Moist Deciduous where *Shorea robusta* is the principal species and is dominant. It has semi-evergreen habit in being deciduous for 15-20 days or less and is a stable Pre-climax in successional status. Lauraceous and meliaceous species are very abundant, but *Terminalias* stand very conspicuous in the top and middle canopy in Shorea forests. There is a belt of Tropical Dry forests described in another chapter as 'Dry-Type' along the foot hills where the crop is mixed and *Shorea* is very few and scattered. Such types also occupy quite a large area in the moist deciduous forests. The author does not consider it of much practical help to classify such forests into various goups, sub-groups, types etc. as done by Champion and Seth; moreover synthetic data do not wholly justify such classifications. The present classification of the plains forests under Pure Shorea Forests, Scattered Shorea Forests, Wet-mixed Forests, Dry-mixed Forests have been found most practical to classify such forests. Shorea forests predominate all over the plains while the Wet-mixed forests are scattered in small and big pockets distributed in such forests. The species are both evergreen and semi-evergreen. Typical evergreen types of forests occasionally occur in the foot hills region which may be termed as 'Wet Evergreen' and this extends over limited areas. Such forests have ***Mesua ferrea, Syzygium*** species, ***Phoebe lanceolata, Phoebe hainesiana, Phoebe attenuata*** and several other evergreen associates. Although Champion and Seth have termed them as Sub-Montane Semi-Evergreen Forests (Under Tropical

Semi-Evergreen Forests) these are better be described under Edaphic Types.

The present author considers the Riverain forests as seral in status and describes them under Edaphic Forests (the author also considers only one Climax i.e. Climatic climax) though somewhat stable in the typical sandy tract. Champion and Seth's Eastern Himalayan Sal, Upper Bhabar Sal, Lower Bhabar Sal, Terai Sal, Moist Mixed Deciduous Forests, Wet Temperate Forests and Mixed Coniferous Forests have been found by the author as mere micro climatic edaphic types as they occupy limited areas and do not signify any general type of vegetation. Similarly, there are some bio-edaphic forests where original better quality forests have suffered a drastic change at the hands of man.

The hills forests get Tropical monsoon rains for 5 to 6 months in a year and have high humidity. Conifer is practically absent which justifies the name of the forests as 'Monsoon Temperate' forests unlike in the Western Himalyas. These monsoon temperate forests cover Champion and Seth's Sub-Montane Wet Hill Forests and 'Montane Wet Temperate Forests'. There are several Edaphic types scattered over them. These forests are very rich in ground vegetation such as mosses, hepatics, lichens and ferns. Sub-Alpine Forests have several tree associates with dense undergrowth of *Arundinaria* species and *Rubus* species.

Based on synthetic survey the following are different types of forests and classifications made by the author:

1. Shorea robusta forests of plains and foot hills

Principal types

(a) Pure Shorea Forests.
(b) Scattered Shorea Forests.
(c) Dry-Mixed Forests.
(d) Dry-Mixed Forests.

Sub-Types

(a) Pure Shorea Forests:
(b) Six Sub-types have been identified:
   (i) *Shorea-Terminalia* association.
   (ii) *Shorea-Terminalia-Schima* association.
   (iii) *Shorea-Terminalia-Lagerstroemia* association.
   (iv) *Shorea-Adina* association.
   (v) *Shorea-Stereospermum* association.
   (vi) *Shorea-Terminalia-Garuga* association.
(c) Wet-mixed Forests:
   Two Sub-types have been identified.
   (i) *Machilus-Artocarpus-Shorea* association.
   (ii) *Machilus-Syzygium-Shorea* association.

(d) Dry-Mixed Forests:
Three Sub-types have been identified.
(i) *Tetrameles-Stereospermum-Terminalia* association
(ii) *Terminalia-Tetrameles-Stereospermum* association.
(iii) *Lagerstroemia-Terminalia-Stereospermum* association.

2. Shroea robusta and mixed forests in the Tista Valley

(a) *Shorea-Terminalia-Tetrameles* association.
(b) *Shorea-Schima-Terminalia* association.
(c) *Terminalia-Schima-Ailanthus* association.

A cross section of the Tista Valley Forests across the river has been shown in the diagram no. 5.

3. Edaphic Forests

In plains
(a) *Shorea robusta* Forests.
(i) *Shorea-Lophopetalum* associes.
(ii) *Shorea-Michelia* associes.
(iii) *Shorea-Ternstroemia-Podocarpus* associes.
(b) Tropical Fresh Water Swamp Forest.
(c) Cane Brakes.
(d) Syzygium Swamp.
(e) Pinanga-Dracaena community.
(f) Bamboo Brakes.
(e) Salix Tetrasperma swamp.

In Foot Hills
(a) *Firmiana colorata* consocies.
(b) *Sapium insigne* consocies.
(c) *Mesua-Phoebe associes.*
(d) *Shorea-Terminalia* associates.

In Middle and Upper Hills
(a) *Betula cylindrostachys* consocies.
(b) *Podocarpus nerifolia* consocies.
(c) *Quercus lamellosa* consocics.
(d) *Tsuga dumosa* consocies.

4. Seral Froests

(h) Riverain Forests:
(i) Pure *Acacia catechu* consocies.
(ii) Pure *Dalbergia sisoo* consocies.
(iii) *Dalbergia-Acacia* associes.
(iv) *Bombax-Albizzia* associes.
(v) *Toona-Gmelina* associes.

(iv) *Trewa-Bischofia* associes.
(vii) *Bauhinia* consocies.
(viii) *Wrightia-Holarrhena* associes.
(ix) *Macaranga* consocies.
(x) Wet Riverain Grassland associes.

5. Bio-edaphic Forests

(a) *Butea-Bombax-Stereospermum-Albizzia* associes.
(b) *Shorea-Terminalia-Syzygium-Lagerstroemia* associes.
(c) Gamble's 'Convolvulus Mictum'.

6. Kalimpong Hills Forests (Montane Sub-tropical and Temperate forests)

Champion and Seth have followed the classification of Cowan (1929) and described the forests under East Himalayan Sub-Tropical Wet-Hills forests and East Himalayan Wet-Temperate Forests under Montane-Subtropical Forests. The analysis of random plots does not show any basic change form the classification made by the aforesaid authors.

The following tree associations have been identified by the author at altitudes mentioned against each form synthetic survey data:

(i) At 300 metres.
Top storey-*Syzygium-Phoebe-Beilschimiedia* association.
Middle storey-*Ostodes-Cinnamomum-Drysoxylum* association.
In Tista valley:
Shorea-Terminalia-Tetrameles
Shorea-Schima-Terminalia
Terminalia-Schima-Ailanthus

(ii)	At 900 metres	*Terminalia-Schima-Bauhinia* association.
(iii)	At 1,400 metres	*Castanopsis-Engelhardtia* association.
(iv)	At 1,600 metres	*Castanopsis-Schima* association. *Castanopsis-Nyssa-Acer* association.
(v)	At 1,800 metres	*Michelia* (including *Alcimandra*)-*Castanopsis-Engelhardtia* association.
(vi)	At 2,000 metres	*Castanopsis-Schima-Acer* association.
(vii)	At 2,200-2,400 metres.	*Michelia* (including *Alcimandra*)-*Acer-Engelhardtia* association. *Machilus-Michelia* association.
(viii)	At 2,400 metres	*Michilus-Quercus* association.
(ix)	At 2,600 metres	*Tsuga-Abies-Rhododendron* association.

Figure No. 1 shows the altitudinal distribution of different tree associations in Jaldacca valley and Inner valleys. Figure No. 3 shows the distribution of *Meliaceae* and *Lauraceae* in the Kalimpong, Kurseong and Darjeeling Hills. Figure No. 4 shows predominant shrubs and herbs on the gorund. It is difficult to establish a definite relationshop of such herbs and shrubs with the tree assocation as the vegetation is very much mixed and except bamboos, some ferns and grasses, others are not well differentiated.

7. Darjeeling and Kurseong Hills Forests (Montane Sub-tropical and Temperate forests)

Principal associations derived from the analysis of random plots and transects are as follows:

(i)	At 300 metres	*Shorea-Terminalia-Schima.*
(ii)	At 900 metres	*Castanopsis-Schima* association.
(iii)	At 1,400 metres	*Castanopsis-Engelhardtia-Beilschmiedia* association.
(iv)	At 1,600 metres	*Castanopsis-Engelhardtia* association.
(v)	At 1,800 metres	*Castanopsis-Nyssa-Garuga* association.
(vi)	At 2,000 metres	*Acer-Machilus-Quercus* association.
(vii)	At 2,200 metres	*Castanopsis-Garuga-Nyssa* association.
(viii)	At 2,400 metres	*Acer-Michelia-Machilus-Quercus* association.
(ix)	At 2,600 metres	*Machilus-Michelia-Acer-Quercus* association.
(x)	At 2,800 metres	*Tsuga-Abies-Rhododendron* association.

8. Sub-alpine Forests

Principal associations are:

(a) Pure *Rhododendron* consociation.
(b) *Pure Juniperus* consociation.
(c) *Tsuga-Abies* association.
(d) *Betula-Rhododendron* association.

D. OTHER GENERAL AND CRITICAL OBERVATIONS

1. General

- The study area extends over bhabar, terai and alluvial flats in the plains and Himalayan mountains upto an altitude of 3500 m in the districts of Jalpaiguri and Darjeeling.
- Heavy monsoon rain in the High hills has resulted in a special type of temperate forests (monsoon temperate forests) which are primarily broad leaved.

- The temperate and sub-alpine forests are rich in flowering herbs and cryptogamic vegetation.
- The physiognomy of the forest is diverse.
- Economically important forest species are mostly middle aged to mature, and there is dearth of established regeneration. Non-commercial species dominate young girth classes.
- Flowering and fruiting of plants continue round the year, but peak period is April-May and September-Oct. Fruiting is at its peak in October-November.
- Herb layer is vey dense in the moist temperate hill forests; it is rather open in the plains forests. The shrub layer is, however, dense in the plains forests.
- Scitaminae is a conspicuous family in herb/shrub layer in the plains; ferns and grasses and other conspicuous ground flora. Tree regeneration forms at about 50% of the total vegetation in shrub layer.
- Simple alternate leaves predominate followed by compound alternate.
- Ephiphytic orchids, ferns and vascular epiphytes form conspicuous physiognomy of the forest; ephiphytic ferns occupy a very important position in the temperate forests where they assume several forms and biological niche.
- Climbers and Lianas tupes of 'life form' and biotic associations are remarkable features of the forests.
- The origin of *Shorea* forests is in savannah like grasslands; riverain forests also spread over large areas and the principal species are *Acacia catechu* and *Delbergia sissoo.*
- Dry-mixed forest is dominated by deciduous and semi-deciduous tree species which occur along the foot hills; the principal species are *Schima wallichii, Lagerstroemia parviflora, Dillenia pentagyna, Sterculia villosa* etc.
- Wet mixed or evergreen type forest occur in pockets and is dominated by *Artocarpus chaplasha, Sloanea sterculiaceus, Castanopsis tribuloides, Szygium* sp. *Amrora wallichii* and *Aphanamixis polystachia.*
- Riverain tracts have low lying savannah like formations which are rich in tall grass and abode of wild animals including rare one horned rhinoceros.
- Principal species in the plains and foot hills forests is *Shorea robusta* (moist deciduous forests of Champion and Seth's Classification of Forest types).
- In middle hills (1200 m-1600 m) *Sizygium* spp. *Engelhardtia spicata, Beilschmiedia* sp. *Phoebe* sp., *Terminalia* sp., *Castanopsis* sp., *Schima wallichii* and several other species

predominate and form district three storeyed forests.

- In upper hill forests (1600 m-2400 m) *Nyssa javanica, Castanopsis* sp., *Michelia* sp., *Acer* sp., *Quercus* spp. *Machilus* sp. predominate.
- Above 2400 m *Quercus* and *Rhododendron* sp. predominate with sprinklings of *Taxus baccata* and *Tsuga dumosa.*

2. Observations on Shorea Forests

Shorea robusta forests attain its diverse and varied ecological form in this small study area not encountered elesewhere in moist deciduous forests which covers more than 25 million ha. in India. This study area has therefore a special significance in ecological studies.

There are clear evidence in the forests that the *Shorea* forests had their origin in Savannah like formation. The *Shorea* forests have their best growth and development in the bhabar tract and also in the deep loamy well drained soil in Tondu and Moraghat forests. The best *Shorea* forests have a few indicator grasses, herb and shrubs viz. *Microstergeum ciliatum, Desmodium* sp. *Flemingia* sp. *Morinda citrifolia, Coffea bengalensis* and *Ageratum conyzoides* while poor types of Shorea in the foot hills are indicated by the presence of *Capillipedium assimele* and *Arundinella* spp. Regeneration of *Shorea* does not get established anywhere. Tall grasses viz. *Saccharum* spp. and *Narenga porphyrocoma* have patchy occurrence in good Sal forests. *Narenga,* however, holds extensive ground as pure crop in repeatedly burnt *Shorea* forests.

Natural regeneration of broad leaved species was found in plentiful. In fact 50% of the ground vegetation in shrub layer has been found to contain such natural regeneration. *Aphanamxis* and *Dysoxylum* species have maximum incidence of occurrence in such forests. The riverain tracts have plenty of regeneration of *Acacia catechu* and *Dalbergia sissoo* each of which grow either in pure gregarious crop or in mixture where *Bombax ceiba* and *Albizzia* spp. also predominate. Lowlying Savannah formation with *Saccharum spontaneum,* S. *procerum, Neyreudia renaudiana, Arundo donax Phragnites karka* occur. Pure *Shorea robusta* forests at good sites have as many as 450 trees per ha. against 220 in other forests. Dry mixed *Shorea* forests has few individuals in ground layer which is comparatively open. The hardy forms of *Schima wallichi, Lagerstroemia parviflora, Dillenia pentagyna, Sterculia villosa, Careya aborea, Terminalia belerica* occur besides other characteristic species, viz. *Capillepedum assimele, Ervatamia divericata, Coffea benegalensis* and *Clerodendrum viscosum.*

In such riverain tract conspicuous species are *Gmelina arborea, Toona ciliata, Bischofia javanica, Trewia nudiflora, Dillenia pentagyna* and *Wrightia tomentosa.*

In general the *Shorea* forests other than wet mixed ones have *Terminalias, Lagerstroemia parviflora, Dillenia pentagyna, Sterculia villosa, Careya arborea* as most conspicuous and widely occurring associates. In wet mixed forests *Shorea* has *Syzygium cumini, Jambosa praecox, Meliosma simplicifoila. M. pinnata, Turpinia pomifera, Dysoxylum, binectrariferum, D. hamiltonii, Aphanamixis polystachia, Elaeocarpus varunua* and *Premna* spp. as conspicuous and widely occurring associates. *Characteristic* ground flora consist of *Calamus sp., Phrynium capitatum, Lasia heterophylla, Forestia glabrata, Oplismenus compositus, Setaria palmaefolia* and *Stenoclaena palustris* (ferns.)

Macaranga Forests:-Pure formations of *Macaranga denticulata* were found wide over the study zone on burnt ground or on abandoned village or cultivation sites.

Plateau Forests:-Limited areas in Buxa hills have dense growth of *Schima wallichii* and Scattered growth of *Shorea robusta, Tetrameles nudiflora, Duabanga grandiflora, Lagerstroemia parviflora* and *Sterculia villosa.*

Evergreen Forests:-Evergreen forests were found extending over limited areas on moist and shady situations in the foot hills. Such forests have *Mesua ferrea, Sloanea sterculacea, Sterculia Alata, Artocarpus chaplasa* and *Canarium sikkimense, Cinnamomum cecedodapahne, Amoora wallichii, Syzygium formosum Duabanga grandiflora* and *Castanopsis tribuloides.* Undergrowth is dense with *Calamus* sp., *Pinanga gracilis* and very many other herbs and shrubs. *Aesculus assamica* occurs on stream bank.

3. Observation on wet-mixed Forests

This type has approximately 25 tree species comprising of about 226 trees per hectare. Evergreen species are predominant. Apart from the fact that most of the evergreen species invaded *Shorea* forests after fire protection was enforced in reserved forests, there were many moist pockets or lowlying areas where evergreen species were already existing earlier. These pockets have some endemic species confined to a small area. A few examples are as follows:-

Podocarpus neriifolia is confined to Dalkajhar and Moraghat forests; *Vatica lancaefolia* in Jamguri forests (Damanpur forests); *Michelia montana* in Lower Tondu and *Eriolaena wallichii* in Rajabhatkhawa forests. *Lauraceous* and *Meliaceous* species are very well distributed and is found regenerating in such areas. *Machilus villosa, M. gamblei, Litsaea polyantha, Aphanamixis polystachia, Amoora wallichii, Macaranga denticulata, Meliosma simplicifolia, Jambosa praecox, Callicarpa arborea, Dillenia indica, D. pentagyna, Premna bengalensis, Elaeocarpus varunua* are typical species of such areas.

There are large number of herbs, shurbs and climbers and they have been enumerated in *transects* and described. *Oplismenus, Cyrtococoum, Digitaria and Setaria* are common grasses, *Dioscorea Zingibar, Curcuma, Alpinia Smilax* common monocots. *Lygodium, Abacopteris* and *Cyclosorus* are the commonest ferns. Regeneration of most of the upper canopy trees (except *Shorea*) are found all over the high forest; *Shorea* regeneration, however, does not establish itself.

Crown classes of the tree species have been determined; Density, frequency and percentage of species have been calculated and analysed. There are three distinct tree canopy layers. Some findings are as under:

Frequency:-
- i) *Callicarpa arborea, Premna bengalensis* and *Macaranga denticulata* have 100 % frequency;
- ii) *Shorea robusta, Olea diocia, Litsaea polyantha, Dillenia pentagyna, Careya arborea* and *Cinnamomum obtusefolium* have 85%.
- iii) Other calculations have been shown in frequency diagram No. 1.

The following are seven species having highest percentage (Histogram-1) of occurrence:

Machilus gamblei	-9.4%	*Premna* species	3.5%
Meliosma simplicifolia	6.4 %	*Shorea robusta*	3.4 %
Macaranga denticulata	5.3%	*Litsaea polyantha*	3.1%
Aphanamixis polystachia	2.2%		

The following are the analysis of (vertical) canopy layers:-

Canopy layer	*Number of species*	*Percentage*
Top storey	34	25.8
Middle storey	64	48.4
Lower storey	34	25.8
	132	100.0%

Histogram Nos. 1-7 and Statement-1 show percentage of different species in different types of forests. Histogram Nos. 12-18 show the canopy class distribution of predominant species (%) and also distribution (%) of Dipterocarpaceae, Meliaceae and Lauraceae—the three top families in different forest types. Statement-VI shows density per hectare and crown class of 30 common species.

Statement-VI
Density per Hectare of Thirty Common Genera and Species of the Plains Forest

Genera and Species	*Forests Types*					*Canopy Class*
	Wet Mixed	*Scattered Shorea*	*Pure Shorea*	*Dry mixed*	*Density*	
Aphanamixis polystachia	2.2	4.1	1.9	6.7	3.7	M.S.
Amoora wallichii	1.3	0.9	0.2	0.2	0.7	T.S.
Albizzia lucida	1.3	0.1	-	0.4	0.4	T.S.
Callicarpa arborea	0.9	0.6	0.6	2.4	1.1	L.S.
Careya arborea	0.6	0.9	0.6	2.4	1.1	L.S.
Castanopsis Tribuloides	2.1	0.8	0.1	0.2	0.8	M.S.
Dillenia pentagyna	0.3	1.6	3.0	4.5	2.3	M.S.
Dysoxylum Binnectariferum	-	0.7	0.6	1.0	0.6	M.S.
Elaeocarpus varunua	1.7	2.9	0.1	0.2	1.2	T.S.
Ficus species	2.0	0.2	0.1	0.6	0.7	T.S.
Gmelina arborea	0.2	0.9	0.6	0.7	0.6	M.S.
Garuga pinnata	2.6	0.1	0.1	1.4	1.1	M.S.
Litsaea monopetala	3.1	0.6	1.7	5.8	2.8	M.S.
Lagerstroemia parviflora	0.4	0.5	1.0	3.0	1.2	T.S.
Macaranga denticulata	5.2	5.0	1.8	1.6	3.4	L.S.
Mallotus philippinensis	0.3	0.3	0.6	3.9	1.3	M.S.
Meliosma simplicifolia	6.4	4.6	2.7	6.7	5.1	L.S.
Machilus villosa	2.2	6.2	2.4	1.5	3.0	T.S.
Machilus gamblei	9.4	2.5	2.1	0.1	3.5	T.S.
Olea dioica	0.9	0.2	-	-	0.3	M.S.
Oroxylum indicum	1.5	0.4	0.4	2.5	1.2	M.S.
Premna bengalensis	2.5	0.4	2.6	2.1	2.1	M.S.
Premna integrifolia	0.2	1.4	1.7	5.4	2.2	M.S.
Stereospermum species	0.5	0.6	-	2.3	0.9	T.S.
Shorea robusta	3.4	22.4	56.8	2.0	21.1	T.S.

Genera and Species	*Forests Types*				*Density*	*Canopy Class*
	Wet Mixed	*Scattered Shorea*	*Pure Shorea*	*Dry mixed*		
Schima wallichii	0.5	6.0	0.2	1.3	2.0	T.S.
Syzygium cumini	1.3	1.4	0.8	3.1	1.6	M.S.
Terminalia belerica	0.6	1.7	2.4	5.6	2.6	T.S.
Toona ciliata	-	0.5	0.1	0.2	0.2	T.S.
Sterculia villosa	1.0	0.8	1.4	2.4	1.4	T.S.

N.B. T.S. = Top Storey
M.S.= Middle Storey
L.S. = Lower Storey.

4: Observation on Scattered and Pure Shorea Forests

This type is transtitional between the wet type and pure *Shorea* forests. *Shorea* forms is about 25% with heavy mixtures of Meliaceae and Lauraceae. There are lesser number of evergreen species that in wet type. The wet mixed pockets do not have the characteristic species like *Vatica, Ternstroemia, Michelia montana* etc. as in this type.

There are three canopy classes of tree species in the following proportions:-

Canopy layer	*Number of species*	*Percentage of total tree species*
Top storey	27	26.5
Middle storey	47	46.0
Lower storey	28	27.5
Total	102	100%

Frequency: In the Diagram No. 3 frequency of the species of this type has been given. *Shorea robusta* and *Amoora wallichii* have 100% frequency; *Syzygium cumini, Terminalia belerica, Premna begalensis, Meliosma simplicifolia* and *Dillenia pentagyna* have 85% frequency.

The following are few species with high percentage of occurrence:- Top Storey—*Shorea robusta* 22.4%, *Schima wallichii* 6.0%, *Terminalia belerica* 1.7%; Middle storey-*Premna* species, *Syzygium cuminii* have 1.4% *Aphenanixis polystachia*, 4.1% *machilus* species and *Dillenia pentagypa* have 1.6%. Lower storey-*Meliosma simplicifolia* 4.6% *Macaranga denticulata* 5.0%.

Histogram Nos. 1-8 and statement-1 show percentage of different species in different types of forests. Histograms 12-18 show Canopy class distribution of predominant species (in %) and also distribution (in %) of Dipterocarpaceae, Meliaceae & Lauraceae the three top families in different types. Frequency of species are as follows:-

Shorea robusta and *Aphanamixis polystachia* have 100% frequency; *Terminalia belerica, Premna bengalensis, Premna integrifolia* and *Dillenia pentagyna* have 85% frequency each.

Canopy layer—In the top *Shorea robusta* has 56.8% *Terminalia belierica* has 2.4% *Machilus gamblei* 2.1% and *Machilus villosa* 2.4%.

In the middle storey *Premna bengalensis* has 2.6% and *Aphanamixis Polystachia* has 1.9%.

In the lower storey *Dillenia pentagyna* has 3.0% and *Meliosma simplicifolia* has 2.7%.

The species and their percentage distribution in three canopy classes are as follows:-

Canopy layers	*Number of species*	*Percentage of total tree species*
Top storey	14	18.2
Middle storey	45	58.4
Lower storey	17	23.4
Total	77	100.0

The ground flora has been analysed and the analysis is given in the appropriate chapter in details. *Clerodendrum viscosum, Smilax* species, *Dioscorea* species, *Bridelia* species, *leguminous* species, regeneration of *Litsaea* and species of *Machilus, Setaria, Oplismenus* and *Abacopteris* are common and universally distributed. In the fire burnt *Shorea* forests pure undergrowth of *Curcuma* and *Zingibar* can be found in many forests of Borodighi, Champsari, Tondu and Media blocks.

5. Shorea in Kurseong Hill Forest

The zone is essentially *Shorea* with *Terminalia tomentosa* and *Terminalia belerica* forming close associate in the top storey. *Adina cordifolia* adds economic value to such valuable forests. These forests have sprinkling of *Dalbergia latifolia* which occurs only to the west of Tista. *Ougenia oojeinensis* which is a close associate of *Shorea* in Uttar Pradesh reaches its eastern limit in the *Shorea* forests of Khairbani. *Homalium zeylanicum* a plant of Malabar evergreen forests is found to occur with *Shorea* in Bamanpokri. In the middle storey *Dillenia, Aphanamixis, Meliosma, Macaranga, Mallotus Dysoxylum* species and Lauraceous species are very prominent and abundant.

6. Special Observations

The ground vegetation and various canopy layers have been found to give refuge to wildlife species. There is abundance of food for various species of wildlife. The wildlife species play a very important role in dispersal of fruits and seeds and keeping down unusual luxuriance of fodder crops. Vast number of plant species, their leaves, stems, fruits and seeds harbour innumerable insect species which form an exhaustible source of food for various insectivorous birds. Seasonal changes in flower, fruit and leaf flush influence the food habit of wildlife. Diagrams 1 &2 represent abundance of fruiting species throughout the year. A study of the fruit size (classified into four size classes) shows that nature has sustained frugivorous birds (classified into four size classes) in the same proportion as the sizes of fruits. Similarly, classification of insects into four sizes shows equivalent proportion of the sizes of insectivorous birds in the study area. The abundance of various insect species goes parallel with abundance of plant species in the lower hill. In such areas there is abundance of lizards reptile and amphibian species. Among the game fishes (Masheer, Katli, Trout, Goonch); Katlis are fond of fruits *(Tinospora cordifolia)* but Masheers primarily feed on planktonic algal flora. Root decoction of *Miltellia pachycarpa,*crushed fruits of *Randia longispira,* crushed bark of *Schima wallichii,* crushed roots of *Derris spp.* are poisonous and have been used extensively to kill fishes. Various species of the families *Lauraceae, Meliacieae, Euphorbiaceae* yield fruits edible to wildlife species. Speciesl of the genra *Litsaea, Machilus. Cinnamomum, Lindera, Beilschemidia, Dysoxylum, Aphanamixis, Amoora, Chisocheton, Glochidion, Bridelia, Bischofia, Symplocos, Castonopsis, Quercus, Pygeum Picrasma,* etc. yield fruits which are liked by frugivourous birds.

Small sized frugivorous birds also feed on the fruits of *Glochidion multioculare, Ardisia himile, Micromelum pubescens, Cocculus* sp. *Stephania* sp., *Setaria* sp., *Paspalum* sp. and on other species. In the hills however, seeds and fruits of *Viburnum* sp., *Rubus* sp., *Berberis* sp., *Cotoneaster* sp. are associatied with frugivorous birds. Among the arillate seeds species of the families, *Celastraceae, Euphorbiaceae, Myrtaceae* are notable. On the whole, the study area is rich in food material for various species of birds, herbivorous animals which obviously harbour carnivores, and other animals. Twenty five species of grasses provide bulk fodder for the herbivores; *Microstegeum ciliatum* is commonest of them. *Tinospora cordifolia, Acacia pennata, Artocarpus* sp., *Saccharum* sp., *Imperata cylindrica* etc. are of obvious choice. The fruit and leaves of *Castanopsis tribuloides, Musa* sp., *Quercus* sp., *Caryota urens, Nyssa javanica, Calamus, Ficus* sp., are liked by various species of monkey and langurs. Malayan squirrel is very much fond of the fruits of *Hodgsonia*, Roots and rhizomes of aroids, *Asparagas, Curcuma, Lasia* and *Zingibar* are favourites of the pigs. The leaves of *Chukrasia, Toona, Bombax, Gmelina*, and *Alpinia* are liked by other herbivores.

A careful study to the quantitative data revealed that many species occurring in the forests are not represented in the plots. Some are not commonly found in the forests. Some species have disappeared owing to natures own process of evloutionary changes. Some got extinct because of man's ignorance and indiscriminate exploitation and destruction of habitat. *Ruwollfia serpentina* has disappeared from the area during last twenty years of ruthless exploitation; other spp. are *Bauhinia anguina, Dioscorea deltoidea, D. prazeri, Swertia chirata, Aconitum* sp., *Nardostachys jatamunsii, Cycas pectinata, Podophyllum* sp. It has been found that several species were very heavily exploited in the past which are as follows:

Fern and ferm allies—*Cyathea* sp. *Botrychium virginianum* and *Angiopleris evecta*

Orchids - *Cymbidium marcohizon, Arundina graminifolia, Cyprepidium elegans, C. himalaicum, Arachnanthe catcheartii, A. Clarkei, Dendrobium densiflorum.*

Other - *Helwingia himalalica, Rhododendron dalhousiae,* other *Rhododendron* sp. in general

Some rare species of the study area need protection, viz. *Gnetum scandens, Ulmus lancaefoila, Populus gamblie, Homalium zeylanicum, Podocarpus nerrifolia, Ternstroemia gymnanthera, Cycas pectinata* etc.

7. Observations on Monocotypedons

Several monocots have been recorded from various plots analysed, but many species occurring sporadically got excluded from such list. Some observations are as under:-

Palms - *Phoenix rupicola* and *P. acullis* are found in Tista and Rangitt valley's. *Licuala peltata* and *Livistona jenkinsiana* have very sporadic or of rare occurrence. *Caryota urens* and *Wallichia distichia* are sporadic in the foot hills but very conspicuous. *Trachycarpus* spp. is also sporadic or rare.

Aroids - *Pothos, Rhaphidophora, Scindapsus, Lasia, Gonatanthus* and *Arisaema* are conspicuous aroids of temperate and tropical forests.

Water & marsh plants - Floating submerged and amphibious, genera are *Hydrilla, Vallisneria, Ottelia, Blyxa, Typha, Pistia, Typhonia, Cruptocoryne, Lemna, Woolffia, Spriodela,*

Alisma, Sagittaria, Monochria, Eichornia and varicus grasses and sedges.

E. SUCCESSIONAL TREND IN THE FORESTS

Various successional trend in the grasses and grassland has been detailed in sections D and E. Such trends have also been indicated as in other forests in Sections B and C of this chapter.

Biotic factors have primarily brought changes in the chain of succession. Vast areas have been felled and converted into man made plantations. Flood in the rivers have brought devastation in the forests and vast areas have been silted or covered by sand, swamps have lost their existence and grazing and fire have stopped the trend of regeneration and succession. A few diagrams produced under Serial No. 40 to 43 will reveal the situation in (i) Deciduous forests (ii) Burnt plantations (iii) Grassland communities (iv) Wet mixed forests (v) Sal forests after clear felling. Two diagrams at S. No. 44 & 45 represent successional stage of vegetation from river bed to foot hills and a cross section of foot hill vegetation from Jaldacca in the Central Duars to the Balason river in the Western Boundary. Sl. No. 46 indcates the present state of vegetation as in Ranjit valley in the east to Terai in the west.

It has been adequately highlighted that plantations have fewer species than the high forest. Persistent grazing and frequent fires have diluted the diversity of genera and species resulting in fewer and coarse type of vegetation. Species composition, frequency, density etc., have been seriously affected owing to large scale felling in the hills and conversion into plantations of a few selected species.

Appendix

SYNTHETIC ANALYTICAL DATA

Table - A	-	Belt transect in wet mixed forests.
Table - B	-	Belt transect in scattered Shorea forests.
Table - C	-	Belt transect in pyre shorea forests.
Table - D	-	Belt transect in dry mixed forests.
Table - E	-	Grid plot analysis in shorea forests.
Table - F	-	Grid plot analysis in dry mixed forests.
Table - G	-	Grid plot analysis of ground vegetation in plains.
Table - H	-	Belt transect and grid plot analysis in foot hills of Kurseong forests.
Table - I	-	Belt transect and grid plot analysis in Kalimpong forests.
Table - J	-	Belt transect and grid plot analysis in Darjeeling forests.
Table - K	-	Grid plot analysis in the hills.

Special Note on Appendix

Tables A, B, C and D

Substantial areas of the forests have been converted into plantation by felling of forests. Forest villages have also been estabilished inside the reserved forests. This created problems as to how to deal with plots falling in such areas.

This was obviated by aligning a belt transect in the vicinity along fertility gradient in the same type of forest and laying plots or the transect at intervals.

Table E

Substantial number of plots falling in Nilpara, Chilapata, Moraghat, Tondu, Buxaduar, Dalkajhar, Rajabhatkhawa and Damanpur forests have been described under Sal forests where various types of forests are mingled together.

Table F

No remark.

Table G

For assessment of ground vegetation in plains, Sal forests were considered as a broad type owing to the fact that the ground vegetation did not differ materially in pure Sal, scattered Sal and wet mixed types. However, Riverain, Dry mixed and Plantation areas were considered separately for such count. Each of the numericals 1 to 10 in Sal forests reflect the findings in 40 plots; in other forests each of the numericals 1 to 5 reflect findings in 20 plots.

Table H

In Terai and Foot hill forests, total number of plots falling in plantations, forests villages and other blank areas were counted by laying a belt transect in the same type of forest. Equal number of plots were laid out at regular interval in the transect.

Table I and J

Similar method as in Tables A, B, C, and D was adopted and belt transects were laid in areas where plots fell in plantations and forest villages. Inaccessible areas and areas biotically effected were not attended to.

Table K

Ground vegetation in hills was identified in three zones viz. lower hill (altitude from 100 m. to 1000 m.), middle hill (1001 m. to 1800 m.) and upper hills (1801 m. to 2600 m.). In each zone results have been expressed on 10 sites involving about 150 plots.

SYNTHETIC SURVEY
FLORISTIC COMPOSITION OF TREE VEGETATION IN PLAINS: BELT TRANSECTS

(i) WET-MIXED FORESTS

REFERENCE TO TABLE A

COLUMN	1-NUMBER OF TREES
COLUMN	2-PERCENTAGE OF SPECIES
COLUMN	3-DENSITY
F%	-FREQUENCY
D	-DENSITY PER 0.4 HECTARE (OR ONE ACRE)
DOMINANCE	-U.S. (UPPER STOREY) -M.S. (MIDDLE STOREY) -L.S. (LOWER STOREY)

Table - A. Wet Mixed Forests.

Sl. No.	SPECIES	MORAGHAT 1.	MORAGHAT 2.	MORAGHAT 3.	LATAGURI 1.	LATAGURI 2.	LATAGURI 3.	DAMANPUR 1.	DAMANPUR 2.	DAMANPUR 3.
1.	*Actinodaphne obovata*	8	0.8	4	4	0.4	0.4	-	-	-
2.	*Actinodaphne angustifolia*	30	3.0	3.0	-	-	-	-	-	
3.	*Aesculus assamica*	-	-	-	18	2.0	1.8	-	-	-
4.	*Aralia* species	-	-	-	-	-	-	-	-	-
5.	*Albizzia lucida*	-	-	-	-	0.5	0.4	2	0.2	0.2
6.	*Alstonia scholaris*	2	0.2	0.2	-	-	-	-	-	-
7.	*Aphanamixis polystachia*	3	0.3	0.3	44	5.0	4.4	4	0.4	0.4
8.	*Amoora wallichii*	33	3.3	3.3	2.4	2.7	2.4	-	-	-
9.	*Anthocephalus chinensis*	2	0.2	0.2	2	0.2	0.2	-	-	-
10.	*Aporosa roxburghii*	3	0.3	0.3	-	-	-	-	-	-
11.	*Artocarpus chaplasa*	45	4.5	4.5	2	2.7	2.4	-	-	-
12.	*Actinodaphne* species	6	0.6	0.6	-	-	-	-	-	-
13.	*Artocarpus lakoocha*	2	0.2	0.2	5	0.6	0.5	-	-	-
14.	*Albizzia odoratissima*	-	-	-	-	-	-	-	-	-
15.	*Albizzia lebbek*	-	-	-	-	-	-	-	-	-
16.	*Baccaurea sapida*	-	-	-	-	-	-	-	-	-
17.	*Bauhinia purpurea*	-	-	-	-	-	-	-	-	-
18.	*Beilschmiedia brandisi*	-	-	-	4	0.5	0.4	-	-	-
19.	*Bombax ceiba*	2	0.2	0.2	-	-	-	-	-	-
20.	*Bridelia montana*	-	-	-	-	-	-	2	0.2	0.2
21.	*B. retusa*	-	-	-	5	0.6	0.5	-	-	-

Sl. No.	RAJABHATKHAWA 1.	2.	3.	APALCHAND 1.	2.	3.	UDLABARI 1.	2.	3.	DOMINANCE	FREQUENCY	DENSITY
1.	38	0.5	3.6	10	1.3	1.0	-	-	-	L.S.	68	1.0
2.	-	-	-	-	-	-	-	-	-	L.S.	17	0.5
3.	8	0.7	0.8	20	2.5	2.0	-	-	-	L.S.	51	0.09
4.	2	0.2	0.2	-	-	-	-	-	-	L.S.	17	0.03
5.	50	4.6	5.0	10	1.3	1.0	10	1.2	1.0	T.S.	84	1.3
6.	2	0.2	0.2	-	-	-	-	-	-	M.S.	33	0.03
7.	80	7.3	8.0	-	-	-	-	-	-	M.S.	68	2.2
8.	18	1.6	1.8	-	-	-	-	-	-	T.S.	51	1.3
9.	-	-	-	-	-	-	-	-	-	M.S.	33	0.06
10.	-	-	-	-	-	-	-	-	-	L.S.	17	0.05
11.	2	0.2	0.2	-	-	-	-	-	-	T.S	51	1.2
12.	-	-	-	-	-	-	-	-	-	L.S.	17	0.1
13.	4	0.4	0.4	-	-	-	-	-	-	M.S.	51	0.2
14.	8	0.7	0.8	-	-	-	-	-	-	M.S.	17	0.1
15.	4	0.4	0.4	-	-	-	-	-	-	M.S.	17	0.07
16.	6	0.5	0.6	10	1.3	1.0	-	-	-	L.S.	33	0.6
17.	48	4.4	4.8	20	2.5	2.0	-	-	-	L.S.	33	1.15
18.	-	-	-	-	-	-	-	-	-		17	0.1
19.	4	0.4	0.4	-	-	-	-	-	-	T.S.	33	0.1
20.	-	-	-	-	-	-	-	-	-	M.S.	17	0.03
21.	-	-	-	-	-	-	-	-	-	M.S.	17	0.1

Table - A. Wet Mixed Forests.

Sl. No.	SPECIES	MORAGHAT 1.	MORAGHAT 2.	MORAGHAT 3.	LATAGURI 1.	LATAGURI 2.	LATAGURI 3.	DAMANPUR 1.	DAMANPUR 2.	DAMANPUR 3.
22.	*Bischofia javanica*	-	-	-	2	0.2	0.2	2	0.2	0.2
23.	*Callicarpa arborea*	5	0.5	0.5	4	0.5	0.4	2	0.2	0.2
24.	*Canarium sikkimense*	-	-	-	-	-	-	-	-	-
25.	*Careya arborea*	-	-	-	-	-	-	8	0.9	0.8
26.	*Castanopsis indica*	2	0.2	0.2	44	5.2	4.4	-	-	-
27.	*C. tribuloides*	45	4.5	4.5	28	3.4	2.8	-	-	-
28.	*Chisocheton paninculatus*	3	0.3	0.3	14	1.7	1.4	-	-	-
29.	*Chukrasia tabularis*	2	0.2	0.2	-	-	-	2	0.2	0.2
30.	*Cinnamomum cecidodaphne*	-	-	-	8	1.0	0.8	-	-	-
31.	*C. obtusifolium*	9	0.9	0.9	6	0.7	0.6	20	2.0	2.0
32.	*Crataeva nurvala*	-	-	-	-	-	-	-	-	-
33.	*Cryptocarya amygdalina*	2	0.2	0.2	6	0.7	0.6	24	2.6	2.4
34.	*Casearia vareca*	5	0.5	0.5	2	0.2	0.2	4	0.4	0.4
35.	*Casearia* species	-	-	-	-	-	-	42	4.6	4.2
36.	*Cephalanthus occidentalis*	-	-	-	-	-	-	-	-	-
37.	*Desmodium racemosum*	-	-	-	-	-	-	-	-	-
38.	*D. pentagyna*	6	0.6	0.6	-	-	-	8	0.9	0.8
39.	*Drimycarpus racemosus*	-	-	-	14	1.7	1.4	-	-	-
40.	*Dysoxylum binectariforum*	3	0.3	0.3	-	-	-	-	-	-
41.	*D. hamiltonii*	12	1.2	1.2	6	0.7	0.6	-	-	-
42.	*D. procerum*	5	0.5	0.5	14	1.7	1.4	-	-	-
43.	*Dalbergia assamica*	-	-	-	6	0.7	0.6	-	-	-
44.	*Dillenia indica*	-	-	-	2	0.2	0.2	4	0.4	0.4

Sl.	RAJABHATKHAWA			APALCHAND			UDLABARI			DOMINANCE	FREQUENCY	DENSITY
No.	1.	2.	3.	1.	2.	3.	1.	2.	3.			
22.	16	1.4	1.6	-	-	-	-	-	-	M.S.	51	0.3
23.	14	1.3	1.4	10	1.3	1.0	10	1.3	1.0	L.S.	100	0.9
24.	2	0.2	0.2	-	-	-	-	-	-	T.S.	17	
25.	-	-	-	-	-	-	20	2.5	2.0	L.S.	33	0.6
26.	-	-	-	-	-	-	-	-	-	M.S.	33	0.9
27.	50	4.6	5.0	-	-	-	-	-	-	M.S.	51	2.1
28.	2	0.2	0.2	-	-	-	-	-	-	M.S.	51	0.4
29.	-	-	-	-	-	-	-	-	-	T.S.	33	0.1
30.	-	-	-	-	-	-	-	-	-	T.S.	17	0.2
31.	2	0.2	0.2	20	2.6	2.0	-	-	-	M.S.	85	1.1
32.	54	4.7	5.4	-	-	-	-	-	-	M.S.	17	0.1
33.	10	0.9	1.0	4	0.5	0.4	-	-	-	M.S.	85	0.8
34.	-	-	-	-	-	-	-	-	-	L.S.	51	0.2
35.	-	-	-	-	-	-	-	-	-	L.S.	17	0.8
36.	6	0.5	0.6	-	-	-	-	-	-	M.S.	17	0.1
37.	2	0.2	0.2	-	-	-	-	-	-	L.S.	17	-
38.	50	4.6	5.0	10	1.3	1.0	85	10.6	8.5	M.S.	85	3.0
39.	-	-	-	30	3.8	3.0	-	-	-	M.S.	33	0.9
40.	-	-	-	-	-	-	-	-	-	M.S.	17	-
41.	56	5.2	5.6	-	-	-	-	-	-	M.S.	51	1.2
42.	-	-	-	-	-	-	-	-	-	M.S	33	0.4
43.	-	-	-	-	-	-	-	-	-	T.S	17	0.1
44.	6	0.5	0.6	40	5.2	4.0	-	-	-	M.S.	68	1.1

Table-A. Wet Mixed Forests.

Sl. No.	SPECIES	MORAGHAT 1.	2.	3.	LATAGURI 1.	2.	3.	DAMANPUR 1.	2.	3.
45.	*Elaeocarpus* species	-	-	-	-	-	-	-	-	-
46.	*Sloanea sterculiacea*	-	-	-	4	0.5	0.4	-	-	-
47.	*Ehrotia acuminata*	-	-	-	-	-	-	-	-	-
48.	*E. rugosus (Elaeocarpus)*	-	-	-	66	8.0	6.6	-	-	-
49.	*Elaeocarpus varunua*	20	2.0	2.0	24	2.8	2.4	6	0.7	0.6
50.	*Engelhardtia spicata*	-	-	-	12	1.4	1.2	-	-	-
51.	*Erythrina suberosa*	12	1.2	1.2	-	-	-	-	-	-
52.	*Eurya cavinervis*	-	-	-	-	-	-	-	-	-
53.	*Eurya acuminata*	2	0.2	0.2	-	-	-	-	-	-
54.	*Evodia lunuankenda*	3	0.3	0.3	12	1.4	1.2	-	-	-
55.	*E. meliaefolia*	2	2.0	0.2	2	0.2	0.2	-	-	-
56.	*Erioglossum edule*	3	0.3	0.3	-	-	-	-	-	-
57.	*Eriolaena wallichii*	-	-	-	-	-	-	-	-	-
58.	*Eranthemum indicum*	-	-	-	-	-	-	-	-	-
59.	*Ficus* species	2	0.2	0.2	16	1.9	1.6	6	0.7	0.6
60.	*Garuga pinnata*	-	-	-	5	0.6	0.5	14	1.5	1.4
61.	*G. hirsutum (Glochidion)*	48	4.8	4.8	-	-	-	-	-	-
62.	*Gmelina arborea*	2	0.2	0.2	-	-	-	2	0.2	0.2
63.	*Gynocardia odorata*	-	-	-	2	0.2	0.2	-	-	-
64.	*Glochidion* species	2	0.2	0.2	-	-	-	-	-	-

Sl. No.	RAJABHATKHAWA 1.	2.	3.	APALCHAND 1.	2.	3.	UDLABARI 1.	2.	3.	DOMINANCE	FREQUENCY	DENSITY
45.	12	1.1	1.2	-	-	-	-	-	-	M.S.	17	0.2
46.	16	1.4	1.6	40	5.1	4.0	-	-	-	T.S.	51	1.2
47.	-	-	-	-	-	-	-	-	-	T.S.	17	1.3
48.	-	-	-	-	-	-	10	1.3	1.0	M.S.	17	0.2
49.	22	2.1	2.2	20	2.6	2.0	-	-	-	T.S.	85	1.7
50.	-	-	-	4	0.5	0.4	-	-	-	T.S.	33	0.3
51.	-	-	-	-	-	-	-	-	-	M.S.	17	0.2
52.	8	0.7	0.8	-	-	-	-	-	-	T.S.	33	0.1
53.	12	1.1	1.2	-	-	-	-	-	-	M.S.	17	0.25
54.	-	-	-	-	-	-	-	-	-	L.S.	33	0.3
55.	-	-	-	-	-	-	-	-	-	M.S.	33	0.4
56.	-	-	-	-	-	-	-	-	-	M.S.	17	-
57.	32	3.0	3.2	-	-	-	-	-	-	M.S.	17	0.5
58.	2	0.2	0.2	-	-	-	-	-	-	M.S.	17	-
59.	12	1.1	1.2	60	8.0	6.0	-	-	-	T.S.	85	2.0
60.	2	0.2	0.2	100	13.0	10.0	-	-	-	M.S	68	2.6
61.	-	-	-	-	-	-	-	-	-	M.S	17	0.8
62.	10	1.0	1.0	-	-	-	-	-	-	M.S.	51	0.2
63.	-	-	-	-	-	-	-	-	-	L.S.	17	-
64.	4	0.4	0.4	-	-	-	-	-	-	L.S.	33	0.1

Table - A. Wet Mixed Forests.

Sl. No.	SPECIES	MORAGHAT			LATAGURI			DAMANPUR		
		1.	2.	3.	1.	2.	3.	1.	2.	3.
65.	*Heteropanax fragrans*	-	-	-	4	0.5	0.4	-	-	-
66.	*Holarrhena antidysenterica*	-	-	-	-	-	-	-	-	-
67.	*Hymenodictyon excelsum*	-	-	-	-	-	-	-	-	-
68.	*Ilex godajam*	2	0.2	0.2	-	-	-	-	-	-
69.	*Jambosa formosa*	50	5.0	5.0	16	1.9	1.6	85	9.0	8.5
70.	*J. praecox*	25	2.5	2.5	5	0.6	0.5	40	4.0	4.0
71.	*Knema linifolia*	2	0.2	0.2	-	-	-	-	-	-
72.	*Lophopetalum fimbriatum*	-	-	-	45	5.4	4.5	2	0.2	0.2
73.	*Lagerstroemia parviflora*	-	-	-	8	1.0	0.8	-	-	-
74.	*Leea* species	-	-	-	2	0.2	0.2	-	-	-
75.	*L. monopetala* (Litsaea)	60	6.0	6.0	4	0.5	0.4	2	0.2	0.2
76.	*L. glutinosa*	-	-	-	-	-	-	-	-	-
77.	*Lannea coromendelica*	3	0.3	0.3	5	0.6	0.5	-	-	-
78.	*Litsaea panamonja*	-	-	-	20	2.4	2.0	-	-	-
79.	*Mallolus albus*	-	-	-	-	-	-	-	-	-
80.	*Macaranga denticulata*	30	3.0	3.0	4	0.5	0.4	22	2.0	2.2
81.	*Machilus gamblei*	60	6.0	6.0	4	0.5	0.4	200	22.0	20.0
82.	*M. villosa*	60	6.0	6.0	30	3.6	3.0	-	-	-

Sl. No.	RAJABHATKHAWA 1.	2.	3.	APALCHAND 1.	2.	3.	UDLABARI 1.	2.	3.	DOMINANCE	FREQUENCY	DENSITY
65.	-	-	-	-	-	-	-	-	-	M.S.	17	0.1
66.	-	-	-	4	0.5	0.4	-	-	-	L.S.	17	0.1
67.	-	-	-	4	0.5	0.4	-	-	-	M.S.	17	0.1
68.	-	-	-	-	-	-	10	1.3	1.0	M.S.	33	0.3
69.	4	0.4	0.4	-	-	-	-	-	-	M.S.	68	2.7
70.	-	-	-	110	14.1	11.0	-	-	-	M.S.	68	3.5
71.	2	0.2	0.2	-	-	-	-	-	-	L.S.	33	0.1
72.	-	-	-	-	-	-	-	-	-	T.S.	33	0.9
73.	2	0.2	0.2	10	1.3	1.0	-	-	-	T.S.	51	0.4
74.	-	-	-	-	-	-	-	-	-	L.S.	17	-
75.	10	0.8	1.0	-	-	-	90	11.2	9.0	M.S.	85	3.1
76.	4	0.4	0.4	-	-	-	-	-	-	M.S.	17	-
77.	2	0.2	0.2	-	-	-	-	-	-	T.S.	51	0.2
78.	-	-	-	-	-	-	10	1.3	1.0	M.S.	33	0.6
79.	4	0.4	0.4	-	-	-	-	-	-	L.S.	17	0.1
80.	38	3.5	3.8	120	15.4	12.0	60	7.5	6.0	L.S.	100	5.3
81.	-	-	-	-	-	-	225	28.1	22.5	T.S.	68	9.4
82.	38	3.5	3.8	-	-	-	-	-	-	T.S.	51	2.2

Table - A. Wet Mixed Forests.

Sl. No.	SPECIES	MORAGHAT 1.	MORAGHAT 2.	MORAGHAT 3.	LATAGURI 1.	LATAGURI 2.	LATAGURI 3.	DAMANPUR 1.	DAMANPUR 2.	DAMANPUR 3.
83.	*Mallotus philippinenisis*	6	0.6	0.6	4	0.5	0.4	-	-	-
84.	*Mangifera sylvatica*	27	2.7	2.7	-	-	-	2	0.2	0.2
85.	*Machilus* species	46	4.6	4.6	2	0.2	0.2	-	-	-
86.	*Mastixia arborea*	3	0.3	0.3	2	0.2	0.2	-	-	-
87.	*Meliosma pinnata*	2	0.2	0.2	-	-	-	-	-	-
88.	*M. simplicifolia*	14	1.4	1.4	-	-	-	195	21.7	19.5
89.	*Michelia champaca*	-	-	-	2	0.2	0.2	-	-	-
90.	*M. montana*	-	-	-	32	3.8	3.2	-	-	-
91.	*Morus lavegata*	-	-	-	-	-	-	-	-	-
92.	*Olea dioica*	5	0.5	0.5	25	3.0	2.5	4	).4	0.4
93.	*Oroxylum indicum*	3	0.8	0.8	-	-	-	-	-	-
94.	*Polyalthia semiarum*	-	-	-	4	0.5	0.4	-	-	-
95.	*Picrasma napalensis*	-	-	-	-	-	-	-	-	-
96.	*Premna bengalensis*	10	1.0	1.0	25	3.0	2.5	65	70	6.5
97.	*P. integrifolia*	2	0.2	0.2	-	-	-	-	-	-
98.	*Pygeum acuminatum*	-	-	-	6	0.8	0.6	-	-	-
99.	*Pandanus furcatus*	2	0.2	0.2	-	-	-	-	-	-
100.	*Pterospernum acerifolium*	2	0.2	0.2	-	-	-	-	-	-

Sl. No.	RAJABHATKHAWA 1.	2.	3.	APALCHAND 1.	2.	3.	UDLABARI 1.	2.	3.	DOMINANCE	FREQUENCY	DENSITY
83.	6	0.5	0.6	4	0.5	0.4	-	-	-	M.S.	68	0.3
84.	4	0.4	0.4	-	-	-	-	-	-	T.S.	51	0.6
85.	2	0.2	0.2	-	-	-	-	-	-	T.S.	51	0.1
86.	-	-	-	-	-	-	-	-	-	M.S.	33	0.1
87.	62	5.6	6.2	-	-	-	-	-	-	L.S.	33	0.9
88.	-	-	-	60	7.7	6.0	60	7.5	6.0	L.S.	68	6.4
89.	4	0.4	0.4	-	-	-	-	-	-	T.S.	33	0.1
90.	-	-	-	-	-	-	-	-	-	T.S.	17	0.6
91.	2	0.2	0.2	-	-	-	-	-	-	M.S.	17	-
92.	6	0.5	0.6	-	-	-	10	1.2	1.0	M.S.	85	0.9
93.	9	8.0	0.9	-	-	-	-	-	-	M.S.	33	1.5
94.	-	-	-	-	-	-	-	-	-	M.S.	17	0.1
95.	4	0.4	0.4	-	-	-	-	-	-	M.S.	17	0.1
96.	16	1.4	1.6	40	5.1	4.0	30	3.7	3.0	T.S.	100	3.5
97.	-	-	-	-	-	-	10	1.2	1.0	T.S.	33	0.2
98.	8	0.8	0.8	-	-	-	-	-	-	M.S.	33	0.5
99.	2	0.2	0.2	-	-	-	-	-	-	L.S.	33	0.1
100.	-	-	-	-	-	-	-	-	-	M.S.	17	-

Table - A. Wet Mixed Forests.

Sl. No.	SPECIES	MORAGHAT 1.	MORAGHAT 2.	MORAGHAT 3.	LATAGURI 1.	LATAGURI 2.	LATAGURI 3.	DAMANPUR 1.	DAMANPUR 2.	DAMANPUR 3.
101.	*Quercus spicata*	6	0.6	0.6	-	-	-	-	-	-
102.	*Q. lanceaefolia*	15	1.5	1.5	6	0.7	0.6	-	-	-
103.	*Schefflera alata*	-	-	-	-	-	-	-	-	-
104.	*Salacia roxburghii*	-	-	-	-	-	-	-	-	-
105.	*Syzygium* species	-	-	-	15	1.8	1.5	-	-	-
106.	*Salix tetrasperma*	-	-	-	2	0.2	0.2	-	-	-
107.	*Sapium baccatum*	6	0.6	0.6	2	0.2	0.2	-	-	-
108.	*S. eugeniaefolium*	2	0.2	0.2	-	-	-	-	-	-
109.	*Saurauia roxburghii*	6	0.6	0.6	6	0.7	0.6	-	-	-
110.	*Sterculia alata*	-	-	-	4	0.5	0.4	2	0.2	0.2
111.	*Streospermum chelonoides*	7	0.7	0.7	2	0.2	0.2	10	1.1	1.0
112.	*Schima wallichii*	16	1.6	1.6	12	1.5	1.2	-	-	-
113.	*Shorea robusta*	70	7.0	7.0	6	0.7	0.6	22	2.0	2.2
114.	*Styrax serrulatum*	-	-	-	25	3.0	2.5	-	-	-
115.	*Symplocos spicata*	-	-	-	-	-	-	26	2 1	2.6
116.	*Syzygium balsameum*	-	-	-	2	0.2	0.2	-	-	-
117.	*S. cumini*	-	-	-	15	1.8	1.5	14	1 6	1.4
118.	*Symplocos ramosissima*	7	0.7	0.7	10	1.2	1.0	-	-	-
119.	*Spondias mangifera*	20	2.0	2.0	4	0.5	0.4	-	-	-
120.	*Sterculia villosa*	20	2.0	2.0	16	1.9	1.6	-	-	-
121.	*Streblus asper*	-	-	-	-	-	-	-	-	-
122.	*Symplocos racemosa*	2	0.2	0.2	-	-	-	2	0.2	0.2
123.	*Talauma hodgsoni*	30	3.0	3.0	-	-	-	-	-	-
124.	*Terminalia belerica*	10	1.0	1.0	2	0.2	0.2	-	-	-
125.	*T. chebula*	-	-	-	2	0.2	0.2	-	-	-

Sl. No.	RAJABHATKHAWA 1.	2.	3.	APALCHAND 1.	2.	3.	UDLABARI 1.	2.	3.	DOMINANCE	FREQUENCY	DENSITY
101.	8	0.7	0.8	20	2.6	2.0	-	-	-	M.S.	51	0.6
102.	6	0.5	0.6	-	-	-	-	-	-	M.S.	51	0.4
103.	6	0.5	0.6	-	-	-	-	-	-	L.S.	17	0.1
104.	10	0.9	1.0	-	-	-	-	-	-	L.S.	17	0.1
105.	-	-	-	-	-	-	-	-	-	L.S.	17	-
106.	-	-	-	-	-	-	-	-	-	M.S.	17	0.3
107.	4	0.4	0.4	-	-	-	-	-	-	L.S.	51	0.2
108.	-	-	-	-	-	-	-	-	-	L.S.	17	-
109.	-	-	-	-	-	-	-	-	-	L.S.	33	0.2
110.	-	-	-	-	-	-	-	-	-	T.S.	33	0.1
111.	12	1.0	1.2	-	-	-	-	-	-	T.S.	68	0.5
112.	-	-	-	-	-	-	-	-	-	T.S.	33	0.5
113.	6	0.5	0.6	-	-	-	8	1.0	-	T.S.	85	3.4
114.	20	1.9	2.0	-	-	-	30	3.7	3.0	M.S.	51	1.4
115.	-	-	-	-	-	-	10	1.2	1.0	M.S	33	0.6
116.	-	-	-	-	-	-	-	-	-	-	-	-
117.	4	0.4	0.4	-	-	-	30	3.7	3.0	M.S	68	1.3
118.	-	-	-	-	-	-	-	-	-	M.S.	33	0.3
119.	1	0.1	0.1	-	-	-	-	-	-	M.S.	51	0.4
120.	20	1.9	2.0	-	-	-	-	-	-	T.S.	51	1.0
121.	4	0.4	0.2	-	-	-	8	1.0	0.8	L.S.	33	0.2
122.	2	0.2	0.2	-	-	-	-	-	-	M.S.	51	0.1
123.	4	0.4	0.4	-	-	-	-	-	-	M.S.	33	0.6
124.	14	1.3	1.4	-	-	-	10	1.2	1.0	M.S.	68	0.6
125.	4	0.4	0.4	-	-	-	-	-	-	T.S.	33	0.1

Table - A. Wet Mixed Forests.

Sl. No.	SPECIES	MORAGHAT 1.	MORAGHAT 2.	MORAGHAT 3.	LATAGURI 1.	LATAGURI 2.	LATAGURI 3.	DAMAHPUR 1.	DAMAHPUR 2.	DAMAHPUR 3.
126.	*Tetrameles nudiflora*	-	-	-	-	-	-	-	-	-
127.	*Trema orientalis*	13	1.3	1.3	-	-	-	-	-	-
128.	*Turpinia pomifera*	5	0.5	0.5	35	4.0	3.5	10	1 1	1.0
129.	*Vitex quinata*	-	-	-	-	-	-	-	-	-
130.	*Vatica lancefolia*	-	-	-	-	-	-	30	3.5	3.0
131.	*Wrightia tomentosa*	-	-	-	-	-	-	-	-	
132.	*Xylosma longifolia*	-	-	-	-	-	-	15	1.7	1.5
133.	*Fagara badrunga*	-	-	-	2	0.2	0.2	-	-	-

Sl.	RAJABHATKHAWA			APALCHAND			UDLABARI			DOMINANCE	FREQUENCY	DENSITY
No.	1.	2.	3.	1.	2.	3.	1.	2.	3.			
126.	60	5.0	6.0	-	-	-	-	-	-	T.S.	17	0.1
127.	-	-	-	-	-	-	-	-	-	L.S.	17	0.2
128.	-	-	-	-	-	-	-	-	-	L.S.	51	0.2
129.	8	0.7	0.8	-	-	-	-	-	-	T.S.	17	0.1
130.	-	-	-	-	-	-	-	-	-	M.S.	17	0.6
131.	2	0.2	0.2	-	-	-	-	-	-	L.S.	17	-
132.	-	-	-	-	-	-	-	-	-	L.S.	17	0.3
133.	-	-	-	-	-	-	-	-	-	M.S.	17	-

SYNTHETIC SURVEY
FLORISTIC COMPOSITION OF TREE VEGETATION IN PLAINS:
BELT TRANSECTS

(ii) SCATTERED SHOREA FORESTS

REFERENCE TO TABLE B

COLUMN	1-NUMBER OF TREES IN THE TRANSECT
COLUMN	2-PERCENTAGE OF SPECIES
COLUMN	3-DENSITY
F%	-FREQUENCY
D	-DENSITY PER 0.4 HECTARE (OR ONE ACRE)
DOMINANCE	-U.S. (UPPER STOREY) -M.S. (MIDDLE STOREY) -L.S. (LOWER STOREY)

Table - B. Scattered Sal Forests.

Sl. No.	SPECIES	MORAGHAT 1.	MORAGHAT 2.	MORAGHAT 3.	LATAGURI 1.	LATAGURI 2.	LATAGURI 3.	DAMANPUR 1.	DAMANPUR 2.	DAMANPUR 3.
1.	*Acronychia pedunculata*	-	-	-	-	-	-	-	-	-
2.	*Actinodaphne obovata*	-	-	-	-	-	-	-	-	-
3.	*Actinodaphne angustifolia*	6	0.6	0.6	-	-	-	-	-	-
4.	*Adina cordifolia*	-	-	-	-	-	-	-	-	-
5.	*Aesculus assamica*	-	-	-	-	-	-	-	-	-
6.	*Ailanthus grandis*	-	-	-	-	-	-	-	-	-
7.	*Albizzia lucida*	-	-	-	-	-	-	-	-	-
8.	*Aphanamyxis polystachia*	-	-	-	-	-	-	7	0.5	0.7
9.	*Amoora wallichii*	6	0.6	0.6	4	0.6	0.4	7	0.5	0.7
10.	*Anthocephalus chinensis*	-	-	-	-	-	-	-	-	-
11.	*Aporosa roxburghii*	-	-	-	4	0.6	0.4	14	1.0	1.4
12.	*Artocarpus chaplasa*	3	0.3	0.3	16	2.4	1.6	-	-	-
13.	*Albizzia odoratissima*	-	-	-	-	-	-	-	-	-
14.	*Bauhinia malaberica*	-	-	-	-	-	-	-	-	-
15.	*Baccaurea sapida*	-	-	-	-	-	-	-	-	-
16.	*Bauhinia purpurea*	-	-	-	-	-	-	-	-	-
17.	*Bombax ceiba*	12	1.2	1.2	-	-	-	-	-	-
18.	*Bridelia retusa*	-	-	-	-	-	-	-	-	-
19.	*B. stipularis*									
20.	*B. tomentosa*									
21.	*Biochofia javanica*	-	-	-	-	-	-	-	-	-
22.	*Callicarpa arborea*	-	-	-	4	0.6	0.4	7	0.5	0.7
23.	*Careya arborea*	33	4.0	3.3	-	-	-	-	-	-
24.	*C. tribuloides* (Castanopsis)	-	-	-	16	2.4	1.6	-	-	-

Sl. No.	RAJABHATKHAWA 1.	2.	3.	JAINTI 1.	2.	3.	BENGDURI 1.	2.	3.	DOMINANCE	FREQUENCY	DENSITY
1.	-	-	-	-	-	-	76	5.2	7.6	L.S.	17	0.9
2.	25	2.1	2.5	-	-	-	4	0.3	0.4	L.S.	33	0.4
3.	-	-	-	-	-	-	12	0.9	1.2	L.S.	33	0.3
4.	-	-	-	-	-	-	-	-	-	L.S.	-	-
5.	-	-	-	-	-	-	-	-	-	T.S.	-	-
6.	-	-	-	2	0.1	0.2	4	0.3	0.4	T.S.	33	0.1
7.	5	0.4	0.5	-	-	-	-	-	-	T.S.	17	0.1
8.	110	9.2	11.0	170	13.1	17.0	28	2.0	2.8	M.S	68	4.1
9.	20	1.7	2.0	5	0.3	0.5	20	1.4	2.0	T.S.	100	0.9
10.	-	-	-	2	0.1	0.2	-	-	-	M.S.	17	-
11.	-	-	-	-	-	-	4	0.3	0.4	L.S.	51	0.3
12.	-	-	-	-	-	-	4	0.3	0.4	T.S.	51	0.5
13.	-	-	-	3	0.2	0.3	-	-	-	M.S.	17	-
14.	5	0.4	0.5	2	0.1	0.2	-	-	-	M.S.	33	0.1
15.	10	0.9	1.0	2	0.1	0.2	-	-	-	L.S.	33	0.2
16.	40	3.3	4.0	6	0.4	0.6	-	-	-	L.S.	33	0.6
17.	-	-	-	6	0.4	0.6	-	-	-	T.S.	33	0.3
18.	5	0.4	0.5	14	1.1	1.4	-	-	-	M.S.	33	0.3
19.	-	-	-	-	-	-	-	-	-	-	-	-
20.	-	-	-	-	-	-	-	-	-	-	-	-
21.	-	-	-	-	-	-	-	-	-	-	-	-
22.	5	0.4	0.5	15	1.2	1.5	-	-	-	L.S.	68	0.6
23.	-	-	-	18	1.4	1.8	-	-	-	L.S.	33	0.9
24.	10	0.9	1.0	2	0.1	0.2	20	1.3	2.0	M.S.	68	0.8

Table - B. Scattered Sal Forests.

Sl. No.	SPECIES	MORAGHAT 1.	2.	3.	LATAGURI 1.	2.	3.	DAMANPUR 1.	2.	3.
25.	*Toona ciliata*	6	0.7	0.6	-	-	-	21	1.5	2.1
26.	*Chisocheton paniculatus*	3	0.3	0.3	-	-	-	-	-	-
27.	*Chukrasia tabularis*	-	-	-	-	-	-	-	-	-
28.	*Cinnamomum obtusifolium*	-	-	-	20	3.0	2.0	23	2.2	2.8
29.	*Crataeva nurvala*	-	-	-	-	-	-	-	-	-
30.	*Casearia vareca*	-	-	-	-	-	-	50	4.0	5.0
31.	*Cephalenthus occidentalis*	-	-	-	-	-	-	-	-	-
32.	*D. pentagyna (Dillenia)*	18	2.0	1.8	4	0.6	0.4	15	1.2	1.5
33.	*Drimycarpus racemosus*	-	-	-	-	-	-	-	-	-
34.	*Drypetes assamica*	-	-	-	-	-	-	-	-	-
35.	*Dysoxylum binectariferum*	-	-	-	-	-	-	42	3.2	4.2
36.	*Embilca officinalis*	-	-	-	-	-	-	-	-	-
37.	*Elaeocarpus sp.*	-	-	-	-	-	-	-	-	-
38.	*E. sterculiaceus*	-	-	-	-	-	-	-	-	-
39.	*Elaeocarpus varunua*	-	-	-	12	1.8	1.2	7	0.5	0.7
40.	*Endospermum chinense*	-	-	-	-	-	-	-	-	-
41.	*Engelherdtia spicata*	-	-	-	-	-	-	-	-	-
42.	*Eurya acuminata*	-	-	-	4	0.6	0.4	-	-	-
43.	*E. carvinervis*	-	-	-	12	1.8	1.2	-	-	-
44.	*Evodia meliaefolla*	3	0.4	0.3	4	0.6	0.4	-	-	-
45.	*Eranthemum indicum*	-	-	-	-	-	-	-	-	-
46.	*Ficus* species	-	-	-	4	0.6	0.4	-	-	-
47.	*Garcinia cowa*	-	-	-	-	-	-	7	0.5	0.7
48.	*Garuga pinnata*	10	1.2	1.0	-	-	-	7	0.5	0.7

Sl. No.	RAJABHATKHAWA 1.	2.	3.	JAINTI 1.	2.	3.	BENGDURI 1.	2.	3.	DOMINANCE	FREQUENCY	DENSITY
25.	-	-	-	6	0.4	0.6	4	0.3	0.4	T.S.	68	0.5
26.	-	-	-	-	-	-	-	-	-	M.S.	17	-
27.	5	0.4	0.5	6	0.4	0.6	-	-	-	T.S.	33	0.1
28.	-	-	-	-	-	-	-	-	-	M.S.	33	0.4
29.	5	0.4	0.5	-	-	-	-	-	-	M.S.	17	0.1
30.	5	0.4	0.5	-	-	-	-	-	-	L.S.	33	0.1
31.	-	-	-	5	0.3	0.5	-	-	-	M.S.	17	-
32.	-	-	-	60	4.6	6.0	16	1.2	1.6	M.S.	85	1.6
33.	80	6.6	8.0	-	-	-	-	-	-	M.S.	17	1.1
34.	-	-	-	-	-	-	-	-	-	-	-	-
35.	5	0.4	0.5	6	0.4	0.6	-	-	-	M.S.	51	0.7
36.	5	0.4	0.5	-	-	-	-	-	-	L.S.	17	-
37.	10	0.9	1.0	-	-	-	4	0.3	0.4	M.S.	33	1.2
38.	-	-	-	-	-	-	4	0.3	0.4	T.S.	17	0.2
39.	185	14.2	18.5	-	-	-	12	0.9	1.2	T.S.	68	2.9
40.	-	-	-	-	-	-	4	0.3	0.4	T.S.	17	-
41.	-	-	-	-	-	-	4	0.3	0.4	T.S.	17	-
42.	-	-	-	-	-	-	30	2.0	3.0	M.S.	33	0.4
43.	20	1.6	2.0	-	-	-	-	-	-	L.S.	33	0.6
44.	-	-	-	-	-	-	-	-	-	M.S.	33	0.2
45.	10	0.9	1.0	-	-	-	-	-	-	M.S.	17	0.15
46.	5	0.4	0.5	-	-	-	4	0.3	0.4	T.S.	33	0.2
47.	-	-	-	-	-	-	4	0.3	0.4	M.S.	33	0.1
48.	10	0.9	1.0	-	-	-	-	-	-	M.S.	51	0.4

Table - B. Scattered Sal Forests.

Sl. No.	SPECIES	MORAGHAT 1.	MORAGHAT 2.	MORAGHAT 3.	LATAGURI 1.	LATAGURI 2.	LATAGURI 3.	DAMANPUR 1.	DAMANPUR 2.	DAMANPUR 3.
49.	*G. hirsutum*	3	0.4	0.3	-	-	-	-	-	-
50.	*Gmelina arborea*	10	1.2	1.0	4	0.6	0.4	42	3.2	4.2
51.	*Glochidion* species	-	-	-	-	-	-	7	0.5	0.7
52.	*Holarrhena antidysenterica*	15	1.8	1.5	-	-	-	-	-	-
53.	*Ilex godajam*	3	0.4	0.3	12	1.8	1.2	14	1.1	1.4
54.	*Jambosa formosa*	-	-	-	-	-	-	7	0.5	0.7
55.	*J. praecox*	-	-	-	4	0.6	0.4	7	0.5	0.7
56.	*Kydia calycina*	3	0.4	0.3	-	-	-	7	0.5	0.7
57.	*Lophopetalum fimbriatum*	-	-	-	50	7.2	5.0	-	-	-
58.	*Lagerstroemia parviflora*	3	0.4	0.3	5	0.8	0.5	-	-	-
59.	*Leea* species	-	-	-	5	0.8	0.5	-	-	-
60.	*Litsaea monopetala*	10	1.2	1.0	12	1.8	1.2	-	-	-
61.	*L. glutinosa*	60	7.5	6.0	4	0.6	0.4	-	-	-
62.	*Lannea coromendelica*	45	5.6	4.5	-	-	-	-	-	-
63.	*Litsaea panamonja*	-	-	-	16	2.4	1.6	-	-	-
64.	*Macaranga denticulata*	180	22.5	18.0	45	6.7	4.5	-	-	-
65.	*Machilus gamblei*	12	1.2	1.2	12	1.8	1.2	130	10.0	13.0
66.	*M. villosa*	10	1.3	1.0	50	8.2	5.0	-	-	-
67.	*Mallotus philippinensis*	10	1.0	1.0	-	-	-	-	-	-
68.	*Marlea chinense*	6	0.7	0.6	-	-	-	-	-	-
69.	*Mastixia arborea*	-	-	-	12	1.8	1.2	-	-	-
70.	*Meliosma simplicifolia*	3	0.4	0.3	12	1.8	1.2	300	23.1	30.0
71.	*Michelia champaca*	-	-	-	-	-	-	-	-	-
72.	*M. montana*	-	-	-	-	-	-	-	-	-

Sl. No.	RAJABHATKHAWA 1.	2.	3.	JAINTI 1.	2.	3.	BENGDURI 1.	2.	3.	DOMINANCE	FREQUENCY	DENSITY
49.	-	-	-	-	-	-	-	-	-	M.S.	17	-
50.	-	-	-	10	0.8	1.0	-	-	-	M.S	68	0.9
51.	-	-	-	-	-	-	4	0.3	0.4	M.S.	33	0.1
52.	-	-	-	-	-	-	-	-	-	L.S.	17	0.3
53.	-	-	-	-	-	-	60	4.0	6.0	M.S.	68	1.2
54.	-	-	-	-	-	-	-	-	-	M.S.	17	0.1
55.	-	-	-	-	-	-	16	1.2	1.6	M.S.	51	0.2
56.	-	-	-	-	-	-	-	-	-	L.S.	33	0.15
57.	-	-	-	-	-	-	20	1.4	2.0	T.S.	33	1.4
58.	20	1.6	2.0	-	-	-	-	-	-	T.S.	51	0.5
59.	-	-	-	-	-	-	12	0.9	1.2	L.S.	33	0.3
60.	-	-	-	6	0.4	0.6	-	-	-	M.S.	51	0.6
61.	10	0.9	1.0	-	-	-	-	-	-	M.S.	51	1.5
62.	-	-	-	18	1.4	1.8	-	-	-	T.S.	33	1.2
63.	-	-	-	-	-	-	-	-	-	M.S.	17	0.4
64.	-	-	-	-	-	-	5	0.3	0.5	L.S.	51	5.0
65.	-	-	-	-	-	-	25	1.7	2.5	T.S.	68	2.5
66.	115	9.6	11.5	-	-	-	260	18.0	26.0	T.S.	68	6.2
67.	10	0.9	1.0	-	-	-	-	-	-	M.S.	33	0.3
68.	-	-	-	-	-	-	-	-	-	L.S.	17	0.1
69.	-	-	-	-	-	-	16	1.2	1.6	M.S.	33	0.5
70.	20	1.7	2.0	-	-	-	8	0.6	0.8	L.S.	85	4.6
71.	-	-	-	-	-	-	4	0.3	0.4	T.S.	17	-
72.	-	-	-	-	-	-	8	0.6	0.8	T.S.	17	0.1

Table - B. Scattered Sal Forests.

Sl. No.	SPECIES	MORAGHAT			LATAGURI			DAMANPUR		
		1.	2.	3.	1.	2.	3.	1.	2.	3.
73.	*Micromelum integerrimum*	-	-	-	-	-	-	-	-	-
74.	*Olea dioica*	-	-	-	8	1.2	0.8	-	-	-
75.	*Oroxylwm indicum*	-	-	-	-	-	-	-	-	-
76.	*Photinia integrifolia*	-	-	-	-	-	-	-	-	-
77.	*Phoebe attenuata*	-	-	-	4	0.6	0.4	-	-	-
78.	*P. lanceolata*	-	-	-	-	-	-	-	-	-
79.	*Podocarpus neriifolia*	-	-	-	-	-	-	-	-	-
80.	*Premna bengalensis*	35	4.3	3.5	16	2.4	1.6	80	6.2	8.0
81.	*P. integrifolia*	6	0.7	0.6	-	-	-	-	-	-
82.	*Quercus spicata*	-	-	-	-	-	-	-	-	-
83.	*Salacia roxburghii*	-	-	-	-	-	-	-	-	-
84.	*Sapium baccatum*	-	-	-	-	-	-	-	-	-
85.	*Saurauia roxburghii*	6	0.7	0.6	20	3.0	2.0	30	2.3	3.0
86.	*Sterculia alata*	-	-	-	-	-	-	-	-	-
87.	*Streospermum cheloneides*	-	-	-	4	0.6	0.4	7	0.5	0.7
88.	*Schima wallichii*	6	0.7	0.6	-	-	-	-	-	-
89.	*Shorea robusta*	200	25.0	20.0	190	28.4	19.0	340	26.7	34.0
90.	*Styrax serrulatum*	-	-	-	4	0.6	0.4	-	-	-
91.	*Symplocos spicata*	-	-	-	4	0.6	0.4	-	-	-
92.	*S. cumini (Syzigium)*	6	0.7	0.6	-	-	-	25	1.9	2.5
93.	*Symplocos ramosissima*	-	-	-	12	1.8	1.2	-	-	-
94.	*Spondias mangifera*	12	1.4	1.2	10	1.5	1.0	-	-	
95.	*Sterculia villosa*	6	0.7	0.6	12	1.8	1.2	-	-	-
96.	*Symplocos ferrugenia*	-	-	-	-	-	-	-	-	-

Sl.	RAJABHATKHAWA			JAINTI			BENGDURI			DOMINANCE	FREQUENCY	DENSITY
No.	1.	2.	3.	1.	2.	3.	1.	2.	3.			
73.	-	-	-	-	-	-	4	0.3	0.4	L.S.	17	-
74.	-	-	-	-	-	-	-	-	-	M.S.	17	0.2
75.	15	1.2	1.5	15	1.2	1.5	-	-	-	M.S.	33	0.4
76.	20	1.7	2.0	-	-	-	-	-	-	L.S.	17	0.3
77.	-	-	-	-	-	-	-	-	-	M.S.	17	0.1
78.	-	-	-	5	0.4	0.5	-	-	-	L.S.	17	0.1
79.	-	-	-	-	-	-	8	0.6	0.8	M.S.	17	0.1
80.	-	-	-	6	0.4	0.6	16	1.2	1.6	M.S.	85	2.4
81.	-	-	-	95	7.4	9.5	-	-	-	M.S.	33	1.4
82.	15	1.2	1.5	-	-	-	-	-	-	M.S.	17	0.2
83.	5	0.4	0.5	-	-	-	-	-	-	L.S.	17	0.1
84.	-	-	-	2	0.1	0.2	-	-	-	L.S.	17	-
85.	-	-	-	-	-	-	16	1.2	1.6	L.S.	68	1.2
86.	-	-	-	-	-	-	8	0.6	0.8	T.S.	17	0.1
87.	5	0.4	0.5	27	2.1	2.7	-	-	-	T.S.	68	1.6
88.	110	9.2	11.0	100	7.7	10.0	265	18.2	26.5	T.S.	68	6.0
89.	115	9.6	11.5	410	31.5	41.0	200	13.8	20.0	T.S.	100	22.4
90.	10	0.9	1.0	3	0.2	3.0	-	-	-	M.S.	51	0.3
91.	-	-	-	-	-	-	20	1.4	2.0	M.S.	33	0.3
92.	10	0.9	1.0	51	4.0	5.1	16	1.2	1.6	M.S.	85	1.4
93.	-	-	-	-	-	-	-	-	-	M.S.	17	0.3
94.	5	0.4	0.5	-	-	-	-	-	-	M.S.	51	0.6
95.	-	-	-	23	1.8	2.3	8	0.6	0.8	T.S.	68	0.8
96.	-	-	-	-	-	-	41	2.8	4.1	M.S.	17	0.5

Table - B. Scattered Sal and Wet - Dry Mixed Forests.

Sl.	*S P E C I E S*	*MORAGHAT*			*LATAGURI*			*DAMANPUR*		
No.		*1.*	*2.*	*3.*	*1.*	*2.*	*3.*	*1.*	*2.*	*3.*
97.	*S. racemosa*	-	-	-	-	-	-	-	-	-
98.	*Talauma hodgsoni*	-	-	-	4	0.6	0.4	-	-	-
99.	*Terminalia belerica*	18	2.0	1.8	-	-	-	21	1.5	2.1
100.	*T. chebula*	-	-	-	4	0.6	0.4	7	0.5	0.7
101.	*Ternstroemia gymnenthera*	-	-	-	-	-	-	-	-	-
102.	*Tetrameles nudiflora*	-	-	-	-	-	-	7	0.5	0.7
103.	*Trema orientalis*	10	1.2	1.0	-	-	-	-	-	-
104.	*Turpinia pomifera*	-	-	-	25	4.0	2.5	50	3.8	5.0
105.	*Vitex quinata*	-	-	-	-	-	-	-	-	-
106.	*Wrightia tomentosa*	-	-	-	-	-	-	-	-	-
107.	*Xylosma longifolia*	-	-	-	-	-	-	7	0.5	0.7
108.	*Fagara badrunga*	-	-	-	-	-	-	-	-	-
109.	Unknown	-	-	-	-	-	-	-	-	-

Sl.	RAJABHATKHAWA			JAINTI			BENGDURI			DOMINANCE	FREQUENCY	DENSITY
No.	1.	2.	3.	1.	2.	3.	1.	2.	3.			
97.	10	0.9	1.0	-	-	-	-	-	-	M.S.	17	0.1
98.	-	-	-	-	-	-	56	3.9	5.6	M.S.	17	0.7
99.	20	1.6	2.0	60	4.6	6.0	8	0.6	0.8	T.S.	85	1.7
100.	-	-	-	-	-	-	4	0.3	0.4	T.S.	51	0.2
101.	-	-	-	-	-	-	60	3.9	6.0	M.S.	17	0.7
102.	5	0.4	0.5	2	0.1	0.2	-	-	-	T.S.	51	0.2
103.	-	-	-	2	0.1	0.2	-	-	-	T.S.	33	0.2
104.	-	-	-	-	-	-	16	1.2	1.6	L.S.	51	1.5
105.	100	8.4	10.0	-	-	-	-	-	-	T.S.	17	1.4
106.	-	-	-	45	3.5	4.5	4	0.3	0.4	L.S.	33	0.6
107.	-	-	-	-	-	-	-	-	-	L.S.	17	0.1
108.	-	-	-	35	2.7	3.5	4	0.3	0.4	M.S.	33	0.5
109.	-	-	-	60	4.7	4.6	-	-	-	-	-	0.7

SYNTHETIC SURVEY
FLORISTIC COMPOSITION OF TREE VEGETATION IN PLAINS: BELT TRANSECTS

(iii) PURE SHOREA FORESTS

REFERENCE TO TABLE C

COLUMN	1-NUMBER OF TREES
COLUMN	2-PERCENTAGE OF SPECIES
COLUMN	3-DENSITY
F%	-FREQUENCY
D	-DENSITY PER 0.4 HECTARE (OR ONE ACRE)
DOMINANCE	-U.S. (UPPER STOREY) -M.S. (MIDDLE STOREY) -L.S. (LOWER STOREY)

Table - C. Pure Sal Forests.

Sl. No.	SPECIES	MORAGHAT 1.	MORAGHAT 2.	MORAGHAT 3.	LATAGURI 1.	LATAGURI 2.	LATAGURI 3.	DAMANPUR 1.	DAMANPUR 2.	DAMANPUR 3.
1.	*Actinodaphne obovata*	4	0.3	0.4	-	-	-	-	-	-
2.	*A. angustifolia*	26	1.7	2.6	-	-	-	-	-	-
3.	*Ailanthus grandis*	-	-	-	-	-	-	-	-	-
4.	*Aralia* species	-	-	-	-	-	-	-	-	-
5.	*Albizzia lucida*	-	-	-	-	-	-	-	-	-
6.	*Alstonia scholaris*	-	-	-	3	0.3	0.3	14	0.9	1.4
7.	*Aphanamyxis ploystachia*	4	0.3	0.4	6	0.7	0.6	7	0.5	0.7
8.	*Amoora wallichii*	12	0.8	1.2	3	0.3	0.3	-	-	-
9.	*Anthocophalus chinensis*	-	-	-	3	0.3	0.3	-	-	-
10.	*Aporosa roxburghii*	-	-	-	9	1.0	0.9	14	0.9	1.4
11.	*Artocarpus chaplasa*	22	1.5	2.2	9	1.0	0.9	-	-	-
12.	*Artocarpus lakoocha*	6	0.4	0.6	-	-	-	-	-	-
13.	*Albizzia odoratissima*	-	0.4	0.6	-	-	-	-	-	-
14.	*Baccaurea sapida*	4	0.3	0.4	-	-	-	-	-	-
15.	*Bauhinia purpurea*	4	0.3	0.4	-	-	-	-	-	-
16.	*Bridelia retusa*	2	0.1	0.2	-	-	-	-	-	-
17.	*Callicarpa arborea*	30	2.0	3.0	3	0.3	0.3	-	-	-
18.	*Canarium sikkimense*	-	-	-	-	-	-	7	0.5	0.7
19.	*Careya arborea*	20	1.4	2.0	3	0.3	0.3	15	0.9	1.5
20.	*Castanopsis tribuloides*	2	0.1	0.2	6	0.6	0.6	-	-	-
21.	*Toona ciliata*	-	-	-	-	-	-	20	1.2	2.0
22.	*Chisocheton paniculatus*	8	0.5	0.8	-	-	-	-	-	-
23.	*Cinnamomum cecidodaphne*	-	-	-	-	-	-	20	1.2	2.0
24.	*C. obtusefolium*	10	0.7	1.0	3	0.3	0.3	50	3.1	5.0

Sl. No.	APALCHAND 1.	2.	3.	UDLABARI 1.	2.	3.	BOIRANTI 1.	2.	3.	DOMINANCE	FREQUENCY	DENSITY
1.	-	-	-	-	-	-	-	-	-	L.S.	17	-
2.	-	-	-	-	-	-	-	-	-	L.S.	17	0.3
3.	-	-	-	-	-	-	-	-	-	L.S.	2	-
4.	-	-	-	-	-	-	-	-	-	M.S.	9	-
5.	-	-	-	-	-	-	-	-	-	M.S.	-	-
6.	-	-	-	-	-	-	-	-	-	M.S.	33	0.2
7.	25	6.2	2.5	10	1.2	1.0	36	2.3	3.6	M.S.	100	1.9
8.	-	-	-	-	-	-	-	-	-	T.S.	33	0.2
9.	-	-	-	-	-	-	-	-	-	M.S.	17	-
10.	-	-	-	3	0.3	0.3	-	-	-	M.S.	51	0.4
11.	-	-	-	-	-	-	-	-	-	-	33	0.4
12.	-	-	-	-	-	-	-	-	-	M.S.	17	0.1
13.	-	-	-	3	0.3	0.3	-	-	-	M.S.	17	-
14.	-	-	-	-	-	-	-	-	-	M.S.	17	-
15.	30	7.5	3.0	-	-	-	36	2.3	3.6	M.S.	51	1.7
16.	-	-	-	3	0.3	0.3	-	-	-	M.S.	33	0.1
17.	-	-	-	3	0.3	0.3	12	0.7	1.2	L.S.	68	0.6
18.	-	-	-	-	-	-	-	-	-	T.S.	17	0.1
19.	-	-	-	3	0.3	0.3	-	-	-	L.S.	68	0.5
20.	-	-	-	-	-	-	-	-	-	M.S.	33	0.1
21.	-	-	-	-	-	-	35	2.2	3.5	T.S.	33	0.6
22.	-	-	-	-	-	-	-	-	-	M.S.	17	0.1
23.	-	-	-	-	-	-	-	-	-	T.S.	17	0.2
24.	-	-	-	-	-	-	-	-	-	M.S.	51	0.7

Table - C. Pure Sal Forests.

Sl. No.	SPECIES	MORAGHAT 1.	MORAGHAT 2.	MORAGHAT 3.	LATAGURI 1.	LATAGURI 2.	LATAGURI 3.	DAMANFUR 1.	DAMANFUR 2.	DAMANFUR 3.
25.	*Cryptocarya amygdalina*	-	-	-	-	-	-	35	2.2	3.5
26.	*Casearia vareca*	6	0.4	0.6	-	-	-	-	-	-
27.	*Casearia* species	-	-	-	-	-	-	-	-	-
28.	*Dillenia pentagyna*	35	2.5	3.5	18	1.8	1.8	15	0.9	1.5
29.	*Dysoxylum binectariferum*	4	0.3	0.4	-	-	-	-	-	-
30.	*D. hamiltonii*	2	0.1	0.2	-	-	-	-	-	-
31.	*Elaeocarpus varunua*	4	0.3	0.4	3	0.3	0.3	-	-	-
32.	*Eurya acuminata*	8	0.5	0.8	-	-	-	15	0.9	1.5
33.	*Evodia meliaefolia*	2	0.1	0.2	-	-	-	-	-	-
34.	*Ficus* species	6	0.4	0.6	-	-	-	7	0.5	0.7
35.	*Garuga pinnata*	-	-	-	6	0.6	0.6	-	-	-
36.	*Glochidion acuminatum*	10	0.7	1.0	-	-	-	-	-	-
37.	*Gmelina arborea*	2	0.1	0.2	6	0.6	0.6	-	-	-
38.	*Holarrhena antidysenterica*	12	0.8	1.2	-	-	-	-	-	-
39.	*Ilex godajam*	10	0.7	1.0	-	-	-	7	0.5	0.7
40.	*Jambosa formosa*	-	-	-	-	-	-	20	1.2	2.0
41.	*Kydia calycina*	-	-	-	-	-	-	-	-	-
42.	*Lophopetalum fimbriatum*	-	-	-	3	0.3	0.3	-	-	-
43.	*Lagerstroemia parviflora*	10	0.7	1.0	-	-	-	-	-	-
44.	*Litsaea monopetala*	8	0.6	0.8	12	1.2	1.2	-	-	-
45.	*L. glutinosa*	-	-	-	3	0.3	0.3	-	-	-
46.	*Lannea coromendelica*	-	-	-	3	0.3	0.3	-	-	-
47.	*Litsaea panamonja*	-	-	-	3	0.3	0.3	-	-	-
48.	*Macaranga denticulata*	25	1.8	2.5	90	9.0	9.0	-	-	-

Sl. No.	APALCHAND 1.	2.	3.	UDLABARI 1.	2.	3.	BOIRANTI 1.	2.	3.	DOMINANCE	FREQUENCY	DENSITY
25.	6	1.5	0.6	3	0.3	0.3	-	-	-	M.S.	51	0.7
26.	-	-	-	-	-	-	-	-	-	L.S.	17	0.1
27.	-	-	-	12	1.4	1.2	-	-	-	L.S.	17	1.2
28.	-	-	-	100	11.8	10.0	12	0.7	1.2	M.S.	85	3.0
29.	6	1.5	0.6	3	0.3	0.3	25	1.6	2.5	M.S.	68	0.6
30.	-	-	-	-	-	-	-	-	-	M.S.	17	-
31.	-	-	-	-	-	-	-	-	-	M.S	33	0.1
32.	-	-	-	-	-	-	-	-	-	M.S.	33	0.2
33.	-	-	-	-	-	-	-	-	-	M.S.	17	-
34.	-	-	-	-	-	-	-	-	-	M.S.	33	0.15
35.	-	-	-	-	-	-	-	-	-	M.S.	17	0.1
36.	-	-	-	-	-	-	-	-	-	M.S.	17	0.1
37.	6	1.5	0.6	-	-	-	25	1.5	2.5	M.S.	68	0.6
38.	-	-	-	10	1.1	1.0	12	0.7	1.2	L.S.	51	0.5
39.	-	-	-	6	0.7	0.6	-	-	-	M.S.	51	0.3
40.	-	-	-	-	-	-	-	-	-	M.S.	17	0.2
41.	-	-	-	3	0.3	0.3	-	-	-	L.S.	17	-
42.	-	-	-	-	-	-	-	-	-	T.S.	17	-
43.	12	3.0	1.2	20	2.4	2.0	-	-	-	T.S.	51	1.0
44.	-	-	-	15	1.7	1.5	110	6.8	11.0	M.S.	68	1.7
45.	-	-	-	9	1.0	0.9	25	1.6	2.5	M.S.	51	0.5
46.	6	1.5	0.6	35	4.0	3.5	-	-	-	M.S.	51	1.0
47.	-	-	-	-	-	-	-	-	-	MS.	17	-
48.	-	-	-	-	-	-	-	-	-	L.S.	33	1.8

Table - C. Pure Sal Forests.

Sl. No.	SPECIES	MORAGHAT 1.	MORAGHAT 2.	MORAGHAT 3.	LATAGURI 1.	LATAGURI 2.	LATAGURI 3.	DAMANPUR 1.	DAMANPUR 2.	DAMANPUR 3.
49.	*M. gamblei*	120	8.3	12.0	18	1.8	1.8	20	1.2	2.0
50.	*M. villosa*	140	9.7	14.0	25	2.5	2.5	-	-	-
51.	*Mallotus philippinensis*	4	0.3	0.4	-	-	-	-	-	-
52.	*Machilus* species	105	7.2	10.5	-	-	-	-	-	-
53.	*Meliosma simplicifolia*	6	0.5	0.6	15	1.5	1.5	220	14.0	22.0
54.	*Michelia montana*	-	-	-	3	0.3	0.3	-	-	-
55.	*Oroxylum indicum*	-	-	-	3	0.3	0.3	-	-	-
56.	*Polyalthia semiarum*	-	-	-	3	0.3	0.3	-	-	-
57.	*Premna bengalensis*	36	2.4	3.6	9	0.9	0.9	85	5.3	8.5
58.	*P. integrifolia*	2	0.2	0.2	3	0.3	0.3	-	-	-
59.	*Pygeum acuminatum*	8	0.5	0.8	9	0.9	0.9	-	-	-
60.	*Pithecolobium angulatum*	-	-	-	-	-	-	-	-	-
61.	*Quereus spicata*	6	0.6	0.6	-	-	-	-	-	-
62.	*Sapium baccatum*	2	0.1	0.2	-	-	-	-	-	-
63.	*Saurauia roxburghii*	10	0.7	1.0	18	1.8	1.8	-	-	
64.	*Stereospermum chelonoides*	4	0.3	0.4	-	-	-	-	-	-
65.	*Schima wallichii*	10	0.7	1.0	3	0.3	0.3	-	-	-
66.	*Shorea robusta*	630	43.4	63.0	680	68.0	68.0	1000	62.5	103.0
67.	*Styrax serrulatum*	10	0.7	1.0	-	-	-	-	-	-
68.	*Syzygium balsameum*	10	0.7	1.0	4	0.4	0.4	-	-	-
69.	*S. cumini*	25	1.7	2.5	-	-	-	-	-	-
70.	*Symplocos ramosissima*	4	0.3	0.4	3	0.3	0.3	-	-	-
71.	*Spondias mangifera*	8	0.6	0.8	-	-	-	-	-	-
72.	*Sterculia villosa*	8	0.6	0.8	-	-	-	-	-	-

Sl.	APALCHAND			UDLABARI			BOIRANTI			DOMINANCE	FREQUENCY	DENSITY
No.	1.	2.	3.	1.	2.	3.	1.	2.	3.			
49.	20	1.2	2.0	6	1.5	0.6	-	-	-	M.S.	68	2.1
50.	-	-	-	6	1.5	0.6	6	0.7	0.6	M.S.	68	2.4
51.	6	1.5	0.6	-	-	-	-	-	-	L.S.	33	0.6
52.	-	-	-	12	1.4	1.2	-	-	-	M.S.	33	1.5
53.	-	-	-	-	-	-	-	-	-	L.S.	51	2.7
54.	-	-	-	-	-	-	-	-	-			
55.	6	1.5	0.6	-	-	-	12	0.7	1.2	M.S.	51	0.4
56.	-	-	-	-	-	-	-	-	-	M.S.	17	-
57.	25	6.2	2.5	6	0.7	0.6	-	-	-	M.S.	85	2.6
58.	12	3.0	1.2	30	3.5	3.0	55	3.4	5.5	M.S.	85	1.7
59.	-	-	-	-	-	-	-	-	-	L.S.	33	0.3
60.	-	-	-	-	-	-	-	-	-	-	-	-
61.	-	-	-	-	-	-	-	-	-	M.S.	17	-
62.	-	-	-	-	-	-		-	-	L.S.	17	-
63.	-	-	-	-	-	-	-	-	-	L.S.	33	0.4
64.	-	-	-	-	-	-	-		-	T.S.	17	-
65.	-	-	-	-	-	-	-	-	-	T.S.	33	0.2
66.	200	50.0	20.0	420	50.0	42.0	970	60.6	97.0	T.S.	100	56.8
67.	-	-	-	7	0.8	0.7	12	0.7	1.2	M.S.	51	0.4
68.	-	-	-	5	0.6	0.5	50	3.1	5.0	M.S.	68	0.8
69.	-	-	-	-	-	-	-	-	-	-	-	-
70.	-	-	-	-	-	-	-	-	-	M.S.	33	0.1
71.	30	7.5	3.0	3	0.3	0.3	-	-	-	M.S.	51	1.4
72.	-	-	-	-	-	-	-	-	-	T.S.	17	0.1

Table - C. Pure Sal Forests.

Sl. No.	SPECIES	MORAGHAT 1.	MORAGHAT 2.	MORAGHAT 3.	LATAGURI 1.	LATAGURI 2.	LATAGURI 3.	DAMANPUR 1.	DAMANPUR 2.	DAMANPUR 3.
73.	*Semicarpus anacardium*	-	-	-	6	0.6	0.6	-	-	-
74.	*Terminalia belerica*	18	1.3	1.8	-	-	-	9	0.5	0.9
75.	*Tetrameles nudiflora*	-	-	-	-	-	-	-	-	-
76.	*Turpinia pomifera*	-	-	-	3	0.3	0.3	20	1.2	2.0
77.	*Wrightia tomentosa*	-	-	-	-	-	-	-	-	
78.	*Fagara badrunga*	-	-	-	-	-	-	-	-	-
79.	*Emblica officinalis*	-	-	-	-	-	-	-	-	-

Sl. No.	APALCHAND 1.	APALCHAND 2.	APALCHAND 3.	UDLABARI 1.	UDLABARI 2.	UDLABARI 3.	BOIRANTI 1.	BOIRANTI 2.	BOIRANTI 3.	DOMINANCE	FREQUENCY	DENSITY PER HECT.
73.	-	-	-	-	-	-	-	-	-	L.S.	17	0.1
74.	12	3.0	1.2	70	8.2	7.0	26	1.6	2.6	M.S.	85	2.4
75.	6	1.5	0.6	-	-	-	-	-	-	M.S.	17	0.3
76.	-	-	-	-	-	-	-	-	-	L.S.	33	0.3
77.	-	-	-	50	6.0	5.0	100	6.2	10.0	L.S.	33	2.0
78.	-	-	-	-	-	-	12	1.4	1.2	M.S.	17	0.2
79.	-	-	-	-	-	-	33	2.1	3.5	L.S.	17	0.4

SYNTHETIC SURVEY
FLORISTIC COMPOSITION OF TREE VEGETATION IN PLAINS: BELT TRANSECTS

(iv) DRY-MIXED FORESTS

REFERENCE TO TABLE D

COLUMN	1-NUMBER OF TREES
COLUMN	2-PERCENTAGE OF SPECIES
COLUMN	3-DENSITY
F%	-FREQUENCY
D	-DENSITY PER 0.4 HECTARE (OR ONE ACRE)
DOMINANCE	-U.S. (UPPER STOREY) -M.S. (MIDDLE STOREY) -L.S. (LOWER STOREY)

Table - D. Dry Mixed Forests.

Sl.	SPECIES	CHAPRAMARI			JAINTI			UDLABARI		
No.		1.	2.	3.	1.	2.	3.	1.	2.	3.
1.	*Actinodaphne obovata*	8	0.9	0.8	-	-	-	-	-	-
2.	*A. angustifolia*	-	-	-	-	-	-	-	-	-
3.	*Adina cordifolia*	-	-	-	-	-	-	-	-	-
4.	*Aesculus assamica*	-	-	-	-	-	-	-	-	-
5.	*Ailanthus grandis*	2	0.2	0.2	-	-	-	-	-	-
6.	*Albizzia lucida*	4	0.5	0.4	-	-	-	-	-	-
7.	*Alstonia scholaris*	6	0.7	0.6	-	-	-	-	-	-
8.	*Aphanamixis polystachia*	105	11.7	10.5	335	30.5	33.5	-		
9.	*Amoora wallichii*	6	0.7	0.6	-	-	-	-	-	-
10.	*Aporosa roxburghii*	-	-	-	-	-	-	-	-	-
11.	*Artocarpus chaplasa*	2	0.2	0.2	-	-	-	-	-	-
12.	*Artocarpus lakoocha*	-	-	-	-	-	-	-	-	-
13.	*Albizzia adoratissima*	-	-	-	5	0.5	0.5	18	1.6	1.8
14.	*Albizzia procera*	-	-	-	10	0.9	1.0	-	-	-
15.	*Acacia catechu*	-	-	-	-	-	-	-	-	-
16.	*Albizzia lebbek*	-	-	-	-	-	-	9	0.8	0.9
17.	*Bauhinia malabarica*	-	-	-	10	0.9	1.0	-	-	-
18.	*Baccaurea sapida*	2	0.2	0.2	-	-	-	-	-	-
19.	*Bauhinia purpurea*	40	4.4	4.0	-	-	-	-	-	-
20.	*Beilschmiedia roxburghiana*	-	-	-	-	-	-	-	-	-
21.	*Bombax ceiba*	-	-	-	-	-	-	-	-	-
22.	*Bridelia retusa*	4	0.5	0.4	5	0.4	0.5	-	-	-
23.	*Bischofia javanica*	-	-	-	-	-	-	-	-	-
24.	*Callicarpa arborea*	-	-	-	15	1.5	1.5	55	5.0	5.5

Sl. No.	BHUTRI 1.	BHUTRI 2.	BHUTRI 3.	DHUMCHI 1.	DHUMCHI 2.	DHUMCHI 3.	KAIRBARI 1.	KAIRBARI 2.	KAIRBARI 3.	BOIRANTI 1.	BOIRANTI 2.	BOIRANTI 3.	DOMINANCE	FREQUENCY	DENSITY
1.	-	-	-	12	0.9	1.2	-	-	-	-	-	-	L.S.	28	0.3
2.	-	-	-	-	-	-	-	-	-	-	-	-	L.S.	-	-
3.	-	-	-	-	-	-	-	-	-	-	-	-	T.S.	5	-
4.	-	-	-	-	-	-	-	-	-	-	-	-	M.S.	-	-
5.	-	-	-	-	-	-	-	-	-	-	-	-	T.S.	14	-
6.	-	-	-	5	0.5	0.5	8	1.3	0.8	-	-	-	T.S.	42	0.4
7.	2	0.2	0.2	13	1.0	1.3	-	-	-	-	-	-	M.S.	42	0.3
8.	29	3.6	2.9	156	12.0	15.6	-	-	-	200	12.5	20.0	M.S.	70	6.7
9.	1	0.1	0.1	8	0.6	0.8	-	-	-	-	-	-	T.S.	42	0.2
10.	1	0.1	0.1	4	0.3	0.4	-	-	-	-	-	-	L.S.	28	0.1
11.	-	-	-	27	2.2	2.7	-	-	-	-	-	-	T.S.	28	0.1
12.	-	-	-	5	0.4	0.5	-	-	-	-	-	-	M.S.	14	0.1
13.	-	-	-	-	-	-	2	0.3	0.2	-	-	-	M.S.	42	0.4
14.	-	-	-	-	-	-	3	0.5	0.3	15	0.9	1.5	M.S.	42	0.4
15.	-	-	-	1	0.1	0.1	15	2.7	1.6	3	0.2	0.3	M.S.	42	0.5
16.	-	-	-	-	-	-	-	-	-	-	-	-	M.S.	14	0.1
17.	-	-	-	-	-	-	-	-	-	-	-	-	M.S.	14	0.1
18.	-	-	-	-	-	-	-	-	-	-	-	-	L.S.	14	0.1
19.	-	-	-	33	2.6	3.3	2	0.3	0.2	95	5.3	9.5	M.S.	42	2.1
20.	-	-	-	30	2.4	3.0	1	0.1	0.1	-	-	-	M.S.	28	0.4
21.	6	0.7	0.6	10	0.9	1.0	9	1.5	0.9	-	-	-	M.S.	42	0.5
22.	5	0.6	0.5	5	0.4	0.5	-	-	-	5	0.3	0.5	M.S.	70	0.4
23.	-	-	-	5	0.4	0.5	35	5.8	3.5	-	-	-	M.S.	28	1.0
24.	44	5.4	4.4	23	1.8	2.3	17	2.8	1.7	9	0.6	0.9	L.S.	85	2.4

Table - D. Dry Mixed Forests.

Sl.	SPECIES	CHAPRAMARI			JAINTI			UDLABARI		
No.		1.	2.	3.	1.	2.	3.	1.	2.	3.
25.	*Carallia lucida*	-	-	-	-	-	-	-	-	-
26.	*Careya arborea*	22	2.5	2.2	20	1.8	2.0	70	5.4	7.0
27.	*Cassia fistula*	-	-	-	20	1.8	2.0	-	-	-
28.	*Castanopsis tribuloides*	4	0.5	0.4	-	-	-	-	-	-
29.	*Toona ciliata*	-	-	-	5	0.5	0.5	9	0.8	0.9
30.	*Celtis tetrandra*	-	-	-	-	-	-	-	-	-
31.	*Chisocheton paniculatus*	4	0.5	0.4	-	-	-	-	-	-
32.	*Chukrasia tubularis*	-	-	-	5	0.5	0.5	-	-	-
33.	*Cordia myxa*	-	-	-	-	-	-	-	-	-
34.	*Cryptocarya amygdalina*	-	-	-	-	-	-	-	-	-
35.	*Casearea species*	4	0.5	0.4	-	-	-	45	4.1	4.5
36.	*Duabanga grandiflora*	-	-	-	-	-	-	-	-	-
37.	*Dillenia pentagyna*	30	3.3	3.0	15	1.5	1.5	235	21.4	23.5
38.	*Drimycarpus racemosus*	2	0.2	0.2	-	-	-	-	-	-
39.	*Dysoxylum binectariferum*	6	0.6	0.6	5	0.5	0.5	9	0.8	0.9
40.	*D. hamiltonii*	2	0.2	0.2	-	-	-	-	-	-
41.	*D. procerum*	16	1.8	1.6	-	-	-	-	-	-
42.	*Dalbergia assamica*	-	-	-	-	-	-	-	-	-
43.	*Emblica officinalis*	2	0.2	0.2	-	-	-	-	-	-
44.	*Ehretia acuminata*	-	-	-	-	-	-	-	-	-
45.	*Elaeocarpus varunua*	14	1.6	1.4	-	-	-	-	-	-
46.	*Engelhardtia spicata*	2	0.2	0.2	-	-	-	-	-	-
47.	*Erythrina suberosa*	-	-	-	-	-	-	-	-	-
48.	*Evodia meliaefolla*	2	0.2	0.2	-	-	-	-	-	-

Sl.	BHUTRI			DHUMCHI			KAIRBARI			BOIRANTI			DOMINANCE	FREQUENCY	DENSITY
No.	1.	2.	3.	1.	2.	3.	1.	2.	3.	1.	2.	3.			
25.	1	0.1	0.1	-	-	-	-	-	-	-	-	-	M.S.	14	-
26.	19	2.4	1.9	-	-	-	15	2.5	1.5	35	2.2	3.5	L.S.	85	2.5
27.	-	-	-	-	-	-	-	-	-	-	-	-	L.S.	14	0.2
28.	-	-	-	11	0.8	1.1	-	-	-	-	-	-	M.S.	28	0.2
29.	35	4.3	3.5	16	1.3	1.6	1	0.1	0.1	-	-	-	T.S.	70	1.0
30.	2	0.2	0.2	-	-	-	-	-	-	-	-	-	M.S.	14	-
31.	-	-	-	2	0.2	0.2	-	-	-	-	-	-	M.S.	28	0.1
32.	2	0.2	0.2	-	-	-	-	-	-	3	0.2	[illegible]	T.S.	42	0.1
33.	7	0.8	0.7	-	-	-	7	1.2	0.7	5	0.3	0.5	M.S.	42	0.3
34.	-	-	-	5	0.5	0.5	9	1.5	0.9	-	-	-	M.S.	28	0.3
35.	-	-	-	-	-	-	-	-	-	-	-	-	L.S.	28	0.3
36.	-	-	-	1	0.1	0.1	-	-	-	-	-	-	T.S.	14	-
37.	8	1.0	0.8	-	-	-	16	2.7	1.6	28	1.8	2.8	M.S.	85	4.5
38.	-	-	-	1	0.1	0.1	-	-	-	-	-	-	L.S.	28	-
39.	9	1.1	0.9	33	2.6	3.3	5	0.8	0.5	8	0.5	0.8	M.S.	100	1.0
40.	-	-	-	-	-	-	-	-	-	-	-	-	M.S.	14	-
41.	-	-	-	-	-	-	-	-	-	-	-	-	M.S	14	0.2
42.	-	-	-	2	0.2	0.2	-	-	-	-	-	-	T.S.	14	-
43.	-	-	-	-	-	-	-	-	-	-	-	-	L.S.	14	-
44.	-	-	-	-	-	-	2	0.3	0.2	3	0.2	0.3	M.S.	28	0.1
45.	1	0.1	0.1	6	0.5	0.6	-	-	-	-	-	-	T.S.	42	0.2
46.	-	-	-	10	0.9	1.0	-	-	-	-	-	-	T.S.	28	0.2
47.	-	-	-	1	0.1	0.1	-	-	-	-	-	-	M.S.	14	-
48.	-	-	-	10	0.9	1.0	-	-	-	-	-	-	M.S.	28	0.2

Table - D. Dry Mixed Forests

Sl. No.	SPECIES	CHAPRAMARI 1.	2.	3.	JAINTI 1.	2.	3.	UDLABARI 1.	2.	3.
49.	*Ficus* species	-	-	-	-	-	-	18	1.7	1.8
50.	*Garuga pinnata*	4	0.5	0.4	5	0.5	0.5	9	0.8	0.9
51.	*Gmelina arborea*	4	0.5	0.4	10	0.9	1.0	18	1.6	1.8
52.	*Grewia* species	-	-	-	-	-	-	-	-	-
53.	*Glochidion* species	2	0.2	0.2	-	-	-	-	-	-
54.	*Holarrhena antidysenterica*	16	1.8	1.6	20	1.8	2.0	9	0.7	0.9
55.	*Hymenodictyon excelsum*	2	0.2	0.2	-	-	-	-	-	-
56.	*Ilex godajam*	-	-	-	-	-	-	-	-	-
57.	*Jambosa formosa*	26	2.9	2.6	-	-	-	-	-	-
58.	*Knema linifolia*	-	-	-	-	-	-	-	-	-
59.	*Kydia calycina*	-	-	-	-	-	-	-	-	-
60.	*Lonicera macrophylla*	2	0.2	0.2	-	-	-	-	-	-
61.	*Lagerstroemia parviflora*	28	3.1	2.8	5	0.4	0.5	65	6.0	6.0
62.	*Leea* species	8	0.9	0.8	-	-	-	-	-	-
63.	*Litsaea citrata*	-	-	-	-	-	-	-	-	-
64.	*L. khasyana*	12	1.4	1.2	-	-	-	-	-	-
65.	*L. monopetala*	-	-	-	45	4.1	4.5	75	6.9	7.5
66.	*L. glutinosa*	18	2.0	1.8	-	-	-	9	0.7	0.9
67.	*Lannea coromendelica*	4	0.5	0.4	15	1.4	1.5	-	-	-
68.	*Litsaea panemonja*	-	-	-	-	-	-	-	-	-
69.	*Mallatus roxburgianus*	8	0.9	0.8	-	-	-	-	-	-
70.	*Macaranga denticulata*	-	-	-	-	-	-	18	1.6	1.8
71.	*Machilus gamblei*	6	0.7	0.6	-	-	-	-	-	-
72.	*M. villosa*	28	3.1	2.8	-	-	-	-	-	-

Sl. No.	BHUTRI 1.	BHUTRI 2.	BHUTRI 3.	DHUMCHI 1.	DHUMCHI 2.	DHUMCHI 3.	KAIRBARI 1.	KAIRBARI 2.	KAIRBARI 3.	BOIRANTI 1.	BOIRANTI 2.	BOIRANTI 3.	DOMINANCE	FREQUENCY	DENSITY
49.	3	0.4	0.3	8	0.6	0.8	4	0.7	0.4	-	-	-	T.S.	56	0.6
50.	6	0.7	0.6	4	0.3	0.4	-	-	-	85	5.3	8.5	T.S.	85	1.4
51.	9	1.1	0.9	-	-	-	-	-	-	-	-	-	T.S.	56	0.7
52.	1	0.1	0.1	2	0.1	0.2	-	-	-	-	-	-	L.S.	28	-
53.	-	-	-	-	-	-	-	-	-	12	0.7	1.2	M.S.	28	0.1
54.	2	0.2	0.2	-	-	-	-	-	-	-	-	-	L.S.	56	0.6
55.	1	0.1	0.1	2	0.1	0.2	-	-	-	30	1.8	3.0	T.S.	56	0.3
56.	-	-	-	-	-	-	4	0.7	0.4	-	-	-	T.S.	14	0.1
57.	-	-	-	-	-	-	-	-	-	-	-	-	T.S.	14	0.4
58.	-	-	-	1	0.8	0.1	-	-	-	-	-	-	T.S.	14	0.1
59.	4	0.5	0.4	-	-	-	-	-	-	-	-	-	M.S.	14	0.1
60.	-	-	-	-	-	-	-	-	-	-	-	-	L.S.	14	-
61.	21	2.6	2.1	4	0.3	0.4	21	3.5	2.1	75	4.7	7.5	T.S.	100	3.0
62.	-	-	-	-	-	-	-	-	-	-	-	-	L.S.	14	0.1
63.	-	-	-	10	0.8	1.0	-	-	-	-	-	-	M.S.	14	0.1
64.	-	-	-	-	-	-	-	-	-	-	-	-	M.S.	14	0.2
65.	145	17.9	14.5	34	2.6	3.4	55	9.1	5.5	3	0.2	0.3	M.S.	85	5.8
66.	43	5.3	4.3	11	0.9	1.1	14	2.3	1.4	8	0.5	0.8	M.S.	85	1.7
67.	5	0.6	0.5	3	0.2	0.3	1	0.2	0.1	35	2.1	3.5	T.S.	85	0.7
68.	-	-	-	3	0.2	0.3	-	-	-	-	-	-	M.S.	14	-
69.	-	-	-	-	-	-	-	-	-	-	-	-	L.S.	14	0.1
70.	-	-	-	1.6	9.0	11.6	2	0.3	0.2	-	-	-	L.S.	42	1.6
71.	-	-	-	2	0.1	0.2	1	0.2	0.1	-	-	-	T.S.	42	0.1
72.	9	1.1	0.9	68	5.2	6.8	7	1.1	0.7	-	-	-	T.S.	56	1.5

Table - D. Dry Mixed Forests.

Sl. No.	SPECIES	CHAPRAMARI 1.	2.	3.	JAINTI 1.	2.	3.	UDLABARI 1.	2.	3.
73.	*Mallotus philippinensis*	40	4.5	4.0	85	7.7	8.5	9	0.8	0.9
74.	*Machilus* species	40	4.5	4.0	-	-	-	-	-	-
75.	*Marlea chinense*	-	-	-	-	-	-	-	-	-
76.	*Mastixia arborea*	-	-	-	-	-	-	-	-	-
77.	*Meliosma simplicifolia*	-	-	-	-	-	-	-	-	-
78.	*Michelia champaca*	4	0.5	0.4	-	-	-	-	-	-
79.	*Olea dioica*	-	-	-	-	-	-	-	-	-
80.	*Oroxylum indicum*	12	1.4	1.2	5	0.4	0.5	36	3.3	3.6
81.	*Photinia integrifolia*	-	-	-	125	11.4	12.5	-	-	-
82.	*Polyalthia semiarum*	12	1.4	1.2	-	-	-	-	-	-
83.	*Phoebe lanceolata*	20	2.2	2.0	-	-	-	-	-	-
84.	*Premna bengalensis*	-	-	-	10	0.9	1.0	55	5.0	5.5
85.	*P. integrifolia*	-	-	-	-	-	-	55	5.0	5.5
86.	*Pygeum acuminatum*	-	-	-	-	-	-	9	0.8	0.9
87.	*Pethecolobium angulatum*	-	-	-	-	-	-	-	-	-
88.	*Sapium engeniaefolium*	-	-	-	-	-	-	-	-	-
89.	*Saurauia roxbourghii*	-	-	-	-	-	-	-	-	-
90.	*Sterculia alata*	-	-	-	-	-	-	-	-	-
91.	*Strespermum chelonoides*	48	5.4	4.8	25	2.3	2.5	-	-	-
92.	*Schima wallichii*	52	5.8	5.2	10	1.0	1.0	-	-	-
93.	*Shorea robsta*	36	4.0	3.6	20	1.8	2.0	27	2.5	2.7
94.	*Styrax serrulatum*	-	-	-	-	-	-	9	0.8	0.9
95.	*Symplocos spicata*	-	-	-	-	-	-	9	0.8	0.9
96.	*Syzygium cumini*	20	2.3	2.0	60	5.5	6.0	45	4.1	4.5

Sl. No.	BHUTRI 1.	2.	3.	DHUMCHI 1.	2.	3.	KAIRBARI 1.	2.	3.	BOIRANTI 1.	2.	3.	DOMINANCE	FREQUENCY	DENSITY
73.	79	9.9	7.9	9	0.6	0.9	1	0.1	0.1	-	-	-	M.S.	85	3.9
74.	-	-	-	3	0.2	0.3	-	-	-	5	0.3	0.5	M.S.	42	0.8
75.	6	0.7	0.6	42	3.3	4.2	3	0.5	0.3	-	-	-	L.S.	42	0.8
76.	-	-	-	-	-	-	-	-	-	-	-	-			
77.	-	-	-	204	15.7	20.4	187	31.1	18.7	-	-	-	L.S.	28	6.7
78.	-	-	-	1	0.1	0.1	-	-	-	-	-	-	T.S.	28	0.1
79.	-	-	-	3	0.2	0.3	-	-	-	-	-	-	M.S.	14	-
80.	39	4.9	3.9	8	0.6	0.8	3	0.5	0.3	65	4.1	6.5	M.S.	100	2.5
81.	-	-	-	-	-	-	-	-	-	-	-	-	L.S.	14	1.9
82.	2	0.2	0.2	-	-	-	-	-	-	-	-	-	M.S.	28	0.3
83.	-	-	-	-	-	-	-	-	-	-	-	-	L.S.	14	0.4
84.	7	0.8	0.7	36	2.8	3.6	17	2.8	1.7	5	0.3	0.5	T.S.	85	2.1
85.	14	13.0	10.4	50	3.9	5.0	31	5.1	3.1	170	10.6	17.0	T.S.	70	5.4
86.	-	-	-	4	0.3	0.4	-	-	-	-	-	-	M.S.	28	6.3
87.	-	-	-	14	1.1	1.4	-	-	-	45	2.8	4.5	L.S.	28	0.7
88.	-	-	-	1	0.1	0.1	-	-	-	-	-	-	L.S.	14	-
89.	-	-	-	2	0.1	0.2	-	-	-	-	-	-	M.S.	14	-
90.	-	-	-	-	-	-	2	0.3	0.2	-	-	-	L.S.	14	-
91.	27	3.3	2.7	31	2.4	3.1	-	-	-	3	0.2	0.3	T.S.	70	2.3
92.	-	-	-	14	1.0	1.4	-	-	-	-	-	-	T.S.	42	1.3
93.	6	0.7	0.6	49	3.6	4.9	10	1.6	1.0	-	-	-	T.S.	85	2.0
94.	21	2.6	2.1	33	2.5	3.3	10	1.6	1.0	-	-	-	T.S.	56	1.1
95.	-	-	-	-	-	-	-	-	-	-	-	-	M.S.	14	0.1
96.	1	0.1	0.1	-	-	-	1	0.1	0.1	157	9.8	15.7	T.S.	85	3.1

Table - D. Dry Mixed Forests.

Sl. No.	SPECIES	CHAPRAMARI 1.	2.	3.	JAINTI 1.	2.	3.	UDLABARI 1.	2.	3.
97.	*Symplocos ramosissima*	2	0.2	0.2	-	-	-	-	-	-
98.	*Spondias mangifera*	--	-	-	-	-	-	-	-	-
99.	*Sterculia villosa*	35	4.0	3.5	20	1.8	2.0	30	2.7	3.0
100.	*Sterblus asper*	-	-	-	-	-	-	-	-	-
101.	*Terminalia belerica*	48	5.4	4.8	60	5.5	6.0	110	10.0	11.0
102.	*T. chebula*	-	-	-	-	-	-	-	-	-
103.	*T. myriocarpa*	28	3.1	2.8	-	-	-	-	-	-
104.	*Trema orientalis*	8	0.9	0.8	5	0.5	0.3	9	0.8	0.9
105.	*Trewia nudiflora*	-	-	-	-	-	-	-	-	-
106.	*Turpinia pomifera*	-	-	-	-	-	-	-	-	-
107.	*Vitex quinata*	-	-	-	-	-	-	-	-	-
108.	*Wrightia tomentosa*	38	4.3	3.8	90	8.0	9.0	27	2.5	2.7
109.	*Xylosma longifolia*	-	-	-	-	-	-	-	-	-
110.	*Fagara badrunga*	-	-	-	15	1.4	1.5	-	-	-

Sl.	BHUTRI			DHUMCHI			KAIRBARI			BOIRANTI			DOMINANCE	FREQUENCY	DENSITY
No.	1.	2.	3.	1.	2.	3.	1.	2.	3.	1.	2.	3.			
97.	-	-	-	4	0.3	0.4	-	-	-	-	-	-	M.S.	28	0.1
98.	-	-	-	1	0.1	0.1	-	-	-	-	-	-	M.S.	14	-
99.	30	3.7	3.0	12	0.9	1.2	16	2.7	1.6	15	0.9	1.5	T.S	*100	2.4
100.	-	-	-	-	-	-	-			65	4.1	6.5	L.S.	14	0.6
101.	50	6.2	5.0	25	1.8	2.5	1	0.1	0.1	160	10.0	16.0	T.S.	100	5.6
102.															
103.	-	-	-	7	0.5	0.7	-	-	-	3	0.2	0.3	T.S.	42	0.5
104.	-	-	-	-	-	-	1	0.1	0.1	-	-	-	L.S.	56	0.4
105.	-	-	-	1	0.1	0.1	32	5.3	3.2	-	-	-	M.S.	28	0.9
106.	-	-	-	15	1.0	1.3	-	-	-	-	-	-	L.S.	14	0.2
107.	7	0.8	0.7	2	0.1	0.2	-	-	-	-	-	-	T.S.	28	0.1
108.	-	-	-	14	1.1	1.4	15	2.5	1.5	250	15.6	25.0	T.S.	85	5.7
109.	-	-	-	-	-	-	6	1.0	0.6	-	-	-	L.S.	14	0.2
110.	-	-	-	1	0.1	0.1	-	-	-	-	-	-	M.S.	14	0.3

SYNTHETIC SURVEY
FLORISTIC COMPOSITION OF TREE VEGETATION IN PLAINS:
RANDOM PLOTS

(i) SHOREA FORESTS INCLUDING SCATTERED SHOREA AND WET-MIXED FORESTS

REFERENCE TO TABLE E

COLUMN	1-PERCENTAGE OF SPECIES
COLUMN	2-DENSITY
F%	-FREQUENCY
D	-DENSITY PER 0.4 HECTARE (OR ONE ACRE)
*	-PRESENT

Table - E. Shorea Forests Including Scattered Shorea and Wet- Mixed Forests.

Sl. No.	SPECIES	NILPARA & CHILAPATA 1.	NILPARA & CHILAPATA 2.	MORAGHAT & TONDU 1.	MORAGHAT & TONDU 2.	BUXADUAR 1.	BUXADUAR 2.
1.	*Adina cordifolia*	-	-	-	-	-	-
2.	*Ailanthus grandis*	*	*	*	*	*	*
3.	*Albizzia lucida*	0.1	*	0.1	*	1.4	0.4
4.	*Alstonia scholaris*	0.9	0.2	0.2	*	0.4	0.3
5.	*Aphanamixis polystachia*	3.8	1.0	3.7	1.1	12.5	0.3
6.	*Amoora wallichii*	7.1	1.9	0.9	0.3	1.8	0.5
7.	*Anthocephalus chinenis*	0.3	0.1	-	-	-	-
8.	*Artocarpus chaplasa*	1.9	0.5	1.0	0.3	*	*
9.	*Albizzia odoratissima*	*	*	-	-	-	-
10.	*Bauhinia* species	*	*	7.4	0.4	-	
11.	*Beilschimiedia roxburghiana*	*	*	*	*	*	*
12.	*Bridelia retusa*	-	-	-	-	-	-
13.	*B. stipularis*	-	-	-	-	0.1	*
14.	*Bischofia javanica*	0.4	0.1	0.4	0.1	-	-
15.	*Canarium silkimense*	0.3	*	-	-	-	-
16.	*Careya arborea*	*	*	0.2	0.1	0.6	0.2
17.	*Cassia fistula*	*	*	-	-	0.6	0.2
18.	*Castanopsis tribuloides*	2.6	0.7	3.1	0.9	0.1	*
19.	*Toona ciliata*	*	*	0.2	0.1	0.1	*
20.	*Chisocheton paniculatus*	-	-	*	*	-	-
21.	*Chukrasia tabularis*	0.5	0.2	*	*	0.1	-
22.	*Cinamomum cecidodaphne*	0.1	-	0.3	0.1	-	-
23.	*Cycas pectinata*	-	-	*	-	-	-
24.	*Dalbergia assamica*	-	-	-	-	-	-

Sl. No.	RAJABHATKHAWA & DAMANPUR 1.	RAJABHATKHAWA & DAMANPUR 2.	DALKAJHAR 1.	DALKAJHAR 2.	FREQUENCY	DENSITY
1.	-	-	1.0	0.3	20	0.2
2.	1.2	0.2	-	-	80	0.2
3.	3.2	0.4	0.2	0.1	100	1.0
4.	0.7	0.1	*	*	100	0.5
5.	6.5	0.8	1.2	0.3	100	5.5
6.	0.5	0.1	0.8	0.2	100	2.2
7.	0.3	*	-	-	60	0.1
8.	0.7	0.1	*	*	100	0.7
9.	-	-	-	-	20	-
10.	-	-	0.5	0.1	40	0.2
11.	0.3	*	*	*	60	-
12.	-	-	0.2	0.1	20	-
13.	-	-	-	-	-	-
14.	3.7	0.5	-	-	60	0.9
15.	-	-	0.1	*	40	0.1
16.	-	-	0.9	0.2	60	0.3
17.	-	-	0.5	0.1	60	0.2
18.	7.5	1.0	1.0	0.3	100	2.9
19.	*	*	-	-	80	-
20.	0.7	0.1	0.4	0.1	60	0.2
21.	1.0	0.1	-	-	80	0.3
22.	0.3	-	-	-	60	0.1
23.	-	-	*	-	40	-
24.	-	-	*	*	20	-

Table - E. Shorea Forests Including Scattered Shorea and Wet - Mixed Forests.

Sl. No.	SPECIES	NILPARA & CHILAPATA 1.	NILPARA & CHILAPATA 2.	MORAGHAT & TONDU 1.	MORAGHAT & TONDU 2.	BUXADUAR 1.	BUXADUAR 2.
25.	*Dillenia indica*	0.2	0.1	0.3	0.1	-	-
26.	*D. pentagyna*	0.4	0.1	2.3	0.7	*	*
27.	*Drimycarpus racemosus*	-	-	-	-	-	-
28.	*Duabanga grandiflora*	0.5	0.2	*	*	0.2	0.1
29.	*Dysoxylum binectariferum*	1.3	0.3	-	-	-	-
30.	*Elaeocarpus* species	0.7	0.2	0.9	0.3	*	*
31.	*Sloanea sterculacea*	*	*	-	-	-	-
32.	*Ehretia acuminata*	-	-	*	*	-	-
33.	*Endospermum chinense*	-	-	-	-	-	-
34.	*Engelhardtia spicata*	0.5	0.1	0.5	-	-	-
35.	*Erythrina suberosa*	-	-	*	*	-	-
36.	*Evodia meliaefolia*	*	*	*	*	-	-
37.	*Ficus* species	0.2	-	*	*	-	-
38.	*Garuga pinnata*	0.6	0.2	0.3	0.1	3.7	0.9
39.	*Gmelina arborea*	0.5	0.1	0.7	0.2	1.0	0.2
40.	*Holarrhena antidysenterica*	0.2	-	-	-	-	-
41.	*Hymenodictyon excelsum*	0.2	0.1	-	-	-	-
42.	*Ilex godajam*	0.1	-	0.5	0.2	-	-
43.	*Lannea coromendelica*	0.7	0.2	0.9	0.3	0.5	0.1
44.	*Lophopetalum fimbriatum*	5.1	1.4	0.4	0.1	-	-
45.	*Lagerstroemia parviflora*	1.3	0.3	0.7	0.2	3.7	1.0
46.	*Litsaea monopetala*	*	-	0.2	*	*	-
47.	*L. panamonja*	*	-	*	-	*	-
48.	*Macaranga denticulata*	-	-	*	*	-	-

Sl. No.	RAJABHATKHAWA & DAMANPUR 1.	2.	DALKAJHAR 1.	2.	FREQUENCY	DENSITY
25.	-	-	-	-	40	0.1
26.	*	*	*	*	40	0.5
27.	-	-	2.3	0.6	20	0.5
28.	0.7	0.1	-	-	80	0.3
29.	-	-	-	-	20	0.3
30.	1.0	0.1	1.1	0.3	80	0.7
31.	-	-	-	-	20	-
32.	-	-	-	-	20	-
33.	-	-	0.5	0.5	20	0.1
34.	0.4	0.1	0.2	0.1	80	0.5
35.	-	-	-	-	20	-
36.	-	-	-	-	40	-
37.	-	-	-	-	40	-
38.	1.2	0.2	5.6	1.4	100	2.3
39.	1.5	0.2	0.8	0.2	100	0.9
40.	0.1	-	-	-	20	-
41.	-	-	-	-	20	-
42.	-	-	0.9	0.2	60	0.3
43.	2.5	0.3	2.4	0.6	100	1.4
44.	0.1	0.1	0.2	0.1	80	1.3
45.	3.7	0.5	2.4	0.6	100	1.3
46.	*	-	0.1	-	40	-
47.	*	-	-	-	80	-
48.	-	-	-	-	20	-

Table - E. Shorea Forests Including Scattered Shorea and Wet - Mixed Forests.

Sl. No.	SPECIES	NILPARA & CHILAPATA 1.	NILPARA & CHILAPATA 2.	MORAGHAT & TONDU 1.	MORAGHAT & TONDU 2.	BUXADUAR 1.	BUXADUAR 2.
49.	*Machilus gamblei*	*	-	*	-	*	-
50.	*M. villosa*	*	-	*	-	*	-
51.	*Mallotus philippinensis*	*	-	*	-	*	-
52.	*Mangifera sylvatica*	-	-	-	-	-	-
53.	*Machilus* species	1.3	0.3	2.1	0.6	1.0	0.3
54.	*Meliosma pinnata*	-	-	-	-	-	-
55.	*M. simplicifolia*	*	*	*	*	-	
56.	*Michelia champaca*	3.2	0.8	*	*	-	-
57.	*M. montana*	-	-	0.6	0.2	-	-
58.	*Ougeinia oojenioides*	-	-	-	-	-	-
59.	*Polyalthia semiarum*	-	-	0.2	0.1	-	-
60.	*Premna bengalensis*	0.7	0.2	1.8	0.5	*	-
61.	*P. integrifolia*	*	-	*	-	*	-
62.	*Pterospermum acerifolium*	-	-	-	-	-	-
63.	*Randia uliginosa*	-	-	-	-	-	-
64.	*Sterculia villosa*	2.0	0.5	2.3	0.7	5.5	1.2
65.	*Schima wallichii*	4.6	1.2	8.3	2.4	6.2	1.6
66.	*Shorea robusta*	47.2	12.5	57.5	17.5	45.5	12.0
67.	*Syzygium* species	4.3	1.2	3.4	1.0	3.7	1.0
68.	*S. cumini*	1.1	0.3	*	-	*	-
69.	*Terminalia tomentosa*	-	-	0.3	0.1	-	-
70.	*Talauma hodgsoni*	-	-	-	-	-	-
71.	*Terminalia belerica*	1.7	0.4	3.0	0.8	9.6	2.5
72.	*T. myriocarpa*	0.4	0.1	-	-	-	-

Sl. No.	*RAJABHATKHAWA & DAMANPUR* 1.	2.	*DALKAJHAR* 1.	2.	*FREQUENCY*	*DENSITY*
49.	*	-	-	-	80	-
50.	*	-	-	-	80	-
51.	*	-	*	-	100	-
52.	0.2	*	*	*	40	-
53.	5.2	0.7	0.9	0.6	100	2.1
54.	-	-	0.1	-	20	-
55.	-	-	-	-	40	-
56.	0.8	0.1	0.8	0.2	80	1.0
57.	-	-	-	-	20	-
58.	-	-	0.8	0.2	20	1.0
59.	-	-	-	-	20	-
60.	*	-	0.4	0.1	60	0.6
61.	*	-	*	-	100	-
62.	*	*	0.3	0.1	40	-
63.	0.1	-	1.0	0.3	40	0.2
64.	6.2	0.8	0.9	0.2	100	3.4
65.	2.3	0.3	10.6	2.7	100	6.4
66.	32.7	4.3	50.6	13.0	100	46.7
67.	5.8	0.7	-	-	80	3.4
68.	*	-	*	-	100	-
69.	-	-	0.9	0.2	20	0.2
70.	-	-	1.5	0.4	20	0.3
71.	3.7	0.5	3.2	1.1	100	4.2
72.	-	-	-	-	20	0.1

Table - E. *Shorea* Forests Including Scattered *Shorea* and Wet - Mixed Forests.

Sl. No.	SPECIES	NILPARA & CHILAPATA 1.	NILPARA & CHILAPATA 2.	MORAGHAT & TONDU 1.	MORAGHAT & TONDU 2.	BUXADUAR 1.	BUXADUAR 2.
73.	*Tetrameles nudiflora*	0.5	0.1	0.2	0.1	0.7	0.2
74.	*Trewia nudiflora*	1.7	0.4	-	-	0.2	0.1
75.	*Turpinia pomifera*	*	*	0.3	0.1	-	-
76.	*Trichillia connarioides*	-	-	-	-	-	-
77.	*Vitax quinata*	*	*	-	-	-	-
78.	*Fagura budrunga*	0.6	0.1	-	-	0.5	0.1
79.	*Unknown*	*	*	0.2	*	*	*

Sl. No.	RAJABHATKHAWA & DAMANPUR 1.	RAJABHATKHAWA & DAMANPUR 2.	DALKAJHAR 1.	DALKAJHAR 2.	FREQUENCY	DENSITY
73.	0.1	*	-	-	80	0.3
74.	2.1	0.2	0.2	0.1	80	0.8
75.	-	-	-	-	40	-
76.	-	-	-	-	-	-
77.	-	-	*	*	40	-
78.	1.5	0.2	2.2	0.6	80	1.0
79.	0.3	*	*	*	100	-

SYNTHETIC SURVEY
FLORISTIC COMPOSITION OF TREE VEGETATION IN PLAINS:
RANDOM PLOTS

(ii) DRY-MIXED FORESTS

REFERENCE TO TABLE F

COLUMN	1-PERCENTAGE OF SPECIES
COLUMN	2-DENSITY
F%	-FREQUENCY
D	-DENSITY PER 0.4 HECTARE (OR ONE ACRE)
*	-PRESENTS

Table - F. Dry Mixed Forests

Sl. No.	SPECIES	SIPCHU 1	SIPCHU 2	BHUTRI & PANA 1	BHUTRI & PANA 2	KALABARI & NIPANTA 1	KALABARI & NIPANTA 2	FREQUENCY	DENSITY
1.	*Adina cordifolia*	-	-	-	-	6.8	0.9	33	2.3
2.	*Ailanthus grandis*	0.9	-	0.8	-	-	-	66	0.6
3.	*Albizzia lucida*	-	-	4.2	0.4	2.5	0.3	66	2.2
4.	*Alstonia scholaris*	2.8	0.3	1.5	0.1	0.2	-	100	1.5
5.	*Aphanamixis polystachia*	4.1	0.4	16.8	1.7	-	-	66	7.5
6.	*Amoora wallichii*	2.6	0.3	2.0	8.2	-	-	66	1.5
7.	*Anthocephalus chinensis*	2.8	0.1	0.7	-	-	-	66	0.8
8.	*Artocarpus chaplasa*	-	-	-	-	0.6	-	33	0.2
9.	*Acacia catechu*	-	-	-	-	0.2	-	33	0.1
10.	*Bauhinia* species	2.8	0.3	3.9	0.4	0.5	-	100	2.9
11.	*Bombax ceiba*	-	-	0.3	-	0.3	-	66	0.2
12.	*Bridelia montana*	-	-	1.5	0.1	-	-	33	0.5
13.	*B. retusa*	-	-	-	-	1.1	0.1	33	0.4
14.	*Callicarpa arborea*	-	-	0.5	-	-	-	33	0.2
15.	*Canarium sikkimense*	2.1	0.2	-	-	-	-	33	0.7
16.	*Careya arborea*	0.2	-	-	-	1.6	0.2	66	0.6
17.	*Cassia fistual*	-	-	0.6	-	1.2	0.1	66	0.6
18.	*Castanopsis tribuloides*	2.8	0.3	1.4	0.1	-	-	66	1.4
19.	*Toona ciliata*	0.2	-	3.2	0.3	6.0	0.8	100	3.1
20.	*Chukrasis tabularis*	1.4	0.1	0.8	-	-	-	66	0.7
21.	*Cinnamomum cecidodaphne*	7.5	0.7	2.0	0.2	-	-	66	3.2
22.	*Casearia species*	-	-	-	-	-	-	-	-
23.	*Dalbergia latifolia*	-	-	-	-	0.2	-	33	0.1
24.	*Dalbergia assamica*	-	-	0.6	-	-	-	33	0.2

Table - F. Dry Mixed Forests

Sl. No.	SPECIES	SIPCHU 1	SIPCHU 2	BHUTRI & PANA 1	BHUTRI & PANA 2	KALABARI & NIPANTA 1	KALABARI & NIPANTA 2	FREQUENCY	DENSITY
25.	*Dillenia pentagyna*	-	-	0.1	-	4.0	0.5	66	1.4
26.	*Duabanga grandiflora*	1.3	0.1	0.4	-	-	-	66	0.6
27.	*Dysoxylum procerum*	*	*	*	-	*	-	-	-
28.	*D. binectariferum*	*	-	*	-	*	-	-	-
29.	*Ehretia acuminata*	-	-	0.5	-	-	-	33	0.2
30.	*Eaeocarpus varunua*	0.9	-	0.6	-	-	-	66	0.5
31.	*Engelhardtia spicata*	0.3	-	0.1	-	0.1	-	100	0.2
32.	*Erythrina suberosa*	-	-	0.3	-	-	-	33	-
33.	*Evodia meliaefolia*	-	-	-	-	*	-	-	-
34.	*Firmiana colorata*	-	-	-	-	0.1	-	33	-
35.	*Garuqa pinnata*	-	-	-	1.1	5.1	0.7	66	2.1
36.	*Gmelina arborea*	5.0	0.5	10.1	1.0	1.5	0.2	100	5.5
37.	*Glochidion species*	*	*	*	*	-	-	-	-
38.	*Ilex godajam*	-	-	-	*	-	*	*	*
39.	*Jambosa formosa*	*	-	*	-	*	-	-	-
40.	*J. praecox*	*	-	*	-	*	-	-	-
41.	*Lannea coromendelica*	0.6	-	0.2	-	1.8	0.2	100	0.9
42.	*Lagerstroemia parviflora*	0.1	-	4.0	0.4	19.5	2.1	100	7.9
43.	*Litsaea citrata*	*	-	*	-	*	-	-	-
44.	*L. khasyana*	-	*	-	*	-	-	-	-
45.	*L. kingii*	-	-	-	*	-	-	-	-
46.	*L. monopetala*	1.8	0.1	2.0	0.2	0.7	0.1	100	1.5
47.	*L. salicifolia*	2.6	0.3	*	-	2.5	0.3	66	1.0
48.	*L. glutinosa*	*	-	*	-	*	-	-	-

Table - F. Dry Mixed Forests

Sl. No.	SPECIES	SIPCHU 1	SIPCHU 2	BHUTRI & PANA 1	BHUTRI & PANA 2	KALABARI & NIPANTA 1	KALABARI & NIPANTA 2	FREQUENCY	DENSITY
49.	*Macaranga denticulata*	*	-	*	-	*	-	-	-
50.	*M. villosa*	2.8	0.3	0.2	-	0.8	0.1	100	1.3
51.	*Mallotus philippinensis*	*	-	*	-	1.1	0.1	33	0.4
52.	*Machilus* species	-	-	4.9	0.5	-	-	-	-
53.	*M. simiplicifolia*	*	-	*	-	*	-	-	-
54.	*Olea dioica*	*	-	*	-	*	-	-	-
55.	*Ougeinia oojenioides*	-	-	-	-	*	-	-	-
56.	*Polyalthia semiarum*	-	-	-	*	-	-	-	-
57.	*Phoebe lanceolata*	3.3	0.3	*	-	-	-	33	1.1
58.	*Premna bengalensis*	1.2	0.2	0.5	-	8.7	1.2	100	3.5
59.	*Pterospermum acerifolium*	0.1	-	-	-	-	-	33	-
60.	*Sapium baccatum*	3.8	0.4	0.9	0.1	-	-	66	1.6
61.	*Sterculia villosa*	2.8	0.3	4.5	0.5	6.2	0.9	100	3.4
62.	*Schima wallichii*	3.4	0.3	1.9	0.2	-	-	66	1.3
63.	*Shorea robusta*	-	-	3.6	0.4	-	-	33	0.9
64.	*Syzygium* species	0.3	-	*	-	0.9	0.1	66	0.2
65.	*S. cumini*	2.8	0.3	1.9	0.2	0.9	0.1	100	1.9
66.	*Stereospermum chelonoides*	10.8	1.9	4.0	-	0.9	0.7	100	5.2
67.	*Terminalia tomentosa*	2.8	0.3	0.8	-	2.0	0.2	100	1.4
68.	*Terminalia belerica*	7.5	0.7	9.4	1.0	13.8	1.8	100	7.6
69.	*Tetrameles nudiflora*	13.0	1.3	5.8	0.6	-	-	66	4.9
70.	*Trema orientalis*	-	-	0.1	-	-	-	33	-
71.	*Trewia nudiflora*	-	-	0.6	-	3.1	0.4	66	0.9
72.	*Turpinia pomifera*	3.3	0.3	1.0	0.1	-	-	66	1.1

SYNTHETIC SURVEY
FLORISTIC COMPOSITION OF GROUND VEGETATION IN PLAINS QUADRATS

(i) SHOREA FORESTS **(ii) RIVERAIN FORESTS**
(iii) DRY-MIXED FORESTS **(iv) PLANTATION**

REFERENCE TO TABLE G

IN EACH COLUMN

*	MEAN PRESENT IN MORE THAN 50% OF THE QUADRATS
F%	-FREQUENCY
D	-DENSITY PER 0.4 HECTARE (OR ONE ACRE)

N.B.
NUMERICALS (1,2,3.......) IN EACH COLUMN MEAN LOCATION ANALYSED

UNDER(i) SHOREA FORESTS AT .... 1. MORAGHAT 2. LATAGURI 3. DAMANPUR 4. RAJABHATKHAWA 5. APALCHAND 6. UDLABARI 7. BENGDUBI, 8. DUMCHI 9. JAINTY 10. BOIRANTI

(ii) RIVERAIN FORESTS AT ... DIMA, SURTI, TORSA, RAIMATANG,

(iii) DRY-MIXED FORESTS AT .. 1. BHUTRI 2. JAINTY 3. BAIKUNTHAPUR 4. 7th MILE 5. CHAPRAMARI.

(iv) PLANTATIONS AT 1. APALCHAND 2. SEVOKE 3. CHAPRAMARI 4. LALTONG 5. MORAGHAT

Table - G. Floristic Composition of Ground Vegetation in Plains

Sl. No.	SPECIES	SHOREA FOREST 1	2	3	4	5	6	7	8	9	10	F%	D
1.	*Setaria palmaefolia*	*	*	*	*	-	*	*	-	-	-	60	1.8
2.	*Cyrtocoecum oxyphyllum*	-	-	*	*	*	*	-	-	*	*	60	1.6
3.	*Isachne albens*	-	-	*	-	-	-	*	-	-	*	30	1.5
4.	*Imperata arundinacea*	*	-	*	*	-	-	-	-	-	*	40	1.5
5.	*Paspalum* species	*	*	-	*	*	*	*	-	*	*	80	2.0
6.	*Digitaria* species	*	*	-	*	*	*	-	*	-	-	60	0.6
7.	*Axonopus cimicinus*	*	-	-	*	-	-	-	*	*	*	30	0.8
8.	*Microstegium ciliatum*	-	-	*	*	-	-	-	-	*	-	30	0.8
9.	*Oplismenus compositus*	-	*	-	*	*	*	*	*	*	*	80	4.0
10.	*Pogonetherum sacchariodes*	-	-	-	-	-	-	-	-	-	-	-	-
11.	*Ergrostis amabilis*	-	-	-	-	-	-	-	-	-	-	-	-
12.	*Adiantum* species	-	-	-	-	-	-	*	-	-	*	20	-
13.	*Thelepteris* species	*	*	-	*	*	*	-	-	*	-	60	0.4
14.	*Lygodium flexuosum*	*	-	-	*	*	*	-	-	*	-	50	2.5
15.	*Cyclosorus molliusculus*	*	*	*	*	*	*	-	-	-	*	70	1.5
16.	*Other species of fern*	*	-	*	-	-	-	-	*	-	-	30	-
17.	*Diplazium esculentum*	-	-	-	-	-	-	-	-	-	-	-	-
18.	*Equisetum debile*	-	-	-	-	-	-	-	-	-	-	-	-
19.	*Cyperaceae*	-	-	-	-	-	-	-	-	-	-	-	
20.	*Scleria elata*	*	*	-	*	*	-	*	*	*	-	70	1.5
21.	*Aneilema* species	-	*	-	-	-	-	-	*	-	-	20	-
22.	*Smilax* species	*	*	*	*	*	*	*	*	-	*	90	3.0
23.	*Dioscorea* species	*	*	*	*	*	*	-	*	*	*	90	4.0
24.	*Curcuma aromatica*	*	-	-	-	-	*	-	*	*	*	50	1.0

Riverain Forests

Sl. No.	DIANA	REHTI	SURTI	TOSA	RAIMATANG	FREQUENCY	D
1.	-	-	-	-	-	-	-
2.	-	-	-	-	-	-	-
3.	-	-	-	-	-	-	-
4.	-	*	*	-	-	40	0.8
5.	*	-	-	-	-	20	1.0
6.	-	-	-	*	*	40	1.5
7.	*	-	-	-	-	20	2.0
8.	-	*	-	-	-	20	-
9.	*	-	-	-	-	20	2.0
10.	-	-	*	*	*	60	0.8
11.	*	*	-	*	*	80	2.0
12.	-	-	-	-	-	-	-
13.	-	-	-	-	-	-	-
14.	-	-	-	-	-	-	-
15.	*	*	-	-	-	40	4.0
16.	-	-	-	-	-	-	-
17.	*	-	-	-	-	20	3.0
18.	*	*	*	*	*	100	-
19.	-	-	-	-	-	-	-
20.	-	-	*	*	*	60	0.5
21.	*	*	*	-	-	60	1.5
22.	*	*	*	*	*	100	2.0
23.	-	*	-	*	*	60	1.5
24.	-	-	-	-	-	-	-